praise for

Fathoms

Winner of the Carnegie Medal for Excellence in Nonfiction —
Finalist for the Kirkus Prize — Finalist for the PEN/E.O. Wilson
Literary Science Writing Award — Winner of the 2020 Royal Zoological
Society of New South Wales' Whitley Award for Popular Zoology —
Winner of the 2020 Mark & Evette Moran Nib Literary Award for
excellence in research and writing — Finalist for the 2021 Stella Prize

---

"A work of bright and careful genius. Equal parts Rebecca Solnit and
Annie Dillard, Giggs masterfully combines lush prose with conscien-
tious history and boots-on-the-beach reporting. With Giggs leading us
gently by the hand we dive down, and down, and down, into the dark
core of the whale, which, she convincingly reveals, is also the guts of
the world."

—Robert Moor, *New York Times* bestselling
author of *On Trails*

"*Fathoms* took my breath away. Every page is suffused with magic and
meaning. Humanity's relationship with nature has never been more important or vulnerable, and we are truly fortunate that at such a pivotal
moment, a writer of Rebecca Giggs's caliber is here to capture every
beautiful detail, every aching nuance. She is in a league of her own."

—Ed Yong, *New York Times* bestselling
author of *I Contain Multitudes*

"Lyrical . . . Giggs's writing has an old-fashioned lushness and elabo-
rateness of thought. Its finest passages—and they are many—awaken a
sense of wonder."

—*Washington Post*

"[Giggs's] narrative widens the aperture of our attention with a literary style so stunning that the reader may forget to blink. . . . In a story that extends across several continents, Ms. Giggs marshals lapidary language to give the crisis a compelling voice. Her prose, like the oceans in which her subjects roam, is immersive; her sentences submerge us in a sea of sensations. A reader fond of dog-earing choice turns of phrase in *Fathoms* might find, at evening's end, a book pleated like an accordion with an abundance of keepsakes."

—*Wall Street Journal*

"A profound meditation . . . Giggs explores how whales have permeated our lives and the many ways we have invaded and transformed theirs. Each chapter orbits a different aspect of this long and fraught relationship—commodification, pollution, voyeurism, adoration, mythology—swerving wherever Giggs's extensive research and fervent curiosity take her. . . . Giggs's prose is fluid, sensuous, and lyrical. She has a poet's gift for startling and original imagery. . . . The lushness of her sentences and the intensity of her vision inspire frequent rereading—not for clarity, but for sheer pleasure and depth of meaning."

—*Los Angeles Review of Books*

"In *Fathoms*, Rebecca Giggs rips the metaphors off whales and brings us closer than we can usually get to the creatures themselves. Along the way, she shows us how intimately whales are shaping our lives, how they change air quality, and crime, and even our conception of time. I can't stop thinking about the connections she has unearthed, how a whale is connected to a meteor, a mother's breast, a landfill. Under the spell of her deliciously evocative prose, you get the sense that you are truly, finally, glimpsing a whale in full glory. Like the busks she writes about—tiny missives carved into whalebone corsets by sailors—this book leaves an imprint."

—Lulu Miller, author of *Why Fish Don't Exist* and cofounder of NPR's *Invisibilia*

"With lyrical language, Giggs embarks on an underwater journey to uncover the place of whales in the chain of life. Her immersive exploration of varied species of whales illustrates the interconnectedness of all life and the ways human depredations reverberate from the smallest creatures to this largest of Earth's animals."

—*Library Journal*

"Brilliantly full of wonder . . . A series of essays that span aeons and vast amounts of space, from the bottom of the ocean to the far reaches of the solar system."

—*The Economist*

"Dazzlingly well-researched . . . combining reportage, cultural criticism, and poem as a call to action in the spirit of Rachel Carson, Giggs is an assured new voice in narrative nonfiction."

—*The Irish Times*

"As we grapple with our Anthropocene anguish, some of the most alive, inventive writing on the planet is nature writing, and Giggs's *Fathoms* is glorious proof. Ostentatious, mythic and strange, this is the kind of book that swallows you whole. Entirely fitting for its subject."

—Beejay Silcox, *The Guardian*

Fathoms

the world
in the whale

Rebecca
Giggs

Simon & Schuster Paperbacks

New York   London   Toronto   Sydney   New Delhi

Simon & Schuster Paperbacks
An Imprint of Simon & Schuster, Inc.
1230 Avenue of the Americas
New York, NY 10020

First Simon & Schuster trade paperback edition July 2021

SIMON & SCHUSTER PAPERBACKS and colophon are
registered trademarks of Simon & Schuster, Inc.

For information about special discounts for bulk purchases,
please contact Simon & Schuster Special Sales at 1-866-506-1949
or business@simonandschuster.com.

The Simon & Schuster Speakers Bureau can bring authors to
your live event. For more information or to book an event, contact
the Simon & Schuster Speakers Bureau at 1-866-248-3049
or visit our website at www.simonspeakers.com.

Interior design by Lewelin Polanco

Manufactured in the United States of America

1 3 5 7 9 10 8 6 4 2

Library of Congress Cataloging-in-Publication Data has been applied for.

ISBN 978-1-9821-2069-6
ISBN 978-1-9821-2070-2 (pbk)
ISBN 978-1-9821-2071-9 (ebook)

For Leanne and Tony

[FATHOM] 1. Anachronism: a six-foot quantification of depth or breadth, originally indexed to a fingertip-to-fingertip measurement (or "arm-span"), and accounting for spools of cordage, cables, cloth, or other materials; commonly used to demarcate the extent of a water column; 2. An attempt to understand: a metaphor for reaching out to make sense of the unknown. Ariel's song in Shakespeare's *The Tempest* (1610–1611) begins: "Full fathom five thy father lies; of his bones are coral made; those are pearls that were his eyes." Scholars have identified this passage as the origin of the expression "a sea-change"—a profound reversal of fortune, perspective, or circumstance from which there is no return.

Contents

PROLOGUE

Whalefall

A FEW YEARS AGO, I HELPED PUSH A BEACHED HUMPBACK WHALE back out into the sea, only to witness it return and expire under its own weight on the shoreline. For the three days that it died, the whale was a public attraction. Locals brought their children down to see it. Then out-of-towners came too. People would stand in the surf and wave babies in pastel rompers over the whale, as if to catch the drift of an evaporating myth. The whale was black like piano wood and, because it was still young, it was pink in the joints under its fins. Waves burst behind it, sending spray over its back. Every few minutes, the whale slammed its flukes against the wet sand and exhaled loudly—a tantrum or leverage. Its soft chest turned slack, concertinaed by the pull of the swell.

At first the mood was festive. People cheered each time the whale wrestled in the breakers. Efforts made to free it from a sandbar in the morning had been aided by the tide. That the whale had restranded, this time higher up the beach, did not portend well for its survival, but so astonished were the people in the crowd, and such a marvel was the animal, that hope proved difficult to quash. What the whale inspired was wonderment, a dilation of the ordinary. Everyone was talking about it, on buses and in corner stores. Dogs on the beach, held back by their own-ers, swept flat quarter circles in the sand with their tails. A few had their hackles up. How the dogs imagined the whale—predator, prey, or distant relation—was anyone's guess, but they seemed keen to get a closer look. At sunset, armfuls of grease-blotted butcher paper—chips and battered hake—were passed around. The local surf lifesavers distributed zip-up hoodies. Wildlife officers, who had been standoffish with the gathering crowd, relaxed and delivered some lessons on whale physiology.

"Whales are mammals," they began, "as we are too."

This surprised those people who were accustomed to thinking of all marine animals as types of fish. They raised their eyebrows and nodded along. *Cetacean*—from the ancient Greek *kētos*, made Latinate as *cetus*: an order of mammals that includes whales, dolphins, and porpoises.

"Under its skin the whale is wrapped in a subcutaneous envelope of fat called blubber," a man in khaki said, cupping his hands.

Trying to imagine the properties of blubber, I could only conjure the agar desserts sold in Korean supermarkets: opaque, calorie-rich, and possessed of a curiously unimpressionable tactility. While in the ocean, the whale's blubber insulates it and allows the animal to maintain a constant inner temperature. Out of the ocean, the blubber smothers it.

"This whale has the opposite problem to hypothermia," the wildlife officer explained. Though we were shivering, the whale, only yards away, was boiling alive in the kettle of itself.

That night, a group of us slept lightly in the dunes, arrayed like question marks and commas on the white sand. Our minds cast to the cetacean huffing beyond the swale, then swooped back into cloudier visions. Surfers arrived in the early hours. Bouncing down to the water's edge, they stood watching. I woke and brushed a second skin of pearly sand off my cheek, my shoulder, one thigh. Were those sharks, raiding a lux channel tipped up by the moon? Hard to tell. We resolved that the whale had been washed too high on the beach for any shark to reach it.

Rinsed by pewter light, every detail was particular and peculiar. Ridges in the sand. Plants like handfuls of knives. It felt cold. It felt cold, to us.

By sunrise, a part of the whale that ought not to be outside of it was outside of it. A digestive organ, frilled and bluish in the foam. The whale's billiard-ball eyes tumbled in its head, and its breathing sounded labored. The sharks slid into vapor, a squinting rumor. No blood on the tideline. People stayed back from the water's edge nonetheless. Swept slantwise, shallow waves smoothed, oversmoothed, smoothed. I palmed an unremarkable shell that sat for months afterward, furred with dust on a ledge in my room, until it was lost. A cordon was set up. Seagulls flew down to peck avian hieroglyphs in the whale's back, their inscriptions legible to yet more skyward gulls that dove to elaborate the wounds. At every nip the whale flinched, still intensely alive.

Walking off some agitation I'd accrued watching the birds, I found one of the wildlife officers crouched a way down the beach. A blocky guy wearing wraparound sunglasses, his jaw was set tight. The whale's central nervous system was so large and complex, he explained, that euthanizing it in the manner that one might kill a cow or an old horse was impossible. A bolt through the brain would take too long for the heart to register it. A shock to the heart wouldn't transmit immediate death to the brain. Exsanguination (opening the animal's arteries and leaving it to bleed out) could take many hours. A day, even. The volume of blood spread across the beach would be gory, epically so.

Talking with the wildlife officer, I began to think of the whale's body as a sort of setting in which dying could take place at multiple sites, over different durations. The animal, alive on a great scale, didn't die in an instant. Only parts of it did. The humpback's death wasn't, in a word, global. This was the kind of death people call "a death of a thousand cuts." The humpback's face—so much as any whale can be said to have *a face*, its eyes on either side of its huge head, its nostrils in its crown—did not agonize, grimace, or wince. Neither did the animal cry out in anguish. People on the beach took this for a dignified stoicism, though we were only familiar with the human cosmology of pain. It was dawning on me that, because a whale's body is attuned to its oceanic environment, and because it occupies such immense, physical dimensions, it might suffer uniquely, according to senses I then knew little about.

The wildlife officer told me there could come a point when strapping the whale with dynamite would prove the most humane option. The cleanup afterward—which needed to be thorough and hygienic if a whale had run aground on a popular public beach—was expensive. (How expensive? In time I'd look it up. Another humpback, found dead nearby, a few seasons hence, cost $188,000 Australian dollars (AUD) to remove. Biological "contaminants" sieved out of the sand had to be incinerated. Wires, chains, crane straps, and tarpaulins purchased for the task of transporting the dead whale were all thrown away. The local council and the state department of fisheries disputed which government authority should foot the bill: their remits extended to different varieties of calamity. "Because it's a mammal, not a fish, they believe it's not their jurisdiction," the mayor said.)

The wildlife officer and I stared out to the horizon, the sea mouthing our shoes. Then we walked up to his van. He wanted me to see the only other mercy he could bring to hand: an injection.

"It's called the green dream," he said.

The needle was near to a foot long, and as thick as a car antenna. A rubber tube ran to a pump container. The whole apparatus was reminiscent of something you might use to administer herbicide. A vivid green liquid swilled inside the plastic canister: the trademarked color of Fairy detergent and Nickelodeon slime. It might work, he speculated, because the whale was small, only a yearling. But you wouldn't want to get the dosage wrong.

If administered, the fatal chemicals would linger in the humpback's carcass long after death, and imperil the survival, too, of any scavenger that came to dismantle the whale and gather what could be picked off the bones. Spiny nibblers and jellied dabs that crawl. Feral carrion feeders, slunk in from nearby parklands. In one recorded case, a dog (the breed was Australian shepherd) fell into a coma after digging up and consuming a scrap of blubber from a whale killed twenty-three days previous, so enduring was the barbiturate drug in the euthanizing injection delivered to that cetacean.

The lesson here, the way I grasped it, is that what instinctively feels like compassion toward one creature can prove poisonous in the orbit of small and smaller organisms left lying out on the beach after we leave.

The officer let me hold the green dream for a minute, this ghastly prop, heavier than it looked. Whose was the dream? I wondered. I pictured the whale's many veins and arteries, which, if you could unpick them, would lead off more than three hundred feet down the beach— thinning to capillaries in the distance, like the red thread from a smashed thermometer.

I asked him, "Is it you who makes the decision?" I suspected he could get a gun instead and use that. I had heard he was empowered by certain regulations to fire on a suffering whale, as though it were a chassis on chocks in a paddock.

He held one hand, crablike, on the wet sand and said nothing. The whale weakly lifted its tail and dropped it again.

What would happen afterward: I wanted details, the process. The wildlife officer sighed. He described two mechanical bobcats assigned to collect the carcass. *Beach and bundle*, he called it, *the policy*. The whale would be chainsawed in half, it would be quartered and trucked to the Tamala Park landfill site, in Perth's Mindarie, to decay. I envisioned it jumbled in with household waste, amid defunct white goods and bags of trash; the skull, an upturned trough of spoil. After death, the whale's putrefaction would generate yet more heat, scorching its bones and turning its organs black within the tight bind of its innards. If no one cut the body open, it might explode. Other whales had before. Gases puff up cavities inside the carcass, straining against the fat. Did the council worry a whale's remains, towed back beyond the shallows, could bring thresher sharks and hammerheads out of the deeps to loiter where swimmers would, after a time, return? I was confused as to why the animal was destined for the junkyard, even if it didn't end up being given the death-dealing injection.

"This whale is malnourished," the wildlife officer offered, apropos of a question he was more routinely called to answer. "We're not sure why it stranded. Maybe it's sick; maybe the mother didn't feed it right as a calf. Maybe the whale ate something it shouldn't have, or it's got parasites, or it's too tired and ill to survive."

He cleaned salt spots from his sunglasses. I saw fatigue pleated around his eyes. "Killer whales pick off the weak ones," he went on.

The problem of the hour—what was killing the whale, now that it had beached—was gravity. The wildlife officer suggested I visualize the whale as see-through. He said to notice how the heaviest bones lay in the whale's topside; the big, leaden vertebrae, its ribs thickest where they met the spine. Buoyant in the ocean, it was no problem for the whale to be built this way. Even diving to great pressure, its weight distribution didn't trouble it. On land, though, its largest bones exerted a downward force on the animal's soft underside, causing crush injuries we couldn't see. "The chest wall caves in," the wildlife officer begun, but stopped himself from finishing the thought. Fragments of sea-foam spotted the whale, trapped in its tonnage on the wrong side of its known world. One final thing he would say on the matter: "There's an argument, a conservation argument, not to put a whale that's been weeded out back in again."

My mother, Leanne, grew up in a township on the southwest coast of Australia where mass strandings of smaller whale species were a feature of the local lore. With our many uncles and aunts, we had often holidayed on the white, overcast beaches of her girlhood; places where pods of pilot whales—both long- and short-finned species—were known to maroon themselves. One hundred and fifty such whales fetched up on the shore of Hamelin Bay together in early 2018, and all but six individuals died. I saw the whales, roughly half the size of humpbacks, thrashing and rigid, and seemingly beset with despair, in footage a cousin posted to our family group chat. Soundless videos; whales shrunk to fit by the dozen into the palm of my hand. Their dark shapes recalled fingerprints inked onto a rap sheet.

The name "pilot whale" comes from the notion that these animals are steered by a leader, though whether this is true has never been indubitably proven. Around Australia such whales are thought to be nomadic, rather than migrating with the seasons, as other species do. No one knows why they might be predisposed to come ashore on this stretch of coastline specifically. Eighty pilot whales stranded in the same region in 2009, when several were hefted into dampened slings and driven to a neighboring bay, in the belief this might reroute them back on their marine wandering. At least a third of those rescued promptly returned to the sand and died. The most daunting number, 320 pilot whales, beached in 1996—though more than 80 percent of this group survived, having been pushed into the sea quickly, by a team of volunteers at high tide.

Such an abundance: *three hundred and twenty whales*. An event like that you couldn't help but see as sacrificial or ominous. Malevolent even, in the way of a curse long since passed into rumor, carrying over to afflict a successive generation. The pilot whales brought to mind Renaissance frescoes of a world corrupted. The so-called "fall of man."

Why do the whales do it? There seemed no rhyme or reason in the conditions that preceded significant numbers of whales casting themselves onto land in the southwest. It happened some years, not others. Many of these animals died soon after they emerged from the ocean or kept on restranding with a seemingly fatalistic zeal; others, having been pulled out into the shallows, turned fast and purposeful for open waters. Biologists could not identify any discernible difference between the whales that insisted on survival and those that gave in and collapsed.

Beneath a brightening sky on the Perth shoreline people were posing for photographs in front of the beached whale. A mother stretched
the elastic strap of a sun hat beneath the fat, folded chin of her rankled
infant. "Get my neck," said a girl to her friend, who was absorbed in
dribbling lotion onto her own thigh. Then a group of teenagers came
down from the dunes with a wreath of plaited seagrasses and pink pig-
face flowers, and proposed laying it over the whale's forehead.

The spectators fostered their own suspicions as to why this young
humpback whale had drawn up on the sand. Hadn't a shooting star flared
icily over Rottnest Island last week? Astral debris was said to have sprinkled down over the Goldfields. Comets and meteorites were believed, by
many, to be connected to whale beachings, though few could say why—
maybe the animals confused night for day when stars fell, or changes in
the stellar positions led whales to misreckon their nearness to land. And
what *was* happening in the sea? The weather was undeniably weird, all
the time—wasn't that the truth? One woman's brother, "a serious man,"
had let slip mention of clandestine naval operations offshore. Military
sonar terrified whales. Oh no, its effects were *physical*. Thumped by in-
frasonic noise, whales bleed from their ears. (The humpback's ears? No
one on the beach could identify them.) Sour mention was made next of
whalers out of Japan, still hunting whales and flouting the consensus
that these animals warranted protection. Thousands of whales butchered
each year—thousands, truly—far from the public eye. Since the humpbacks summered within range of the hunt, someone pondered aloud if
this whale might be lost because a maternal bond had been broken: the
mother killed, perhaps. (Two tourists agreed this was wrongheaded; the
Japanese pursued smaller species of whale.) Besides which, didn't humpbacks feature in the ancestral Noongar stories? The yearling, thrown
from its element, would be a matter of concern to elders, surely. What
was happening was anomalous. A bad business for the region.

I distrusted the inflection in these voices even as I, too, brimmed
with guesswork, troubled by the whale from afar. Offered in candor,
these theories were nonetheless conspiratorial, being premised on the
assumption that deeper streams of logic undercut the frail authority of
science, and the wildlife officers' superintendence of the whale. The official narratives could prove no whale-beaching hypothesis to the crowd's
satisfaction. Their loyalty was to the unverifiable hunch, to intuited

patterns of allegory, augury, or plot. As if the whale itself, in its fleshly presence, testified to hitherto unplumbed dimensions of reality. Or so it seemed, as the sun found its zenith.

Every few minutes the whale emitted a louder rush of air that dried out to a wheeze, rubbly with unseen obstructions. It hurt to listen to; people felt it inside their own chests. A few families turned away. The surfers knelt in vigil or shame, their wetsuits half-peeled to expose tattoos of regional creeds and constellations, the hair on the backs of their heads like wicker. One woman broke free from the crowd and strode into the water with the wilting wreath in her fists overhead. She sung, bell-clear. Her skin was tanned. It took three wildlife officers to pull her off the side of the whale, kicking. She had spiritual reasons, she said. She had spiritual *skills*. Her fury wasn't dignified. It was incandescent. The whale never wore the sodden wreath.

Back then, though I, for one, had been hesitant to link the whale's demise to origins that were cosmic in scale or diabolical in character, the logic of the crowd's explanations struck me as important. That such immense and enigmatic creatures as whales should be governed by forces of equal mystery seemed apt. Apt, or perhaps the better word is "proportionate." Stood before the humpback, this otherworldly arrival, who among us could renounce, offhand, the probability of inscrutable and stupendous powers, active in the sea? Brought down before its time, the humpback vouchsafed, however tacitly, the existence of other phenomena beyond our ken. But later, when I searched online for the reasons whales decline and die, I saw that the top-ranking results were neither unfathomable in the way of star fields nor motivated by barbarous intent, per whaling fleets. Nothing extraordinary was returned. Nothing covert. Instead, I found myself scrolling through a list of quotidian conveniences, overlooked residues and debris; a spill of lowly, forgotten things, sundry as junk mail.

That was how I first learned about the sperm whale, washed up dead on the Spanish coastline with a greenhouse—an entire greenhouse—in its belly. In the flattened greenhouse—from a hydroponics business in Almería—were enclosed tarps, hosepipes and ropes, flowerpots, a spray canister, and bits of synthetic burlap. It had once sheltered off-season tomatoes, grown for export to Britain. High winds likely collapsed the structure, bundling it from dry land into the ocean. Flash flooding and storms of eerie, unexpected power were gaining in the region, known as the "salad bowl" of Europe. An inventory of the animal's gut itemized other indigestible objects too; alarmingly, goods of comfort and leisure. In addition to the greenhouse, the sperm whale had swallowed parts of a mattress, a coat hanger, a "dishwater plastic pot," and an ice-cream tub. Like a chamber furnished for a prophet or a castaway, these stomach contents recalled stories of people surviving inside whales, I thought. Old parables, anyone can tell, in which such traces are taken for proof of life. Here, though, the midden amounted to a cause of death. Domestic products were newly dangerous, seen from this angle. The banality of household items belied their potential for a gruesome afterlife. Abandoned fishing gear—drift nets, longlines, fish traps, and oyster racks—had long been identified as hazards for sea creatures, but whoever guessed that whales could, or would, eat bits of bedding and kitchenware?

Reading on, I came to understand that marine impedimenta, as a category, had expanded across the early twenty-first century to draw in consumer goods and the refuse of terrestrial agriculture in manifold forms. Whales assayed, revealed the extent. Because whales are so well insulated by their thick layer of blubber, they attract fat-soluble toxicants, absorbing molecular heavy metals and inorganic compounds that comprise pesticides, fertilizers, and other pollutants that have come to powder the modern sea. The body of a whale is a magnifier for these chemicals, both because cetaceans live a long time and because many species accrue a toxic ballast from the organisms they consume. Whales also lack a key gene that, in land-living mammals, functions akin to an antioxidant to neutralize low concentrations of organophosphate—run off from croplands and collected by the animals' tissues—and, unlike some seabirds, whales cannot shunt chemical burdens into feathers to be shed during a molt. Fractional exposure builds up over multiple seasons, making some whales more polluted than their environment. Which is so different to how we conceive of pollution ordinarily, I think, as pervading a landscape or being an atmosphere through which, and within which, animals move—each loaded with *less* malignancy than their surroundings. To view animals *as* pollution is both worrisome and novel.

Estuarine beluga in Canada had been discovered to be so noxious that their carcasses were classed as toxic waste for disposal. Scientists declared Earth's most toxified animals to be killer whales living in Washington's Puget Sound—a place starfish were presently being rent apart by a disease that induced their arms to crawl *off* their bodies ("Some locations saw complete mortality of sea stars," reported *The Seattle Times*). Residents of the Chukotka district of the Russian Federation worried over what they deemed "stinky whales": gray whales hauled as part of a traditional hunt that reeked when carved and jointed, and that triggered numbness in those who consumed the meat. A biotoxin was thought to be to blame, most probably. These gray whales, caught near the Bering Strait, had apparently begun to eat seaweed and decomposing fish from the seafloor, maybe because populations of their customary food, amphipods—bumblebee-small crustaceans that dot-dot-dash the water column—were waning. (Though, too, some scientists pointed at spillages from wells and tankers in the Alaskan oil fields as a possible source of chemicals in the whales' malodorous flesh.) "It has a medical smell, like iodine," reported the Russian deputy commissioner to the International Whaling Commission (IWC). As "when you enter a pharmacy, for example, but it's of course stronger."

Of the industrial substances corralled in whales' bodies, not all were agrochemicals incipient in seawater and prey. Being surface breathers, whales also inhale airborne carcinogens, including cadmium, chromium, and nickel, emitted by the world's refineries and chrome-plating factories. The biggest cetacean species possess Earth's most colossal lungs, and draw the planet's deepest breaths. They sometimes hold that air for record lengths of time, beyond two hours. Subject to depth pressure underwater, abundant oxygen is pressed out of the respiratory system to saturate whales' muscles—making them especially prone to being permeated by atmospheric contaminants. Chromium in endangered North Atlantic right whales, for instance, had been found to match the levels of factory workers employed in metal dipping.

Prior to learning this, I had supposed that very large animals would be less vulnerable to contaminated air than smaller animals. Because these tainting gases are mostly imperceptible, I assumed they lacked potency. I took it for granted that the quantity of pollution afflicting any animal depended only on the chemical profile of its environment (and not how its body functioned to soak up and stockpile the poison). I saw now that this belief was erroneous. Size itself, along with physiology and habitat, turned out to be jeopardizing.

How much amounted to a harmful dose? Were the effects of pollution on whales distinct from its effects on other mammals and sea creatures? Theories varied. A minor reassurance: when contained in blubber, scientists believed that any cache of chemicals (whether inhaled, digested, or absorbed) remained "metabolically inert," which is to say, the toxicants couldn't injure otherwise healthy whales because the substances weren't metabolized and recirculated through the animal's organs. The threat comes when a whale begins to starve and its body reverts to ketosis—breaking down blubber for energy in the absence of food. Released back into the bloodstream, stored toxins then cease to be dormant.

The humpbacks seen from Australia's coastline carry lower accumulations of synthetic chemicals compared to whales that live, year-round, near industrialized ports and along highly trafficked seaways, but because humpbacks tend to fast when they migrate up from Antarctic waters—relying on their blubber as a camel, crossing the desert, does its hump—these whales may be at higher risk of reexposure to adulterants they have acquired: a seasonal poisoning from within. Stranded whales, too, often undergo ketosis. The sublethal impacts of even low levels of industrial pollutants on a whale's health, and behavior, are poorly studied, being difficult to monitor.

In Europe, around the western Mediterranean Sea, the southwestern Iberian Peninsula, and elsewhere, I read that whale species were found to be riddled with an especially wicked class of chemicals, to pernicious consequence. Polychlorinated biphenyls, PCBs—once used in coolants, concrete, paint, light bulbs, and electrical capacitors—are persistent compounds that, having entered the ocean through stormwater and waste, can take many decades to break down into benign molecules. Though these substances were phased out by governments in the 1970s and '80s, PCBs have remained durable in the ecosystem. One site where they have intensified is within the bodies of killer whales—iconic, black-and-white whales, apex predators also known as orca. Research models released in 2018 foretold that all killer whale populations offshore from nations that had used PCBs would die out in thirty to fifty years. The species would only continue to thrive, the report concluded, in the Arctic and some parts of the high seas.

From a detailed article on the development of PCBs and other benzenes, I discovered that these widely used, artificial compounds—eventually extracted, at scale, from coal tar—were first engineered by chemists as isolates from gases rendered out of whale oil. This was during the era when whales were a global commodity and a proto-energy industry—their fat sheared off by whalers and distilled to light lamps, grease machines, process textiles, and fuel the late stages of the industrial revolution. What a cruel and intimate historical loop: whale bodies provided the base chemistry from which the precursors to PCBs were extracted, and now, so many decades later, the legacy elements of these substances came to rest and accumulate in the living animals.

I thought of the humpback in the dump. The whale as landfill. It was a metaphor, and then it wasn't.

Female whales shed some of their toxicity to their calves: during pregnancy, through the placenta, and then in their uncommonly creamy milk. The firstborn calf, most of all, arrives seeded with iotas of human industry because it is subject to the mother's lifetime load; subsequent calves benefit from the birth of the first as though it were a kind of live sequestration. Yet most of the killer whales dying now, off Scotland, Gibraltar, Brazil, Japan, and in the northeast Pacific, had not been born—or were even conceived—when governments acted to ban the production of the persistent pollutants, PCBs, years ago. Oceans bank the emanations of manufacturing long after laws and technologies improve, or our industries move on to more efficient, less noxious methods of fabrication. Substances that have been made illegal, or those that have simply been replaced by innocuous substitutes, in Australia, America, and the wealthier countries of Europe, may remain in production elsewhere; a lag time that results from investment in capital assets such as factory equipment and legal property, including patents over manufacturing processes. In this way, the past is as unevenly distributed as the future.

Though I had started out seeking answers to how and why whales died, what had begun to click into place was this: my entire definition of pollution demanded revision. Some pollutants were still sold by the barrelful, emblazoned with a skull and crossbones; some were vented from smokestacks as haze. Others had long ago been built into the hardscape of cities (like PCBs embedded in ceiling insulation and coating electrical cables). If it was a surprise to find these familiar, albeit insidious, contaminants seeped into the remote environments of whales, then it was a shock to hear of domestic objects (tubs, hoses, coat hangers) infiltrating the animals themselves. Such items were not chemical *by*-products, they were *the* products; end products. Consumables. Hygienic, durable, and disposable: for many people, plastic meant safety. That safety, coupled with low cost, had resulted in plastic filling every hollow of suburbia: cribs, cars, kitchens. Though the qualities that primed plastic for home amenity hadn't materially changed, having departed the orbit of human use to tumble through the world as waste, plastic was recategorized a pollutant. From "consumable" to indigestible. Safe to lethal. Bedeviled by this future, each mundane commodity on the shop floor, however nontoxic, called you to envision not just any pollution spawned by its manufacture but the pollution that it, itself, might in time become. Even perishable produce, stacked in supermarket vegetable bins, relied on unseen plastic to foster its existence: rippling hothouses and ripening sacks tied over vine fruit, irrigation tubes, and transportation cushioning. Where did it all go? That the simple fact of a wintertime tomato radiated complicity in the death of a sperm whale seemed, to me, at once monstrous and bewildering.

Plastic and toxicants found in whales originated, unmistakably, out of the machine-turned world, but the foul flesh of Russian gray whales hinted at more protracted kinds of culpability. That whales could be eating an unnatural, biotoxic diet of rotted fish and plant life indicated a changing food web. Was human activity not also, in a more remote way, accountable there (having influenced the availability of the whale's natural prey, the amphipods, through climate change or otherwise)? Misplaced or misprocessed in an ecosystem, did organic matter, too, rise to the level of a pollutant? Why *had* whales suddenly started eating all the wrong things?

Just as my understanding of what a pollutant was came undone, so, too, did reading about whales challenge my grasp on how pollution was disseminated. A few types of old chemicals didn't dissipate over time but concentrated in places far from where they'd been deposited, transmitting even to unborn animals, untouched by any environment outside their mothers' interiors (this, the curse skipping a generation).

Having never opened their eyes and as yet unbreathing, these fetal animals nonetheless bore the trace of our terrestrial past on a cellular level—more so, even, than their immediate ancestors: those animals exposed to the pollution at the time it was generated. Lately, I doubted the exteriority of pollution altogether, for even in a pristine, natural setting I understood that a toxicant might be reemitted from a place *between* an animal and its habitat: the whale's swaddling blubber. It mattered not just where pollution came from, and how much of it there was, but what sorts of bodies received it, and how they passed it on.

What befell any organism was a function of its distinctive physicality. The gravity of being stranded on land, I knew now, could kill a whale, and air temperature could overheat it: these were natural phenomena. Unnatural presences, also, filtered through animal bodies. To fully comprehend the degree to which any environment was damaged, you needed to consider the ways in which it was damaging from the perspective of other species—from inside the sensorium of animals.

Whales may be subject to pollution, but for people who eat whales, the cycle can sometimes turn backward, bringing what has been lost, forgotten, or prohibited to indwell the human body. Greenland's Inuit women, who seasonally consume whale meat, whale skin, and fat as traditional food, had been warned off eating beluga during pregnancy and advised to stop nursing their babies altogether. Their mammary tissues had become a locus of concentration for the chemical by-products from the whales, because the composition of human breasts, being spongy and replete with estrogen receptors, make these body parts prone to act as dumping grounds for many types of transportable chemicals. The Inuit women may live in some of the most isolated and least industrialized regions on the planet, but sustaining themselves on whales had turned their bodies into habitats of contamination. According to the BBC's *Planet Earth: The Future* (2006): "If her milk was in containers other than her breasts, she would not be allowed to take it over state lines." Nearly all the Inuit tested had levels of mercury and organochlorines that exceeded World Health Organization standards. Their levels proved comparable to those of people living downstream from goldmines in China and South America.

As I absorbed myself in the research, whales were making visible something that had been invisible to me before: how regular human life seeped into the habitus of wildlife, and how wildlife returned back to us, the evidence of our obliviousness. But if whales brought to light the imperishable past and its unforeseen effects, then what bubbled up in my mind now was a question more difficult to quantify. Did whales also have something to teach us, I wondered, about our capacity for change?

During the weeks that followed the humpback beaching in Perth, I found myself unhappily preoccupied. There was an emergency out there—in truth, all of us had heard news of it. The superabundant cyclones that barreled down corridors of unseasonal warmth. Hundred-year storms on annual rotation. Die-offs, dead zones, and reefs rotted to the color of old money. Who hadn't yet seen that abysmal picture of the tin can, spotlit by a submersible in the silt of the ocean's deepest trench, or the other one: the photo of the seahorse latched on to a floating Q-tip? Seascape—the obsession of a golden age of painting, and once *the* saturnine vista with which to dramatize the psyche—had since reverted to kitsch: a mixed-media project, churning found objects. Every thing a foreigner in its own home. What was lost, if you took the time to think about it, was the timelessness the sea had always stood for.

Word had it that the seawater itself had begun to acidify: a change too subtle to taste, smell, or touch, but staged across the breadth of oceans in tandem with rising carbon dioxide (CO_2). As the oceans took in more CO_2 from the air, their baseline chemistry shifted. Marine acidification verified what seemed a very ancient fear: that even as what was coming on promised to assume the dimensions of a vast and totalizing phase shift, it unfurled presently, on a molecular and insensate scale. Would we know it, the moment when it became too late; when the oceans ceased to be infinite?

My mind returned to the stranded whale. All my life, I'd heard the history of whales told as a tale of victory. Notwithstanding the censured Japanese whalers, or those few lawful hunters in First Nations' territories, that these animals had been saved was a celebrated conclusion. Over three decades ago, or longer: "Save the Whales," the faded bumper stickers. See: whaleboats ratcheted into dry docks, harpoons disabled. Whales had since rebounded. Humpbacks and sperm whales were no longer red-listed as endangered animals. In many places (though not for all species) cetacean numbers were on the uptick. Brought back from the brink of vanishing, their populations testified to the dénouement of commercial whaling and the stewardship of conservation groups. Whales buoyed hearts. Whales were a wellspring of awe. How hungry we were, now, for awe! Whales elicited our smallness set against the largess of nature: they proved nature's sovereignty and its resilience. Whales gave people cause to reflect, too, that governments had been known to be benevolent, that industries could be restrained, and that the protection of wonderment was a value shared across the planet. So it was that whale watching surfaced feelings of humility and mastery both, for though it was humbling to be faced with such astonishing animals, that whales existed at all was due to past endeavors thwarting their extinction.

Whales were how the Western environmental movements first learned to tell a story as big as the world. The anti-whaling campaigns of the early 1980s had been predicated on the idea that whales should be viewed as the universal inheritance of all humankind and that the people of the future, regardless of nationality, deserved to live on a planet that hadn't been denuded of its largest animals. But now that the sea had evolved into an amphitheater of hazards far less demarcated than whaling ships, people could no longer view whales the way they once did—a triumph of activism, a thrilling brush with wildness. This, to me, felt like an urgent subject: Beyond the number of whales in the ocean, how did whales live, and how did they die? How did they encounter us? To learn that even whales—these paragons of green devotion—were turning up containing industrial toxicants, plastic, and pesticides, seemed darkly momentous. Apart from the visceral harm done to the animals themselves, their symbolism lay ransacked. A kind of *hope* was being polluted. Whales demanded a new story, I thought. A story that overrode the narrative of success I'd long been told, even as it promised to be just as planet-sized.

For as long as there have been humans, the whale has been a portentous animal. A whale warrants pause—be it for amazement or for mourning. Its appearance, and its disappearance, are significant. On the beach, an individual whale's death may not prove "global" in the way of its body powering down abruptly, like a switch being flicked, but in a different sense the deaths of whales today are global. The decline of a sperm whale—filled with sheeting and ropes, plant pots and hose-pipes—belongs to a class of environmental threat that, over the past few decades, has become dispersed across entire ocean systems, taking on transhemispheric proportions. This whale's body serves as an accounting of the legacies of industry and culture that have not only escaped the limits of our control but now lie outside the range of our sensory perception and, perhaps even more worryingly, beyond technical quantification. We struggle to understand the sprawl of our impact, but there it is, within one cavernous stomach: pollution, climate, animal welfare, wildness, commerce, the future, and the past. Inside the whale, the world.

There is a different story that I caught on the beach that day—one about whales that expire very far out to sea, perhaps of old age or ship strike. More than any other explanation of how whales perished, this rendition would stay with me, transfix me, long after I returned from the shore-front, though the reasons why took a long time to clarify.

Here is how that story goes: If whales that expire mid-ocean are not washed into the shallows by the wind and tides, their massive bodies eventually sink and simultaneously decompose on the descent. This disintegration is called a whalefall. Afloat at the beginning, they are pecked at by seabirds, fish, swimming crabs, and sharks attracted by scent trails. Carrion eaters debride the underside of the carcass. In calm weather, ripples divulge the scavengers' thuggish toil—creating the illusion, perhaps, that the dead whale still trembles. This part takes weeks, a month. Over dusk's shift-change hours, daylight creatures rotate with nighttime meat eaters. Some species of cetacean turn out to be more buoyant than others. Deceased sperm whales will hang off the oil-filled chambers contained in their huge, blocky heads at the surface longer than most, though they are one of the largest and heaviest types of whales. But, in time, any whale will go down, all the way to the seafloor.

A dead whale slips below the depth where epipelagic foragers can feed from it. The whale's mushy body decelerates as it drops, and, where pressure compounds, putrefying gases build up in its softening tissues. It drifts past fish that no longer look like anything we might call fish but resemble instead bottled fireworks, reticulated rigging, and musical instruments turned inside out. The whale enters the abyssopelagic zone. No light has ever shone here, for so long as the world has had water. Entering permanent darkness, the whale passes beyond the range of diurnal time. Purblind hagfish slink: jawless, pale as the liberated internal organs of other animals. Jellyfish tie themselves into knots. The only sound is the scrunch of unseen brittle stars, eating one another alive. Slowly. It is very cold. Hell's gelid analogue on Earth. The hagfish rise to meet the carcass and tunnel in, lathering the passages they make with mucus. They absorb nutrients right through their skin.

The whale body reaches a point where the buoyancy of its meat and organs is only tethered by the force of its falling bones. Methane is released in minuscule bubbles. The ballooning mass scatters skin and sodden flesh below it, upon which grows a carpet of white worms waving upward, like grass on its grave. Then, sometimes, the entire whale skeleton will suddenly burst through the cloud of its carcass. For a time, the skeleton might stay hitched to its parachute of muscle; a macabre marionette, jinking at the spine in the slight currents. Later, it drops, falling quickly to the seafloor, into the plush cemetery of the worms. Gusts of billowing silt roll away. The mantle of the whale's pulpier parts settles over it. Marine snow—anonymous matter, ground to grit in the sun-filtered layers of the sea—sprinkles down ceaselessly. The body is likely to settle far deeper than the depths to which any living whale will ever descend.

Rattails, sea scuds, other kinds of polychaete worms, and eelpouts appear. No one knows from where. Opportunist octopuses bunt between ribs. Sightless, whiskered troglodytes, like ginger tubers, burrow into the surrounding sediment, which is blackened with fat and whale oil. From the dark come red-streamer creatures that flutter all over. Colorless crabs: their delicate gluttony. Life pops. It is as though the whale were a piñata cracked open, flinging bright treasures. On the body gather coin-size mussels, lucinid clams, limpets, and crepitating things that live off sulphate. More than two hundred different species can occupy the frame of one whale carcass. A pink, plumed tube shrinks back into the gothic column of its name, the *Osedax*—Latin for "bone devourer." Mouthless and gutless, the *Osedax* is nonetheless insatiable; it eats through its feet, which extend, like trickling roots, into the marrow.

Some of the organisms that materialize on the whale are called "fugitive species." Some live nowhere else but in dead whales, and a few are so specialized they thrive only within the remains of a single cetacean species. Others are found, very occasionally, at hydrothermal vents or around briny cold seeps on the seafloor—spots where life on Earth is theorized to have first begun, with a plethora of millimeter-high animals inhabiting a thin band of gas-enriched water. After the whale's soft tissues and cartilage are consumed, these tiny organisms broadcast their larvae into the sea to drift in dormancy—infinitesimal, barely perceptible, and hopeful (if the larval can be said to be hopeful) of finding more dead cetaceans. A whale body is, to this glitter splash of biology, a godsend, and an occasion for gene exchange. To think such extremophiles indestructible—too ancient, or too deep to be affected by the impoverishment of the sea above—is to disregard their interaction with the corpse whales, which function as engines of evolution, and stepping-stones for their migration between stringent, oxygen-poor biomes. Without whales, many kinds of detritivores fail to colonize new habitats. When their vents and seeps deplete, their kind will decline. These creatures exist, they have evolved, because of the fall of whales. Whales as transient, decomposing ecosystems that amass, pulse, twitch, and dissolve.

No whalefall had ever been seen by a human eye until 1977, when, in the course of submersible training, a US Navy crew discovered a gray whale skeleton laid out on the stiff clay of an abyss west of Santa Catalina, more than four thousand feet down. After that, scientists deliberately sank cetacean carcasses to track what creaturely communities transitioned on and around them. One estimate holds that there may be as many as 690,000 whalefalls coming undone at this very moment. Over half a million dead whales lying on Earth's basement floor, jiggling with life.

The remaining bones of the dead whale on the seafloor are stripped, hollowed, and then they fluff up with flounces of silver-white bacteria so that it appears as if the skeleton is draped in downy toweling. There the bones stay, lashed softly by microbes. Decades may pass, a hundred years even, before nothing remains—only a dent that holds the dark darker.

In undersea sites bereft of seasons (as we are wont to understand the seasons), a whalefall is tantamount to springtime—a fountain of life; spectacular, then squalid. A whale in the wild goes on enriching our planet, ticktocking with animate energy, long after its demise. So the death of a whale proves meaningful to a vibrant host of dependent creatures, even as it may look senseless from the shore. The story of whalefall: I found it *emotional*. Whalefalls outvied the crowd's loss, describing instead a great, pluripotent detonation of life striking from a whale's demise. In the flatlining of a whale, in the falling apart of its colossal body, this story seeded the rise of organisms more spellbinding and weirder than any I had ever heard of, or dimly pictured before. How little is yet known about the wildness that attends the whale, I realized, and how well the world is built to work without us.

But what message should we, who never venture to these depths, take from the whalefall—what does this story boil down to? What I carry forward is this: Nothing ends without adding vigor to the conditions under which new beginnings are conceived. No state is condemned to be changeless. As would reveal itself to be true of my own contact with the stranded humpback in Perth, the death of a whale can prove not a tragedy but a turning point.

Low tide arrived toward the close: a small group of people persevering. I shuffled in to hear the humpback's irregular gasps. Whales are conscious breathers, which means they have to remember to do it. The whale's eye—the color of midnight, mid-ocean—had no eyelashes and, according to another wildlife officer, no tear ducts (for what would be the point of crying in the ocean?). I could not catch its gaze. Did the whale see where it was? Could it grasp the destiny being prepared for it: Not the plenteous, oceanic dark but the rubbish heap? Despite the whale's prodigious size, we understood that it shared certain traits with people: whales being social animals, affectionate with their young. We'd heard whales possessed brains complex enough to imply the capacity for abstract thought, native concepts of time and self. And, maybe, death?

If it was hard to tell all the ways the whale was hurt, it was harder yet to imagine if, and how, the animal *characterized* its suffering—whether it understood its injuries as a prelude to the end, being capable of anticipating that mortal moment. Was the whale afraid; did it cling to a twist of fate? Did the whale suffer mentally, conceiving the impossibility of a future? No one could know. I still held out for a flash of recognition, some emergency beacon flicked on within the whale's brain. But it showed no sign of apprehending the presence of people, our effort or our grief.

What felt important, in that moment, was seeing this thing through to its resolution. Agreeing not to leave the whale alone. Fealty. Fellowship. Or more than that: *kinship*, I guess, was what we tendered. Who could say if it was more or less welcome than the shot of barbiturate still packed up in the van? The wildlife officer had opted not to use it, or any of the grisly methods in his arsenal. No one clicked a cartridge into a rifle or brandished a stick of explosive. I sensed the thin border between hospitality and hostility; how it wavered but held. Nature, as they say, would run its course. That was a phrase we trusted. We passed it back and forth, hand to hand, between each other. We pressed that line close and thought that, because we were humans, it might still be possible to be humane in ways other species couldn't be.

All throughout those closing hours I was dogged by the uncertainty of the obligation enlivened there, on the Perth beach, and to whom, exactly, that obligation was addressed—to the whale, its kind, its ecosystem (*our* ecosystem)? My unease lingered on and later changed shape into a kind of tension snapped taut between the mercy of the green dream, toxic as it was, and the unintended cruelty of the greenhouse, swallowed whole by the sperm whale. Did we owe whales greater distance or more intervention? And who was the "we" in that sentence? At the very least, I reasoned it included myself and the crowd drawn down onto the beach by wonder. The duty of awe was—wasn't it—care?

Inside the whale it got hotter, though how that was happening proved difficult to envision. People, I think, tend to devise death as a gradual loss of heat; the gleam retracted from every corner, pulled to a wick within, a guttering out. Evicted, the human body turns cold. The whale's descent was different. The whale was burning up, but we could not see the fire.

I started talking to the whale; saying soothing, hackneyed things. I had an idea of each sentence as I spoke it, round and cool as a river stone placed onto the whale. But what did the whale understand by my voice? A germane sound, inlaid with comfort, or just noise, background babble: as the wind speaks in the trees; as dogs bark, being tugged away by their owners on leashes. Do human voices sound as ethereal to the whale as whale voices sound to us? Or do we scratch and irritate the whale, a pin in the ear?

I put one hand briefly on the skin of the humpback and felt its distant heartbeat, an electrical throbbing like a refrigerated truck, sealed tight. I wanted to tap on the outside of the animal and whisper to it: *Are you in there, whale? Neighbor, is that you?* Life on that scale—mammalian life on that scale—so unfamiliar and familiar in turns. Oh, the alien whale. The world-bound whale. A stranger inside. I hated to watch it.

The occasional plosive rush of air, less frequent. Then, the mumbling of the tide in the tiny bays of the sea.

Why we seek contact with animals, and what we believe they give to us: these queries seem, to me, to have gained significance in recent years. Encountering a whale can occasion a rush of awe—suggesting, as it does, novel worlds hidden within our world. Animals have different ways of living with, and within, nature, being equipped with faculties unlike our own and moved by impulses we cannot resolutely apprehend. The limits of human imagination were never more concrete than in the seconds that pass, eye to eye, with another sentient organism. Why *does* the animal do what the animal does? The animal's inner life is a closed book. Yet in attempting to map the varieties of intelligence and ingenuity that make a home on this planet, we enchant ourselves to think that there are more dimensions to this world, and wilder ways to experience it, than we have the scope to fully comprehend. After its lifetime, the unraveling of a whale—its enchainment all the way down to blooms of outlandish life on the seafloor—remains a source of wonder. Animals embiggen our existence; they enliven our sense of mystery.

But awe and wonder are not the only emotions we experience when we face up to the fauna we expect to share a future with. Something else has arisen between us. There is a kind of hauntedness in wild animals today: a specter related to environmental change. In antiquity people corresponded with spirits that took animate form, spirits that came from other realms and dictated moral lessons. But what is most cryptic about animals in this moment, many people suspect, are things we cannot yet gauge about our own impacts on their habitats and their bodies. The ability to identify damage is lapped by the ability to *do* damage. We stand in the wake of something we cannot yet comprehend. What is at the heart of an animal, governing its behavior, may reveal itself in time not to be a compelling mystery but, instead, a shameful familiarity: the rubble of the marketplace (still displaying its trademarks), or the polluted air. Our fear is that the unseen spirits that move in *them* are *ours*. Once more, animals are a moral force. We look to wild animals to see the history of our material intimacy with remote places and the outer edge of our compassion. We look to animals, too, to see how we might survive the world to come and how to cohabit with other creatures there.

This book tackles what it means to pollute not just places but organisms; and then not "just" organisms but *beings*, a category of creatures to which we have granted a central place in our imaginaries—whales, which we have often projected human qualities onto, and personified, but which are also responsive, cognitively sophisticated animals, perhaps even capable of configuring our relation to their social worlds. It is thus an exploration of obligations to another animal, beyond attempts to merely preserve their species: what kind of sensory realities do we want to protect animals from, what sort of lives in the wild do we want to ensure. This book looks at how contact with whales—through tourism, captivity, and via the media—has provided a remedy to people's feelings of malaise about the natural world, and it asks what it means to desire connection, at a time when connection can entail damage to animals, or expose us to grief. The bid this book makes is for the potential of a scientifically literate imagination to allow us to better understand the sensoriums of other species, to gauge the true extent of the changing environment from perspectives other than our own. But it is also a book about where we find hope today, how we control ourselves, where commonalities can be sought with the past, and between cultures, and how to remain compassionately engaged with distant, unmet things.

Across the course of making this book I ranged between the high art of museums and the low culture of selfies; I looked at whales cut into rock faces, whales rendered in exhaustive detail by cameras built for photographing the planets, and whales that scarcely resemble whales at all, hand-drawn on ancient maps. I encountered real whales too. I learned about species abundance, defaunation, endangerment, parasitism, hybridity, evolution, and about how extinctions can be triggered without us ever meeting the disappeared species. I learned about the sorts of whales we never see and why that might be so: I learned of the whale that has no name, the whale with two voices, whales with two pupils in each eye, and whales puppeted by storms on the sun. I discovered that whales have been the subjects of cuisines and conspiracies, that they have housed monsters and do still. I learned that we change the sounds of whales even where we do not make a noise, that humpbacks have pop songs, and that beluga have tried to speak human tongues. I learned about whale vision, biosonar, and memory: human grief, human love, and interspecies recognition. I set out to draw a few lines between myself, the stories I knew about whales, and the science of our changing seas. By the time I came to the end, I understood that these connections were far from esoteric concerns. Whales, I saw, can magnify the better urgings of our nature and renew those parts of us that are drawn, by wonder, to revise our place and our power in the natural world.

01

Petroglyph

Zoomorph at Balls Head — Dune Amplifier — Whale-Bone Shelter
— Dolphin Teeth, a Bride Price — The Whale on a Walrus Tusk —
Bangudae — A Refuge for Ghosts, Is It a Refuge for Hope? — Viking
Chess — Seal Balloons — Harpoon — A Basque Code: Colored Flags,
Burning Straw — Whalers Write Over the Yaburara — Whales, Do They
Travel the Ocean in Veins? — Expanding Regions of Whalelessness
— A Whale-Furnished World — Oil — Boiling the Books — Women
Shrunk — Busk — Whale Piano — A Cure or a Grave — Leviathan
— Did the Other Animals Make It Transparent? — Thom van Dooren
— The Too-Muchness of the Flower — Killing All Around Them —
Concealment — Why Did a Second Surge Occur after Fossil Fuels? —
The Whale Claw — A Listening Prey — Soap and Butter and Bombs
— Stratospheric Whales — Whaling Olympics — Minke-Fed Mink
— Urchin Barrens — Animals in the Atmosphere — Voiding — The
Whale Under the House — Shallow and Getting Shallower

How deep does the history connecting whales and humans run? When I think of what passes between whales and people, of what can come to intervene between a whale and the world, then I want to go first to the outline of a humpback that has been surfacing, for several thousand years, out of the ground within earshot of Sydney Harbour, at Balls Head.

This whale is a petroglyph—an engraving made by nicking a motif into the bedrock using smaller, harder rocks, or shell styluses fashioned for the task. Artworks, not documents, are where the oldest stories of whales are kept; geology, the most persevering canvas. Such depictions go back to before writing, to a time when the landmass known as Australia was not superimposed under that name, though it thrummed with countless languages. Petroglyphs are a kind of rock painting, or, more accurately, because they exist in three dimensions, they are sculptures: land art at fingertip depth. Chipping, scraping, and circling inaugurated the line: the whale at Balls Head was communal work, completed near to where living whales might be spotted. Enclosed within the hoop of the humpback's twenty-foot expanse is a smaller figure: a man who is headless. Faintly, a third creature, a diminutive horse or perhaps a dog, appears next to the whale's right flank. Dry leaves swirl across the tableau. Somewhere, a chain-link fence tinks. The whale in the rock was once the work of the Cammeraygal people, in what is the Eora nation.

So far as we are yet aware, the desire to represent animals is a desire we share with no other animal. To date, no statuettes of people fashioned by nonhuman animals have ever been unearthed; no dens have been excavated on the walls of which appear friezes of us at play. Capturing the likeness of another creature implies an imaginative or emotional relationship that exceeds the exigencies of survival: the instinct to fear a predator, or an appetite that hungers after prey. With no way of telling what emotions stirred in those Cammeraygal craftspeople, did they lower their tools to listen, if tail flukes slapped the sea beyond? From stone blades to finer styluses, adzes and awls, a sequence of implements released the graphite scent of their mineral worktop to glitter and twist around them. The petroglyph would have taken months to lay down, over seasons of fire, then pinwheeling frost. Stars nailed high in the sky, the rock face humming, warm with friction, was it *awe* that stilled the engravers there, when real whales passed by their semblance?

Zoomorphic petroglyphs populate sandstone platforms across the Sydney basin. Hundreds of designs—gestural emus, snakes, kangaroos, possums—range from outcrops in national parks, down the coast, and into the city. Many sacred, some secret, all the petroglyphs are culturally vital in the knowledge systems of Indigenous people. Those petroglyphs that are public crop up on cliff tops, in bushland, between suburban buildings, in golf courses, along rivers and roadsides, and at waterfronts. The biggest animals depicted in a statewide menagerie of petroglyphs, whales feature prominently in the stone gallery.

Their living counterparts—humpbacks, right whales, sperm whales, and killer whales—had a multidimensional function in the lives of the engravers: as food, and part of a totemic exchange. All across the ancient world people are known to have pursued whales for sustenance and raw resources. Fresh or salted, whale meat was sometimes gifted, sometimes swapped for other goods. The Eora tell of ancestors twisting their feet in the fine beach sand to transform a squeaking dune into a towering amplifier, emitting sounds analogous to whale calls. Less a hunt than a trap, this was just one means to draw sick or injured cetaceans to strand—but whether a whale was enticed into the shallows by imitating its voice, or if, by happenstance, it was discovered beached, it was always a relished windfall. Not a tragedy but a boon: a whale was a storehouse of fats and vitamins otherwise scant during wintertime.

Worldwide, both the killing practices and the function of whale parts varied, culture to culture. In the Great Australian Bight there remain vestiges of whale rib bones lashed together to construct beehive-shaped structures, not dissimilar from wigwams. Dugout canoes were used in the Solomon Islands to drive dolphins onto land; dolphin teeth were a bride price on the islands. Sperm whales were speared off Zanzibar during an era of mud houses. In the Pacific Northwest—where tradition mandated a tribal whaler needed to dream the whale before it could be glimpsed awake—lances might be weaponized with blades of elk horn and serrated mussel shell; whales died there having had shards of other animals driven into them. Darts tipped with juice from the lethal root of the monkshood flower killed right whales, slowly, off Kodiak Island. The Nuu-chah-nulth of Canada vowed celibacy before a hunt and took ceremonial baths. Early whalers from Iceland and Norway dispatched minke whales using *dödspiler*, "death arrows" rolled in the necrotic tissues of other dead and moldering animals, causing the whales to die of septicemia. For reasons having more to do with ritual, the Aleut dabbed menstrual blood on their whaling spears and relied on the magnetic allure of amulets, with whale pupils at their core (as did the Chukchi). The skulls of infant bowheads are found, today, in the Arctic ruins of the Thule (their stone-barbed pikes were capable of felling smaller whales, but not full-grown adults), and, less frequently, a human skeleton is exhumed there, too, perhaps a warrior, interred ceremonially between the brackets of a pair of whale jaws.

Among the oldest pictures depicting encounters with whales, many are object lessons on how to bring the animals down, and what to do next, to get them ashore. A three-thousand-year-old walrus tusk, found in Russia's Chukotka, showed scratchings of tiny figures aboard an umiak—an animal-skin boat—unmistakably chasing a whale. Archaeologists noted that whales still come into the freezing beaches there, near to where the artifact was recovered, to scour barnacles off their flanks on the shingle underwater (itself, a kind of tool use).

The earliest incontrovertible evidence of whaling—dating back possibly eight thousand years, to the late Neolithic—comes from South Korea, where, on a shale facade alongside a river in Bangudae, near Ulsan, whale outlines are lassoed by figures standing on a curved pontoon. Other whales, already dead, appear to be kept buoyant by makeshift floats; their backs are marked with dashed lines for division, as on a butcher's chart. The whales have spouts and belly wrinkles. Showing hundreds of animals, the Bangudae rock faces bustle with life: pods of whales are flanked by turtles and porpoises; on the perimeter are wolves, deer, pigs, and fishpounds. Despite the archaic quality of these renderings, most of the whales are identifiably the kind South Koreans call "ghost whales" (*gwisin gorae*)—a name derived from the animals' coloration. Hovering between black and white, they are known elsewhere, in English, as gray whales. Gray whales have disappeared from seas around the Korean peninsula today, though the waters through which the species once migrated remain protected as a whale reserve. A refuge for ghosts. Which is to say, a refuge for hope. The last gray whale seen from South Korea went by in 1977, but some people take, as cause for optimism, news from Namibia and Israel of recent gray whale sightings where these animals have not been recorded for at least two hundred years.

Some petroglyphs in Australia very likely approach the age of the Bangudae relics, though none explicitly illustrate the labor of whaling. At the time the petroglyphs were first inscribed, Indigenous custodians and anthropologists say the impressions were neither mere outlets of creativity nor observations of the physical world as it was. Rather, a petroglyph denoted an intention to generate, or invite, change. A petroglyph spoke to the future. Large aggregations of petroglyphs marked out "increase sites": places where animal intensification rituals would be practiced, luring prey from hiding toward ambush—though this is only a superficial explanation of their significance, eliding the inner articulacy of the animal in dialogue with the person, or people, calling it. Some petroglyphs tied families to animal species in relationships of obligation and reciprocity. Others yet are, perhaps, the obverse of stellar configurations; the night sky's constellations represented as animal forms, navigational aids to daylight travelers. Those responsible for the petroglyphs, their communities and descendants, often iteratively reetched the grooves, and may do so still.

Whether a specific petroglyph bespeaks whale harvesting, whale watching, totemic whale kin, or star charting, the relevancy of these ancient impressions is not limited to a sacrosanct past. The Balls Head whale on its promontory has many symbolic layers; the engraving's history has tracked alongside a more recent lineage of contact between people and whales too. It was first etched into land called Yerroulbine, saddled later with the name "Balls Head" (after a First Fleet lieutenant, Henry Lidgbird Ball). The petroglyph has stayed, moored to the floor of the world, while pockets of Aboriginal whaling across the globe mingled with shoreside whaling fleets and coastal populations of whales fell before a host of transient, zealous whalers. When stocks of whales were depleted close to land and when those seafaring fleets moved out across the high seas and into circumpolar oceans, the sunrise still shone bronzy across the back of the Balls Head whale. The two smaller engravings that accompany the whale—the man, the horse-dog—were placed into, and next to, it during an era in which seaborne methods of disassembling whale bodies were established so that ships no longer needed to return to port to process their haul and whaling could expand to become the first truly globalized extractive industry. The petroglyph was covered up; it was revealed; it was blighted by reflective road paint and cleaned. The Balls Head whale lies now within the fence line of an eco-living initiative, behind a garden wall.

Living whales have been hunted with the weapons of every age that the Balls Head whale has passed through; from hand-knapped axes fused with spruce resin to harpoons fired by gunpowder, swivel guns, small backbreaking bombs and sonar, and remote-controlled lances that deliver electric shocks. Ultimately, the Balls Head whale, writ in rock, outlasted both high tidemarks of commercial whaling: the first, in the mid-nineteenth century, and the second, an arguably more vicious chapter, traversing the fossil-fuel age and the Cold War, rising to a climax in the wake of World War II. To impart a capacity for witness to the Balls Head whale, it has seen—though no eyes are etched into it—the largest cull of any animal order ever perpetrated by our kind. The world has formed and reformed around the petroglyph; and, motionless though it is, the whale has been retouched, over and over, as it divides itself between visible and invisible realms.

————————

As far back as the eighth century Vikings traded whale bones, and the Sami perhaps also joined them (evidence for which: a northern European version of chess, played with remnant fragments of whale skeleton)—but it was sixteenth-century Basque whalers who began, in earnest, the trade of preserved flesh and products from North Atlantic right whales. That trade, and their ventures in the Bay of Biscay (between northern Spain and southern France), industrialized ship-based whaling to a scale that spanned oceans: the Basques would be at the forefront of whaling for more than a hundred years.

Subsistence whale hunting relied on the supply of surrounding environments, as the componentry of archaic whaling weapons attests. Whatever woods grew tall beachside were whittled for spears. In the absence of hewable trees, as in the ice-rimed Arctic, whalers snicked local stones into sharp, leaf-shaped microliths; the leather of native mammals girded their boats. Fortuity, strategy, and, to a lesser extent, taboo determined which whales were hunted by these communities: species that strayed into embayments and estuaries were targets of opportunity, though the designation of any specific animal as huntable accorded with limits placed on the whaling season, and with the exclusion of species, or whales at certain life phases, according to cultural tenets. A kill shot, hand-flung at close proximity, must only have been feasible in the case of porpoises, beluga, and calves, up to a certain measure. If bigger whales were pursued along the coast—or hunted from an ice shelf—it was necessary to harangue the animals over days and nights. Hunters in boats struck at their quarry from above, using lances and bludgeons. Whales were sometimes strung with "drogues," buoyancy devices that slowed their escape and impeded diving (a whale's plunge is a "sounding"). The Inuit are known to have gutted and inflated seals for this purpose, having stitched up the creatures' eyes and orifices to make, in a rudimentary sense, seal balloons. Their harrying technique relied on exhaustion and serial injury to enfeeble and eventually kill an animal, after which snarling it in floats prevented it from sinking.

The plentiful provisioning constituted by a single whale, the precariousness of the weaponry involved, and the fragility of small craft used in pursuit, meant that whales loomed large in mythos and were often believed to have an adversarial or heroic character. Whale hunters were solemnized, having assumed the mantle of challenging volatile beasts in an inhospitable wavescape. In this context, to speak of whales as "supernatural" animals is appealing, were it not to perpetuate the ethnocentric fallacy that these communities recognized clear partitions between the natural and the otherworldly—the whale could not be *super*natural in an environment where nature and the spiritual were so extensively cross-pollinated that their domains did not warrant separation. Charged with seeing through the whale's eyes to predict its movements and hasten the end, some hunters painted whale masks: the subterfuge of laying one animal's face upon another, a forceful magic. There were whale dances, whale prayers. Whale shamans, high up in the hills, suffered in proxy. Among the codes of hunting, whaling especially—though not uniquely—was encircled by ceremony. But as traditional whaling cultures expanded farther out into the sea to connect along whale migration routes, their idiosyncrasies began to aggregate into a unified set of tools, and the unique regional disposition of both whale hunters, and whales as they appeared in culture, was in many parts of the world subsumed.

By the early 1600s, whaling had become a commercial concern. Whales were undergoing a categorical migration from stranded serendipity and fearsome prey to being a for-profit commodity. The species exploited turned from those that were the nearest and easiest to acquire to animals constituting the highest economic value. Whalers became technicians: their skill, less a function of dialing into the mysterious caprice of the whale and more a matter of adroitly wielding a new kind of metal weapon. The word "harpoon" comes from the Basque tongue— its origin *arpoi*, to "take quickly," not always (or, in fact, hardly ever) the tool's effect. The Basques deployed double-pointed harpoons equipped with iron heads and tied to a braided rope, spooled in Flemish coils to be run out at speed. Having struck through the blubber, these weapons proved near impossible for whales to dislodge. The harpoon's flared shape, barbed on the diagonal like an oversize rose thorn, meant the backward tug also cut and stuck. Lacking the propulsion to penetrate the whale's essential organs, the harpoon's impact was not lethal, but its effect was nonetheless grievous: whales, hitched to the boat and hit repeatedly, died from blood loss and flesh wounds as they suffered on the surface.

The Basque industry had begun shoreside. Lookouts installed in stone watchtowers around the Bay of Biscay signaled surfacing whales to boatmen by burning straw or waving colored flags. The watchtowers—mossy, frigid as in a fairy story—foretokened the crow's nests, atop wooden sailships, from which Basque seamen would, in time, spot their cetacean targets as the hunt moved out beyond coves and headlands. Until the invention of sonar in the twentieth century the best vantage to see what moved below the ocean's surface remained high up above it, on a mast. Vessels grew bigger, and crews more numerous, to undertake pelagic (open-ocean) whaling. The inducement was mercenary: whale meat and whale oil were lucrative goods. The inducement was, also, a necessity: between 1530 and 1610, Basque whalers hauled more than forty thousand right whales (many of them slow-swimming pregnant females), decimating the whales' shoreside populations and emptying their nursery waters.

Thereafter, Basque whaling spread out into the far corners of the North Atlantic, to Norway, Newfoundland, and Labrador, reaching down toward the South Atlantic. Nestled in foreign anchorages, the Basques built stations for flensing whale carcasses. Weeds grew in their brick ovens during the down season; forged tools whistled in the wind. The frontier of Basque voyaging frayed to cemeteries, wrecks, and marriages. Sometimes the Basques left behind a cabin boy, under direction to learn the custom of the land. Where Basque seamen encountered local whalers and Indigenous people they absorbed crew members, pidginized trade lingo, and bartered goods. Idioms and jokes that make slender sense today—*How are you? Not as good as the priests!*—flew ear to mouth, mouth to ear, even, inexplicably, in dialects where "priest" was an alien idea.

As per ink dripped onto linen, whaling inched alongside and outward from harbor settlements. Townships were founded near to where whales appeared close to land. In places where the soil was untillable and the climate grinding, whaling's income laid roads and sent houses, churches, and marketplaces rocketing out of the ground. This was unorthodox; excepting the tenacious Indigenous peoples of the Arctic circle, permanent habitancies had, before, demanded a resource base that was cultivated or constant. In time, Basque hirelings conveyed their skill set to Dutch, Danish, and British enterprises, moving into the icy apron of Greenland, following cold-water whales, the bowheads—huge, torpid animals with blowholes the American author Barry Lopez once described as being "volcano shaped," allowing them to surface to breathe between cracks in the freeze. By the late 1700s, with improved knowledge of whale behavior and anatomy, these vessels plunged into waters off South Africa and New Zealand, where they chased humpbacks as well as right whales.

Whaling had flourished in the seas off eastern North America, so by the time of the American Revolution the colonies were supplying Britain almost fourfold what British fleets collected along Greenland. Montauk and Shinnecock people, joining the hunt, got paid in cloth and blubber. From Nantucket, whalers begun to chase immense sperm whales—a stupendous, deep-diving species, found then abundantly in tropical waters—out onto the high seas. Shoreline whaling had been seasonal, but pelagic whaling roved between whale species, picking up migratory whales at all stages of their journeying: the industry could run down the coast and the calendar (a grace note of the restlessness that would come to characterize all capital in centuries to come). Yankee whalers adopted explosive weapons (shoulder guns and grappling hooks), expanding the distance at which a whaler could be lethal to a whale—and they also found ways to deconstruct whales on deck, alleviating the need to return to land to transform their haul into a salable product. Just as living whales roamed oceans, so whale goods became inherently mobile freight: pulled direct from capture point to the highest bidder in an importation market. Whaling's amphitheater widened.

In 1842, Jacob Anderson, a nineteen-year-old African American seaman, disembarked a ship named *Connecticut* to climb a ridgeline on Rosemary Island in Murujuga, an area encompassing the Dampier Archipelago branching from Western Australia. Using a metal barb or pick, Anderson inscribed his own name, the date, the name of his captain, and the ship's particulars atop a basalt rock face that bore a gridded pattern, laid down earlier by local Yaburara people:

> *JACOB ANDERSON*
> *New London*
> *IN THE SHP*
> *CONNECTICUT*

There are more than 260 extant rock-art panels on this tiny island, which lay then within what was known to whalers as "the New Holland Ground"—waters plied by American, British, French, and colonial Australian whaling ships. By the mid-nineteenth century, whaling had become a global trade, and the seaward engravings of Indigenous people in this part of Australia came to be crossed over by the names of North American whalers and whaling ships originating thousands of miles away. Pelagic whaling was America's fifth largest industry, employing some seventy thousand whalemen annually, with more than seven hundred vessels at sea on multiyear voyages (a metalsmith in New Bedford made and sold on average 1,463 new harpoon heads each year). One in six American whalers were recorded as African American.

The north of Western Australia was not yet assailed by the pearlers, sheep pastoralists, and police that would, in the decades to follow, inflict horrific violence on the Yaburara people—the Flying Foam Massacre, in which more than sixty Yaburara were killed in acts of false reprisal by white settlers, hovered some twenty years off in the heat haze. The whalers were, very likely, the first overseas foreigners the Yaburara met, though they were not the earliest voyagers to see the pindan shoreline from the sea. There is no evidence that these encounters were either antagonistic or friendly. The whalers would have observed Yaburara campfires at night and perhaps noticed their wooden canoes traveling between islands during the day. The grid over which the whaler Jacob Anderson scratched his name consists of straight, thin, parallel lines: reminiscent of a net. Mathematical. The cross-hatching is, in some deliberate way, emblematic. Being so geometric, it could not have been mistaken for natural weathering or the functional marks of sharpening a blade. The Yaburara, for their part, must have heeded the seasonal arrival and departure of the whalers' ships and, later, noticed the men's inscriptions.

Stand to one side a moment, to picture it: a seabird dawn, the departing ship faint on the gray horizon. Someone runs their thumb over the newfound text, its little notched serifs, the mystifying, sidewinding *S* in *Shp*. *CONNECT–I–CUT* rolls under their touch and they inquire of the word, its feeling as a sensation; a texture, a sculpture, an animacy. So *Connecticut* makes landfall soundlessly, dressed in the disguise of its anglicization, having crossed an earlier border from Algonquian where once it was "Quinnehtukqut" (meaning "beside a long tidal river"). A touchstone, a dashed track. What the Yaburara thought, or said, on discovering these foreign incisions, is lost—not because of the attrition of time and memory, but as a result of the genocide and generational traumas that followed. Until microscopic analysis revealed otherwise, researchers suspected that the hatching action might have been the Yaburara striking out the whalers' words; that the Indigenous petroglyph was an overwrite. But perhaps it was no epiphany to the Yaburara to hear it was the other way around.

By the time the *Connecticut* returned to its New London port, a little under a year after its name was scratched into the red rocks of the southern hemisphere, the ship carried eighteen hundred barrels of whale oil—a gainful cargo. And these were the takings of one ship alone. Over the course of the century, fleets with crew from assorted nations crisscrossed the animals' migration routes, hauling some 236,000 whales. By 1856, the price of whale oil had hit $1.77 US dollars (USD) per gallon. Whalers spoke of sperm whales traveling the oceans in veins, like gold. A fitting metaphor, not only because whales were a precious lode, but also as the animals were proving to be a finite resource in the way of a mineral deposit. Whaling on this scale had long outpaced the animals' reproductivity. Their numbers, trickling around the world, thinned. In some whaling hot spots, species disappeared entirely. For those individual whales overlooked by whalers, the waning of their kin meant more unsociable oceans; less likelihood of encountering newcomers, and the gradual diminuendo of whale songs from afar. Whales occupied expanding regions of whalelessness. If the animals knew it, or sensed it, it might have felt like falling into an abyss, but without falling at all.

The sound of hammers echoed through the air where the Balls Head whale lay. Saws shuttled and raked. At every corner of Sydney's port, laborers dripped sweat, thickening the decks of sailships by adding a second layer of planks atop the first, to support the weight of cast-iron cauldrons in which whale blubber would be liquified at sea.

People of the nineteenth century—across an array of classes, professions, and life stages—dressed in, slept, and dreamed on the stuff of whales; they cooked with, played with, desired with, and made art from, looked through, healed with, explored via, were disciplined by, disciplined with, and made divinations, out of whales. In the ordinary course of life, they were almost constantly in contact with whale-gleaned products, in much the same way as most people today are never far from plastic objects. The inverse of all our faded shopping, spilt out into the whale's environment; those nineteenth-century forebears lived inside a whale-furnished world.

Basque whalers initially traded whale meat—a food the Roman Catholic Church sanctioned for Fridays and in Lent, when the flesh of terrestrial animals was verboten—but it was oil, rendered from blubber by boiling, that would turn out to be the whale's most bankable extract. Depending on species and storage, whale oil ranged from a flax yellow to a reddish tea color; it retained the odor of its origins and smelled like a sardine cannery after a fire. Thinner in viscosity than modern vegetable oil, whale oil met many needs. It eased mechanical cogs in textile and metalwork factories, aided in the cleansing of wool and the tanning of leather. The oil lit smokeless streetlamps, factory-floor and shop lights, extending trading hours and mercantile business into evening hours (so whales contributed to changing concepts of public space in night-time and were even thought, by some, to reduce crime). Whaling was the context that coaxed industrial manufacturing and commerce into its modern shape—whale oil initiated automation; sped up repetitive, task-based workflow and expanded the working week; it preconditioned the transformation of the natural environment by numerous enterprises, driven to faster and more thorough manufacturing schedules. In agriculture, humpback oil was included as an ingredient in insecticide washes, glossing the leaves of Californian fruit trees and French vines. A melon's shine in a painting—that gleaming spot may betray a smidgen of whale. Whales remained material to culture in other ways too. Printing inks were emulsified with whale oil; in stately houses, whole libraries might be spelled out in alphabets distilled from whales.

Sperm whales were chased for another kind of fat found not in their blubber but in their heads. Spermaceti, a sort of natural wax, resembles solidified coconut cream. A high-grade, long-burning, and noncorrosive substance, spermaceti was prized for use in industrial hardware (looms, trains, guns) and, most particularly, as an illuminant. Spermaceti made such superior candles that New Bedford, the epicenter of sperm whale hunting, became known as "the city that lit the world." As the candles burned, whales evaporated into glow and smoke. But sperm whales were found in equatorial waters, which posed a logistical problem for whalers: the flesh of a bowhead, hauled in the Arctic, was unlikely to rot during the time it took to convey the carcass to a whaling station in Greenland—in warmer climates, dead whales tended to go rancid. For this reason, as well as the incentive to up the portability of oil and spermaceti, Yankee whalers pioneered on-deck processing of dead whales. Using a series of block-and-tackle mounts, stoves and cauldrons, they pulled the whales apart mid-ocean.

Deconstructing any whale was grueling labor. Dead, a whale would be chained to the side of the ship and flensed of scarves of fat about fifteen to twenty feet long, using curved blades like sickles. A status job, the ship's officers straddled the body to make the first cuts, even where the water below was lethally freezing and measled by sea ice (or, worse yet, diced by sharks). Sometimes the flensing was done in a spiral fashion, as you might peel an orange, gradually turning the whale against the boat's hull by way of a system of pulleys. "Blanket strips" of blubber, pared from the whale's exterior, were brought up onto deck using boarding knives and hooks; there, these quilt-sized pieces, each weighing as much as a ton, slid from side to side with the ship's motion, creating a lather that men struggled to avoid slipping in. Brought below-decks to a "blubber room" by way of shovels, the blanket strips were cut down into smaller "horse pieces" atop a wooden "mincing horse"—like a carpenter's saw bench—after which long corrugations were scored in, skin side down, not cutting through the blubber completely but creating squarish segments called "bible leaves." The men, bare-chested in the ship's fetid chamber, likely listened to marine predators bump against the ship's hold, mobbing what they could from the whale's blubberless underside. (Amid Arctic ice, crews feared polar bears charging the ship, which might lead them to unchain and abandon a carcass.) In the blubber room, one whaler, at the end of the line, worked with a "leaning knife" to trim any remaining sinew from the fat—the flesh otherwise tainted the oil, leaving it a muddy shade. Having been sectioned, scored, and tidied of meat, these blubber pieces were then known as "books."

The blubber books were winched up on a cable to swing above a row of iron try-pots (cauldrons) on deck, typically surrounded on three sides by insulating brickwork. Unprofitable scraps of blubber, skin, and tissue—to the workers known as "kreng"—were fed into the flames beneath the try-pots, so that a dead whale fueled its own dismantling. The books were boiled, while a few workers tended to the try-pots, stirring to ensure the oil didn't burn as it was rendered. Using a skimmer, the wrinkled whale skin (the "fritter") was removed from the liquid, after which the oil was bailed into copper cooling casks and then finally siphoned to wooden barrels for storage. It took thirty-six hours to butcher a whale—a mephitic, spattering task. On average, each humpback yielded thirty-seven gallons of stable oil, rich in triglycerides. There was added work to be done excavating the heads of sperm whales for liquid spermaceti, which turned opaque, as if by magic, when it was exposed to the air. In the historical records the crew are merely recorded as "whalers," but mid-ocean they worked as cutters, extractors, hookers, scourers, blubber boilers, gut luggers, gear repairers, blacksmiths, rollers, and loaders; toilers in the animal's red abattoir.

While a whale was being flayed of fat, baleens—from species including the humpback, the right whale, and the bowhead—were also tugged from its mouth to be traded and repurposed as sundry products. Baleens are the bristly combs some whales have in place of teeth (envision a mustache *inside* your upper lip). Pliable as fingernails when ripped fresh from a whale, the baleens—also colloquially known as "whalebone" (one word, distinct from a whale's skeleton)—dried stiff and could be split or remolded by steaming or soaking in hot water.

There, at the beginning of consumer culture, baleens ended up most especially in women's fashions and parasols, but also in hairbrushes, fishing rods and reels, and pastry crimpers; in men's shoehorns, hat rims, and the frames of eyeglasses; in prosthetic limbs, tongue scrapers, throat swabs, and the filamentous cord used for surgical stitching. Baleens became sofa stuffing; they were coiled into furniture, used as springs in pocket watches and for the mechanisms in wind-up toys, and they were braided and bent into what we call now Hula-Hoops. Baleens were hardened into police nightsticks as well as the thin classroom canes that raised welts on bad children's palms (from which the phrase "to whale on" is derived). Baleens were fashioned as whips to spur laggardly ponies, and as Y-shaped divining rods thought to twitch toward freshwater aquifers underground. Persistent beauty standards—the flummoxing hourglass ideal for women's bodies—may have been reinforced via societal duress, but the shape was obtained via the strictures of baleen: whale-buttressed garments, compressing women's rib cages to a cinch. That we still have this body form as an archetype in culture is a legacy of whaling; the shrinking of the feminized figure at its waistline happened in lockstep with the vanishing of whales.

Into whalebone corsets (or "stays") could be secreted what were called "busks"—a kind of love token made from a strip of baleen, to be laced in hard against a woman's bare solar plexus. Where whales were dismembered in the open ocean, shipmen might retain small, cleaned paddles of baleen from the deck, which they polished and then scored with missives of ardor, longing, commitment, or lust. The busks were the product of lethargic hours: erotica for idlers at sea. Chains of hearts, twisting vines, stars, paired initials, love mottos, figures embracing, women (and men) stripped and lascivious, all finely etched and stained with soot or tobacco juice. Whale valentines. And when these women undressed at the end of the day and were released from the clenching columns of their fashion, when they peeled off any busk from the skin that reached from between their cleavage to above their navel: Did they then see there the red imprint of those engravings? Their lover's initials transposed back to front, say. A promise, or a risqué proposition written in reverse, lasting only until it was time to go to bed.

Anything left over, the viscera and inferior bones of whales, were ground for livestock feed, brought back to fertilize croplands, or were, more usually, tipped off the ship. A few good bones might go to a piano company to supply white keys (voiceless in afterlife, whales nonetheless continued to resonate in a major chord). Near the turn of the century it became possible to extract gelatin from whale skin—used, sometimes, in photographic film, so that even in the visual records of whaling, there may be a lick of actual whales.

A whale's highest worth was as a commodity for international trade, but at least one Australian township repurposed local humpbacks as pharmaceutical. Near the turn of the century, a medical hotel for therapeutic "whale cures" sprung up in Eden, on the south coast of New South Wales. The greases and gases of a whale's decomposition were thought, by many residents and people in the surrounding region, to alleviate fatigue and rheumatism, and to remedy accompanying mental maladies. People with chronic pain and ill mood traveled to Eden to spend weeks bathing *inside* dead whales and their extractions.

A handful of archival photographs held by the National Library of Australia illustrate the treatment. In one early image a whaler sits serene as a spa patron within an unlit try-pot positioned on a riverbank, while his brothers tip a bucket of foul sludge over his back. The next picture in the series shows the same group of men: one rubs his brother's trapeziuses, easing knots from his shoulders and neck. His muscled arms are honeyed with oil. A collie dog points its snout into the frame, sniffing the fat-daubed air.

A different picture in the folio shows women in gowns, more stiffly ensconced within the body of a harpooned humpback, pulled up into the Kiah Inlet tryworks. Conical pits have been gouged out of the whale's side where it lies, its skin split and putrefying. Because the prints are black and white, the bloodied cynosure of the whale is not apparent—yet how its flesh must have sung redly against the green and gray bushland! The women are up to their shoulders in the whale, and still wearing ribboned hats. Their faces might be game or revolted; it's hard to make out. They would have been extremely hot given the whale's steaming decay, and the smell was surely rank ("the aftereffects of the cure are that patients give off a horrible dead odor for a week or two"). The hotel's most famous patient, described in the *Pambula Voice*, was a man "obliged to walk with sticks," who "after one trial of the whale treatment, was so much benefited that he was able to walk back to Bega" without sticks—a distance of roughly thirty-three miles. In 1896, the *New York Times* reported on these Australian whale bathers:

> Upon news of a whale being taken, they are rowed over to the works on a boat. The whalers dig a sort of narrow grave in the body, and in this the patient lies for two hours, as in a Turkish bath, the decomposing blubber of the whale closing around his body and acting as a huge poultice. The whalers make no charge for a dip in the whale, and go on with their work on another part of the body while the patient is enjoying his immersion.

A sort of narrow grave. The meaning of a whale, and how it is written of, reveals something about our attitude to its world, I think. The potential of a whale to be both a grave and a cure, say. The way an undersea whalefall is a death plot, but also ignites a life force. The way whales may entomb the history of human enterprise within their bodies—yet simultaneously be a wellspring of wild wonder to us, and once offered, to Eden's patients, a healing experience: a chance to be reborn into health.

In literature, going into whales is a punishing and puritanical business; a story that ascends to the level of theology. Tales about being engulfed by whales are, by and large, morality tales. Salvation and redemption are their overriding themes. The leviathan in the Old

Testament was first a sanctuary from the purpling storm meant to command the prophet Jonah's obedience—but in so many retellings that sheltering stomach becomes, variously, an abyss; a cell; a tribunal; a vise; a grave. Calvin the Protestant finds an infirmary inside the whale, but in the King James Bible Jonah's "treatment" is his torturous compression. Jonah is a prophet who rejects prophesy and flees his fate. He is commanded to foretell the ruin of Nineveh (a great Assyrian city, fallen into renowned wickedness) but fears that, upon hearing of the prediction, the Ninevites will repent and his God will forgive them, thereby rendering the prophecy false. So Jonah runs away: he boards a ship departing in the opposite direction. Then Jonah and the voyagers are hit by a huge storm of celestial force. Jonah ends up in the ocean, where he is consumed by "a great fish," in the belly of which he survives, praying for deliverance even at the cost of being branded a false prophet. Ultimately, what he bitterly fears *does* come to pass—having been released from the castigatory whale onto the shore, Jonah delivers his dire forecast, the Ninevites fast, pray, wear sackcloth, and sit in ashes, and then they are redeemed; Jonah is an object of satire. God's punitive disaster never sweeps over Nineveh. The whale bathers were certainly familiar with this narrative. Perhaps they, too, believed they needed to risk ridicule, to suffer further, to be purged of their conditions.

The two engraved figures that accompany the Balls Head whale—the human inside it, and the leggy creature to its right side—postdate the whales' ancient inscription and were most likely etched in the nineteenth century, as whale oil circulated across the sea and suffused unnumbered workhouses and homes and while Eden's whale bathers stomached their therapeutic interment. Regarding the external animal, its ambiguity: if it is a horse, then the carving has been made after horses, which is to say, after European invasion, begun in 1788. Though, equally, it could be a dingo, a native doglike animal. Why is the man inside the whale? Why is he headless? Does he have, as it seems, a clubfoot? And what is his relation to the third animal: Has he been thrown from the horse and injured, perhaps? Has he been bitten by the dog; is his leg engorged with infection? The figure is like a message, stoppered in a bottle, green leaves flickering in the trees above.

Whether or not the Cammeraygal people had always viewed the Balls Head whale as transparent, or if it became so, when they engraved the man within it is a point of contention. Was the whale, to its engravers, becoming ghostly? Fewer real whales neared the engraving, having been so widely hunted. Moreover, the colonizers, in growing numbers, were stripping the harbor of seafood and edible vegetation the Cammeraygal had once harvested for themselves. Anthropologists have speculated that the man was put inside the whale to call on powers remedying an illness; the interior of the whale, a place of healing. Professor of Indigenous Research, Dennis Foley, posits the man may not be *in* the whale but riding it, surfing the immense animal to shore for a feast, at a time when Eora tribes had begun to starve. In both interpretations—whale as hospital, whale as reprieve from hunger—the animal brings the possibility of renewal.

The Australian philosopher and ethicist Thom van Dooren has an idea I've long found compelling: that when animals die out, the cultural and ecological relationships that furnish their existence can be experienced as a kind of nonstop haunting. Van Dooren is interested in the kinds of stamina that attend extinction and endangerment—how vulnerable animal species (he focuses on birds) don't just disappear in a snap but continue to be revealed *after* their passing in different varieties of fortitude and mourning, both in human societies and in the lives of interdependent organisms. When a creature is gone, its significance may grow: this is the basis of his fieldwork. My impression is that when van Dooren talks of "haunting" he means to refer to both physical and symbolic connections sustained after the vanishing of animals. The way, for example, the ornate shape of a blossom's throat might have coevolved with the ribbon-length tongue of a pollinating bat that has since been killed off. The apparent senselessness and surplus-ness of that flower might then emanate a kind of strangeness. So long as such a feature doesn't steer its proprietor into coextinction (so long as other pollinators can still propagate the plant), there will be a kind of ghostly residue: a physical communication with no visible respondent. A fascinating notion: that the too-muchness of the flower might point out a lack in the world.

Van Dooren also studies communities that have attributed a symbolic quality to an animal and how those cultural narratives are experienced when animal populations ebb, become rarer and perhaps more exotic, or when they are extinguished. Take the huia (not one of van Dooren's subjects, but nonetheless illustrative): a red-cheeked wattlebird, favored by hatters and taxidermists in early twentieth-century New Zealand, which turns up in field recordings of Maori hunters long after the bird's extinction. Mimicry of the huia's call passed from human generation to generation, enclosed within a stalking ritual, despite the real huia's perishing. So people continued to fill the forest with the words of birds, since melted away. An acoustic fossil: like the brand name of a product discontinued since childhood, the imitative call accrues its mournful power from the extinguishment of the bird—though in the Maori tradition it isn't clear whether the profundity of the huia's sound derives from the memory of the living bird, or if its call has become detached from its progenitor to become a wholly human noise. Other languages spoken today reference the calls of some disappeared birds, as "dodo" in English is believed to be mimicry. *Dough-dough*: an imitation of cooing. The dead bird's tongues, spoken from our mouths.

By the beginning of the twentieth century, the Siberian Koryaks near the Bering Sea had ceased hunting bowheads. The Makah along the Pacific Northwest coast gave up on gray whales; there were no more gray whales there to kill. The loss of Yupik whaling contributed to a famine on Saint Lawrence Island, only partially averted by the efforts of Presbyterian missionaries to introduce reindeer. Fanalei villagers on Malaita, in the Solomon Islands, suspended their dolphin drive— though, this was most likely because Christian proselytizers exhorted their conversion and changed the wedding customs. Aboriginal whale hunts may have been suspended, but though the cultural practices that attended them were bereft of their creaturely referents, they did not fade out. Whale masks might have looked even more like monsters to the person who had never seen a whale. Perhaps whale dreams, without real whales to satisfy their inference, were received as dictates from the realms of the ancestral past instead of the animalic future. But however surreal or mythic a whale became, the animal itself was not forgotten.

No species of whale was ever driven to extinction by whaling operations, though the extirpation (localized destruction) of distinct, separate populations, and the removal of so many whales from marine environments, has had long-lasting effects. A useful word exists to describe this today, though until recently, such vocabulary was scant. The word is "defaunation." "The loss of a place's absolute animalness" is how the journalist Brooke Jarvis puts it, a depletion of abundance. A young verb in the scientific literature, defaunation refocuses attention away from the extermination of species, and onto the effects of reducing animal populations—for those animals, their ecosystems, and the cultures that are nourished by them.

Whaling amid the Pacific Islands in the nineteenth century not only led to the defaunation of whales in the ocean but it also altered the composition of terrestrial ecologies. Studies focused on the contents of ship logbooks record that across seventy-four Yankee voyages between 1846 and 1901, at least 5.3 million pounds of meat (other than whale meat) went into the bellies of whaling crews, metabolized for the ergs of energy needed to labor over killing more whales. Hungers were satiated with walruses, ducks, deer, and river fish in the northern hemisphere. Kangaroos are mentioned in the south, along with more numerous porpoises, sunfish, and seals killed with clubs the whalers called "persuaders." Researchers estimate some thirteen thousand Galapagos tortoises were consumed by a mere seventy-nine American whaling vessels in the middle decades of the 1800s. Over time, fleets depleted populations of many edible animals that had the poor fate to live around hunting waters. The ancient people of Un'en'en, in Chukotka, once used walrus tusks as canvases for their artworks, but during the nineteenth century some walrus colonies in polar seas were exterminated entirely, or were so shattered by overexploitation that they never recovered.

The knock-on effects of the reaping of these coastal and island ecosystems are difficult to quantify, but biomass flowed the other way also, onto the land. Rats, off-boarding whaling ships in temperate and tropical climates, set to work consuming the eggs and young of endemic fauna—animals that were often ill-equipped to counteract rodent appetites. Men deserted to build new livelihoods in the undergrowth, and hogs disembarked opportunistically to author other kinds of local damage. Probably we cannot know all the types of island creatures—birds, reptiles, amphibians, insects, and plants alongside them—that disappeared, indirectly, because there were whales in surrounding waters. A few unique species of limited range were likely lost before they were discovered. So food webs, and thus, too, the architecture of the vegetation, the canopy, and the shade that dapples through it, may be shaped in ways that are ill-defined by the arrival of the non-native organisms that the whales' presence first conveyed, and the resultant reshufflings of wildlife and flora.

Whale bones and viscera discarded onto islands by whaling ships mid-journey, or at the terminus of a voyage, changed nutriment levels to the benefit of persisting species too. On craggy atolls, savaged by high weather, the arrival of cetacean leftovers could mean a sudden ballooning of scavengers: crabs, seabirds. Then all the things that eat crabs and gulls, and so on. In some icy parts of the world archaeologists are able to track the settlement histories of traditional whaling communities by aerially surveying vegetation. Rich patches of greenery divulge fertilization by old, buried whale bones. Limnologists (scientists of marshland) examining the layered sediments of ponds on Somerset Island in the Canadian Arctic have identified algae and micro plants that peaked when dead whales leached into pond water. The algae still bloom off whale gristle, though at least four hundred years have passed since carcasses were last flensed there.

Even as whaling pillaged from undersea ecosystems and destroyed hundreds of thousands of individual whales, the movement of whalers and whaleboats chain-linked terrestrial ecologies in new ways, introducing novel resource pressures and evolutionary opportunities. Whaling wasn't simply an economic phenomena, dismembering whales into commodities, commodities into profits, and connecting people through trade. Whaling was also an ecological force. The killing of whales threaded together disparate organisms, picking up and relocating creatures over the sea that would otherwise have remained unconnected in their biomes. Whaling was a transportation, drawing island into island.

The Balls Head whale exists in an ecology of carvings not immediately apparent to the naked eye. The whale triangulates Sydney Harbour along with other petroglyphs of marine creatures found around the water's edge and in the crenulations of Sydney's bays: at Berry Island Reserve, at Mackenzies Point, in Bondi golf course, at Cowan Creek in Ku-ring-gai, and at Jibbon Head in the Royal National Park. These petroglyphs may not be as immense as, say, the Uffington chalk horse, dug along an Oxfordshire hillside during the Bronze Age, or the likenesses of birds and mammals early Peruvians cut into the Nazca Desert, but many of the Australian engravings are connected to one another on a scale that is equally as expansive—perhaps more so. One purpose a petroglyph serves is to dilate a person's perspective outward, to consciousness of a bigger landscape and to relatedness between animals. Etched onto a natural ingress, overhang, or clearing, whale engravings can seem to be set off from clusters of other creatures. At least, this is how it can look to tour groups or casual passersby: the whale solitary, unaddressed to other beings. How easy it is to fail to notice that the whale petroglyph is fleeing a shark petroglyph, located on a headland, miles away.

The networks of knowledge that vibrate between these artworks are detectable, in their full extent, only by certain people according to Indigenous protocol. The inscrutability of the petroglyphs points not just to the depth of their historical provenance but to the cultural securities that surround the stories, embedded into the rock. Untranscribable as those narratives are, it is nonetheless more than conceivable that the specific connotations attached to the Balls Head whale were impacted when, in the early 1920s, a colossal sperm whale petroglyph nearby was dynamited to flatten the ground for a strut of Sydney's iconic Harbour Bridge. This pulverization, across the harbor at Milsons Point, must have been felt all along the traceries of the cultural web of which the petroglyphs are only partial, visible evidence.

One underrepresented truth of the world: things that have been removed from the past exert their pressure on the present moment, just as much as the things that persist.

At an uncertain date in the 1920s the Balls Head whale was buried, remaining hidden from sight for several decades. The whale's ancient outline rested near the main exit route from a coal-loader depot. Major infrastructure projects were a hallmark of the advent of the fossil-fuel age (more than 180,000 tons of coal went into the construction of the Harbour Bridge, used to smelt steel and in the production of cement). Some people say the petroglyph was deliberately concealed with soil and gravel by those who sought to preserve it, before workers came to lay down the thoroughfare to the loader—others say that the whale itself slipped behind a curtain of country, perhaps in apprehension of being destroyed after its sister work, over the water, was blown up. During this time the evocations that attended to the engraving were displaced. Kept company by earthworms, brushed by the colorless roots of orchids, the ground surrounding the Balls Head whale juddered each time a coal truck rumbled past.

It is, perhaps, easy to imagine the enterprise of whaling in an era of drawing rooms, sniffing-salts, and hand-lit streetlamps, but to conceive of the industry undergirding the birth of the automotive city requires surprising effort (let alone to reflect on whaling being coterminous with the space age). Whaling was supposed to have gone the way of parlor séances and the medicinal leech: Victorian preoccupations since debunked by sounder science. Yet from 1900 to 1999, an estimated three million cetaceans were killed and removed from the world's oceans— more whales than had been hauled in all previous centuries. Scientists estimate the total biomass of baleen whales found in the sea surrounding Antarctica was reduced by 85 percent. In 1960, whales had become the single most valuable animal on the planet, worth USD$30,000 each (close to USD$260,000 in today's money). This was in spite of cheaper vegetable oils, spring steel, and petroleum products, including thermoplastics, offering more competitively priced feedstock for items that had once been constituted by whales. Why did whaling not tail off, as it had seemed set to do? Why did whaling, in fact, *accelerate* in the twentieth century?

There are two threads in the explanation that accounts for whaling's resurgence: a technological thread and a geopolitical one. In a nutshell, though new technologies provided substitutes for whale products—which might have been expected to induce a transition away from whaling—innovation also opened up fresh markets for whale oil, and permitted whalers to exploit previously untouched whale stocks in the far southern hemisphere. Additionally, demand for whale oil's burgeoning commodities was exacerbated by international competition and emergent norms around cleanliness and diet. Finally, the whaling industry successfully lobbied to gain short-term footholds in secondary markets for luxury and specialist goods; particularly those that linked whaling to transportation, defense, and health.

During the first whaling peak in the mid-1800s, the oceans had been dominated by Yankee fleets, along with colonial British and French, and other European whalers. But American whaling lost its technological edge in the twentieth century and became a moribund industry there in all but a few outposts. The second surge in whaling happened under more complex investment structures but was led initially by a conglomerate of Norwegian (Norwegian crewed or financed) and British enterprises, and then by Japan and the Soviets. Before World War II, British, Dutch, Norwegian, German, and Japanese fleets—and to a lesser degree South American whalers from Chile and Argentina—could be found pursuing whales in the open ocean—in addition to (and sometimes overlapping with) coastal whaling operations off Australia, New Zealand, Korea, South Africa, Canada, and Peru. Most whaling companies were supported by transnational investment, so that, for example, a Norwegian whaling fleet might be financed by British and German capital, while South American whaling could be backed by Norwegian capital. After the war, the industry bifurcated more distinctly into whaling for meat and whaling for oil: the former was dominated by the Japanese, the latter by the USSR. Whaling ceased to be commercial through the Cold War; it transmogrified into a powerfully political business, all the same.

Had ships under sail remained the predominant means of pursuing a whale, it seems likely that the largest and fastest cetacean species, in the coldest parts of the world, would always have evaded hunting. But the fossil-fuel era saw swifter steam-powered ships running on coal furnaces, and later diesel vessels, enter the oceans. Whaling's second wave entailed the exploitation of species that had, for the main, eluded whalers in earlier decades—blue, fin, and sei whales, as well as the small but quick minke. As the ships sped up, they came not only to be powered by fossil fuels; they also acquired on-board equipment running on diesel, petrol, and electricity—air compressors to pump up whale carcasses and prevent them from sinking, hydraulic tail grabbers and winches ("the whale claw," one was called), pressure cookers, ventilation fans, and refrigerators to preserve whale fats. By 1913, there were twenty-one factory ships—huge, mobile slaughterhouses—in Antarctica and six landings. In the Southern Ocean some species were brought down to populations numbering barely hundreds.

Harpoon technology had also moved on from the guns Yankee whalers had utilized. Mechanized weapons, pioneered by the Norwegians, were bow-mounted on canons; some had explosive tips or deployed fragmentation bombs, and a few were designed to electrocute whales in the water. That these weapons traversed greater distances was a necessary innovation. The ships were so loud, whales ("the listening prey," as they had been known) heard the whalers coming at a far remove, turned tail, and fled. In answer, the hunt expanded along a vertical axis too. Radar, masterminded during World War II, enabled whalers to identify their quarry from a range that was, if not outside the whales' perception, then at a distance at which the ship's noise had not yet spooked the animals. Spotter planes spearheaded the hunt, maneuvering vessels below from on high. Having homed in on a target, it was rare the whales outswam (or out-dived) the whalers.

When the price of whale oil had first begun to plunge (a gallon of whale oil nosedived to USD forty cents in 1896 and fell further to thirty-one cents by 1905), the industry's response had simply been to haul more whales and expedite their processing. By the 1920s, a single large whale could be flensed in under an hour. But growth in supply would have been fruitless had there not also been an evolving demand for whale products. Whale oil proved elemental to the glycerol used for nitroglycerine in explosive munitions in World War I, and was also used to stave off gangrenous trench foot. A single British battalion has been estimated to have gone through ten gallons of whale oil every day. As whale oil was of national interest, a number of European governments subsidized its production over World War I so that the industry did not bear the full expense of pursuing, killing, and processing a whale. That being so, whaling did experience a downturn during the war, when many whaling ships were repurposed as supply bulkers carrying armaments too sizable to be transported otherwise. The interval was not long enough to allow populations of whales to rebound—besides which, aquatic mines blew many marine animals up.

Whale oil also expanded into two cornerstone goods of working-class life: soap and margarine. In the early 1900s, a rise in living standards and a concern for hygiene born of an awareness of germ theory created a growing market for soap. In the 1920s, advances in the technology of hydrogenation made whale oil palatable in margarine: touted as an inexpensive, healthier alternative to butter, margarine sent whale oil straight to the heart of diets in Germany, Scandinavia, and Great Britain.

That some of the world's largest animals should have melted into toast, or become iridescent bubbles to pop midair, seems an almost miraculous act of transubstantiation.

After World War II, whale products became paradoxically more personal and more stratospheric. The industry found niches in products like lipstick, perfume, and colored leather gloves. A hormone gland behind the whale's brain, bioprospected by the pharmaceutical industry, became an active ingredient in drugs treating arthritis (Eden's whale cure turns out to have been prescient, albeit that bathers used the incorrect portion of the whale for a remedy). Vitamin A, from whale livers, appeared in supplements, while whale oil was used in culturing penicillin and as a base for ointments.

Spermaceti, meanwhile, found a place on intercontinental missiles and small, specialized mechanisms on satellites, such as the camera shutters on the Corona missions, flown over Russia and China. General Motors used spermaceti in the transmission fluid of its vehicles up until 1973, when the US Endangered Species Act came into force. Proponents of whale oil worked to modernize the product by associating it with industries that functioned in the public good, like health, or those at the vanguard of technological progress, such as space exploration. That barely a smattering of whale oil droplets ever made their way beyond the stratosphere was beside the point: by the 1960s, whale fats had gone from being a mundane butter substitute to being a part of the space race.

Whales, everywhere they could still be found, were in crisis. A 1960 review of whale populations in the southern hemisphere, staged by an independent committee of biologists, returned a dismal diagnosis: there might be fewer than a thousand blue whales left; humpbacks would likely not rebound for eighty years. Meanwhile, whale oil transitioned to being quantified by the ton rather the gallon, climbing from around USD$118 per ton in 1968 up to USD$460 in 1977, while the rarer spermaceti hit USD$850 that same year.

Established in 1946 in Washington, DC, the International Whaling

Commission (IWC) formed to assume responsibility for regulating whale catches outside of territorial waters but found that their efforts to make whaling a sustainable fishery malfunctioned from the get-go. Seasonal quotas were set with reference to "blue whale units" (BWU): meaning that whalers, if they adhered to the quotas, were supposed to hunt fewer large, high-yield whales, or more small, low-yield whales. A single blue whale—a BWU—was worth two fin whales, two and a half humpbacks, or six sei whales. The quotas were applied across all whaling nations that were IWC members, so at the open of the Antarctic whaling season (which ran each year from December until March), whalers raced to hunt whales, often concentrating on smaller species— many of which were just as endangered as their larger cousins.

Despite all the effort that had gone into improving on-board oil processing, the competitive haste resulted in terrible wastage. Harpooned whales might be abandoned without being hauled if they appeared less fatty than other whales, spotted on the horizon. Some years, whalers only observed the quotas in the Southern Ocean, disregarding the provisos on their return journey and whaling along the migration route. Undersize calves and immature adults were frequently excluded from the reporting (the rogue hunter Aristotle Onassis reputedly took sperm whales so young their baby teeth hadn't come in). Anticipating further reductions in the quotas, or the exhaustion of whale populations entirely, some whalers simply rushed to take as many whales as was possible, believing they would be able to sell their oil reserves at higher prices after heavier restrictions came into force. Yet whaling was beginning to go into a downturn, due not to regulation but because whales had edged into "commercial extinction": it was no longer economic to outlay the expense of a factory ship in Antarctica, weighed against the diminishing returns of the catch. All whale populations south of sixty degrees were showing signs of collapse.

For two nations, the USSR and Japan, the calculations in whaling's continuation proceeded according to different criteria. After World War II, Japanese whaling came to represent an assertion of self-sufficiency. Into the 1960s, Japan experienced a food crisis born of the wartime decimation of supply chains and the Japanese agricultural sector. The Japanese people were starving and crippled by vitamin deficiencies. Dished out to elementary and middle school children, whale meat became a symbol of the nation's resilience. The Japanese hunted mostly minke and disavowed the electrified harpoon because it blistered and toughened the product. Whaling sustained the cultural logic of national recovery and has continued to be subsidized by the Japanese government to this day.

In 1993, post glasnost, the Russian Federation declassified its whale-catch documentation from the twentieth century to reveal that 180,000 more whales had been harvested, beyond what had been reported to the Bureau of International Whaling Statistics. Soviet whaling did not commence until the 1930s, but surged after World War II and through the Cold War. The Soviets needed spermaceti for military programs because its synthetic replacements had been embargoed by Western governments. There was also a moderate appetite for whale meat in the USSR—but production was largely untethered from demand; the animating motivation was nationalistic. The Soviet state had not yet had its "share" of whales, in contrast to European powers and America—and Soviet officials, consistent with bristly Cold War posturing, were eager to build a maritime industry that exceeded the scale and prowess of other nations. With no regard as to the sustainability of whaling, domestic quotas set by the state's five-year plans far exceeded the national use of whale products throughout the 1950s, '60s, and '70s, so that even as the official economic targets put Soviet captains in contravention of the IWC, the national market was glutted with blubber, canned meat, and bonemeal.

Soviet factory ships couldn't process whales as fast as they needed to be hauled to satisfy the quotas; many were summarily peeled of blubber and let fall back into the sea. Some twenty-two thousand humpback whales were harvested in one slice of the Southern Ocean over three consecutive summers from 1959 to 1962. When the ships were returned to port, whale carcasses that were too decayed to be portioned for human consumption were sold to fur farms. Arctic foxes, chinchilla, mink, and sable grew glossy on whale, before being slaughtered for garments. Running my hand down a clothes rack in a second-hand market, I sometimes think about the enfolding moral gravity of a woman's fur coat, or a muff, from Russia in this era: factory-farmed mink, nourished on minke whale. As what lies behind the greenhouse tomato today is copious unseen plastic, the whale, too, merges with the folds of the fur; it is as much there as it is in the whalebone corsets of the nineteenth century.

Did people know it then, the moment when the oceans ceased to be infinite? By the time the anti-whaling movement had begun to coalesce in the late 1970s, the obliteration of whale populations lay decades back in the rear-vision mirror of history. Species and subspecies were dashed in even the most remote and bitter climes. The pivot from whaling as an extractive industry to whales as heralds of a conservation ethic may seem, in retrospect, an unprecedented triumph—but the veer would have been greater had the success been achieved during the period when the animals were not yet a rarity. For activists, there was leverage in the fact that as whales vanished, their charisma intensified—the public's enchantment with whales grew at the threshold of their disappearance.

And yet, the shift in attitudes *was* immense: the anti-whaling coalition required an intensity of transnational effort unlike any that had ever been known, coordinated across multiple spheres of action. For many ordinary people, joining anti-whaling marches meant participating in a globalized environmental citizenry that had never, before, been imaginable. They were called to be responsible for something they rarely, if ever, beheld: an icon of no nation, but of the planet at large. Protecting whales meant more than saving animals for either future resource exploitation or, indeed, for the animals' own sakes; the existence of whales shored up reserves of emotion associated with meeting nature on its most expansive dimensions: emotions like awe, humility, wonder. Geopolitical and ecological conditions may have incited the collapse of twentieth-century whaling, but that we stand now in a moment when the stranding of a single humpback whale occasions tender feelings, and a debate over how best to stage the animal's merciful death, is a turnaround so complete as to be rightly deemed incredible.

Whales—contrary to the gold they'd been compared to in the nineteenth century—are a regenerative resource: finite but, to varying degrees, able to repopulate. As a study in the junctures between economics and culture, twentieth-century whaling offers a germane provocation for the twenty-first century. Whaling, an ecologically untenable industry, was not suppressed by the invisible hand of the market as alternatives to its use became viable—rather, it continued long after it had ceased to be advantageously economic. Our expectation that renewable energy sources will, as a matter of course, supplant fossil fuels by function of expediency runs counter to this history.

Cave painting is an accretive practice, layering pigment over rock, but the making of a petroglyph is subtractive; inch by inch, rock matter has been chinked away. What a petroglyph fundamentally consists of is: a gap. That the effect of a petroglyph is to recall the form of an animal— that it creates the apprehension of *a presence*—is a mental feat, for a petroglyph is an artwork made out of absence.

What presences are left, after something is taken away? Before whaling was an economic force, it shaped culture. After whaling ended, its ecological aftermath continued to accrue. We stand now, many centuries on, but the story of whaling is not beside the point or, more simply, old news. The extent of the damage done to Earth's marine environments, by industrial whaling, is, even now, still being quantified by biologists— and its effects continue to amplify. Kelp forests off southwest Alaska, lofty as any waterless woodlot, are thought to suffer the demolishing attentions of herbivorous sea urchins today because killer whales, lacking for larger whale species to prey on in the lee of industrial whaling, turned to eating sea otters ("like hairy popcorn," as one researcher put it). The otters would otherwise have suppressed the ravening urchin populations, being one of the urchin's few natural predators.

This is an unexpected outcome: fewer bowhead and sperm whales in the Bering Sea, more urchins. At such profusions, these thriving urchins are a scourge; they move in fronts, shredding plant life and exposing, behind them, a seafloor kibbled to lumpy sand, bereft of marine life. These newly terraformed undersea environments are known as "urchin barrens," and though, for a season or two, they might be wandered by smaller, scavenger urchins, like an avant-garde literature made exclusively from punctuation, they sustain no recoverable narrative. A barren rarely returns, without human intervention, to being a thicket of fleshy macroalgae, and a home for rockfish, sea snails, and seals. Decades after commercial whaling was concluded, the influence of whaling persists as a still-creeping wasteland, drawn forward by the appetite inside these spiny globules. The impact of whaling fleets on undersea food webs will likely never be completely described; ecological instabilities introduced continue to cascade and restabilize.

The aftershocks of whaling clearly reverberated in terrestrial and aquatic environments, but perhaps even more bizarrely, we now know that whaling altered the air and changed Earth's atmosphere. In mid-2010, a group of scientists from Flinders University in Adelaide published a remarkable set of findings. They had confirmed that the activity of one species of deep-diving whale—the sperm whale—significantly and quantifiably affected the composition of atmospheric gases worldwide. Later studies found that the same was true of humpbacks.

The researchers explained how. Whales function as nutrient "pumps" in the sea by feeding at depth on squid and krill, and then releasing fecal plumes—long, flocculent excretions, typically orange in color—at a shallower level. In this way large whales move great volumes of organic matter from unstirred or slow-moving lower waters up to the more rapidly mixing photic layers. (Whales seem not to defecate down where they feed, likely because the higher pressure requires them to shut down some of their bodily functions.) In the iron-depleted waters of the cold seas the nutrients needed to fertilize the unicellular organisms and tiny plants that form the base of the oceanic food pyramid are meager. There particularly, whale manure proves a vital catalyst in accelerating plankton abundance. The vertical motion of whales also kicks up the plankton, like some dark force moving through a nebula with stardust clinging to its wake. The turbulence created, called "diapycnal" mixing, brings more plant life toward stronger light and so to greater rates of photosynthesis and growth.

These plankton absorb carbon dioxide and emit oxygen on a planetary scale. Over time they are smattered and egested by zooplankton and fish larvae, dying microscopically, expansively, to sprinkle through the ocean's interior as sediment (marine snow). Planktonic remains drag down atmospheric carbon to settle on the surface of the seafloor, eventually being buried beneath it by yet more settling debris, concentrating and locking carbon there for centuries in silt. This mechanism, the circulation of plankton, is thought to currently account for the absorption and displacement of roughly half the carbon dioxide produced by the burning of fossil fuels—a greater proportion than is sequestered by rainforests and by all other land-based vegetation combined. Whalefalls similarly contribute to the downward travel of carbon; a carcass of forty tonnes carries, on average, two tonnes of carbon to rest beneath the sea. That much carbon would otherwise take two thousand years to accrue on the seafloor. It is well-known that timberlands help to regulate the climate; now we understand that animals do too. Each whale has been calculated to be worth more than a thousand trees in terms of carbon absorption. The *Guardian*'s George Monbiot has called large cetaceans "a benign form of geo-engineering."

The removal of so much whale mass from oceanic ecosystems, and the implication of whales into chains of terrestrial, human consumption, therefore had an indirect, but palpable, effect on the chemical composition of the atmosphere. "A century of whaling equates to burning more than seventy million acres of temperate forest," said Andrew Pershing, a scientist from the Gulf of Maine Research Institute in 2015. Far from being mere passengers or victims carried along by the environmental crises set in motion by climate change, according to these models, the presence (or absence) of whales in the sea continues to shape carbon dioxide levels. More incredible yet, researchers have projected that increased numbers of whales could help offset a measurable quantity of emissions. Joe Roman, a conservation biologist and lead author of one of the scientific papers on whales as carbon sinks, said, to Alaskan radio: "This may be a way of mitigating climate change, if we can restore whale populations throughout the world." A study conducted by the Institute for Capacity Development in the International Monetary Fund (IMF) found that increasing phytoplankton productivity by just 1 percent would have the same effect as the sudden appearance of two billion mature trees.

I had been troubled by the notion of whales as landfill; of cetacean bodies as, in some instances, a type of animate superfund site. But this research recasts whales as a means of renaturalizing the air—not as the end point of atmospheric pollution, but as the mechanism of its remediation. Whales as gardeners in the greenhouse. I wanted to breathe in and believe it.

The whale at Balls Head has not been reengraved in recent history—though that it hasn't been upkept doesn't mean it has been separated from its stories. It mattered, and it matters still, how shade gets caught in the line of the engraved whale, the way rain circuits in the channel, or where water overflows to pool in moonlight, like poured metal; it matters—I cannot know how—that the sunset drags along the cleave, that the last light drains off from this angle, or that. Shallow, and growing ever shallower as a result of erosion, a time will come when the whale will lift out of the rock completely. Lichens stipple its head. Council workers once daubed its outline with road paint to make the whale more visible, and though the paint was scrubbed off in 2008, square holes betray where the posts of a pink wooden fence were, for a time, driven into the rock, penning the whale into a tight rectangle for sightseers (this, at least, prevented people from walking over it). Today small-time vandals have added their own demolishing logos to the whale: a tiny window, several tally marks in paint pen. Or perhaps someone in a high-vis vest will since have come by with a rag and mineral spirits to burnish these marks away.

Elsewhere in the world, the extant art of early cultures has tended to be preserved in sites that are challenging to access: remote sea caves acrid with guano, and dry, windless rock shelters found at dizzying altitude, above the clouds. Yet, the fact that the Australian petroglyphs extend into urban areas—and though they are inscribed, not in subterranean grottos but, more typically, onto open-air platforms—does not mean they are all prominent. Where land meets sea, tides capriciously pull sediment up over the carvings, concealing them until the right hour comes to sluice back, leaving a residue of white sand settled in the grooves. Rumors circulate of mansions with trapdoors and panes of glass inset into basement floors, displaying animals imprinted on the foundation stones. In the midst of a drought, a petroglyph will sometimes swim out from under a recoiling lawn, as if to willfully insist on conscience.

Chancing the solemnity, one day when I visit the Balls Head whale, the gift I am given is this: an iridescent green beetle, zipping along the divot of the line, like a slot car on a racetrack. A bead of impossibly shiny color, it goes around and around, zooming and stopping and zooming again. An animal trapped in the image of an animal. A breeze catches it, and it flies away.

02

The Oooo-ers

Eden — Whale Sneezes: How They Carry — Picture-Perfect — Close,
But No Touching — Humpbacks — On Whales Being Left- or Right-
"Handed" — Hunting with Bubbles — Swimmerets — Krill: A Faunal
Tide — Pink Milk — Do They Eat Rain? — The First Things with
Faces, Being Only Faces — Does a Whale Come When Called? —
Prawn Song — Whale "Footprints" — Spouts and Exhalations
— A Wilderness Without — Minified and Mystified — Drolleries
— Ecotourism — "W" in the First Alphabet — Love — Pandas as
Leverage — Letters Signed "Voices from the Future" — A Soundtrack
to Quartzes and Channeling — Disasters in Common — Wreck Event
— Between Thing and Kin — Sea as Inner Space — Freud and the
Eel — Acidifying the Oceanic Feeling — Whales, They Leap! — Does
Jumping Serve a Purpose? — Mer-Whales Share a Smoke — Whale
Conveyor Belt — Again, Numerousness — Sei Whale Graveyard —
Hand-Feeding — Mugging — Holy Shiver — The Animal You Own —
John Berger — Haruspices — Dry Dreams — Take Only Pictures — A
Sign: The Shape of Australia — *Augenblick*

In a dawn still recognizable as coral-colored, I go looking for living whales.

The small mountain that fills the window when I lift the blind is Eden's Mount Imlay. That higher world, knifed by the sunrise. Amid gullies dripping with tropical ferns, a bat with diamond-shaped ears surely tweaks back into a cleft, powdered in pollen, sated on moths. Every plant up there the staircase to another plant. A seven-hour drive south from Sydney, Eden amply satisfies visions of paradise that detach from the haze on inland highways, though the name does not, as it turns out, nod to any Elysium fantasy—"Eden" is merely an honorific, the surname of a colonialist British Whig.

Over the past two afternoons, it has been possible to spot splashes on the surface of the South Pacific impaneled in the other direction, at the end of Eden's main drag. Fluke slaps and airy uprushes—humpbacks. I have waited on the bitumen at the edge of town, shading my face and squinting for any glimpse of a humpback throwing its tail. If, for a moment, a whale does leap into the sky, it is as small and dark as an insect. Stand ankle-deep on the beach at night, and you hear them. That is, you hear the whales *breathing*. How the sound carries across the water, I do not know. A whale sneezes: you jump. It sounds like a roll-up door slamming.

Eden's whale-cure hotel was demolished long ago. The hotel's golden years passed in the late 1800s, a time when twenty-seven whaleboats waited on the call of "*Rush O*" to bring some two hundred whalers dashing into Twofold Bay, throwing the quick-release hitches on their launches. Humpbacks, southern rights, pilot, and melon-headed whales were hunted here. At the height of Eden's shoreside whaling industry, eight on-land whaling stations bubbled and stank, including the Kiah Inlet tryworks—extracting oil for buyers in Australia, the United States,

Europe, and China. Through the bush, the whalers walked paths paved with yellowing whale vertebrae. But local whaling wound down before the Depression. The descendants of those humpbacks that have since returned are now the foundation of several ecotourism ventures. At the grocer's the day before, I picked up a brochure and read: *Picture-perfect humpbacks. Be delighted and amazed. Be awed and inspired.* The whale-watching catamarans depart early, ahead of the onshore breeze.

Greetings ripple through the group of tourists gathered on the wharf. It is the shoulder season, late October, so most holidayers are either retirees or very dedicated whale watchers. A few young families: kids grousing, canyoning around. A woman applies streaks of sunscreen to her girl's arms. About seasickness and its remedies I hear: "copper bracelet," "Kwells," "scopolamine—the patches," and "vomit off the edge, but take your teeth out first." The whale watchers don fleeces; they adjust the webbed straps and drawcords of backpacks. Anticipating the wind's unwitting thievery, they have brought along the scratched sunglasses from the beach bag or a cheap pair ordinarily relegated to the car's console.

On the boardwalk, this sign: GAZE OUT TO SEA AND BACK IN HISTORY, WHEN BOATS WERE MADE OF WOOD AND MEN WERE MADE OF STEEL!

Before we embark, there's a briefing from one of the crew. Legally, the tour is permitted to get within 330 feet, after which the skipper is obliged to cut the engines and wait to see if the whales are inquisitive enough to nudge closer. A thousand feet out, any vessel that has sighted a whale must decelerate and only putter ahead. No one, we're told— strictly *no one*—is permitted to feed or touch a whale, or to try to. It's hard to see how this might be possible from the deck, but the forceful-ness with which the prohibition is iterated suggests both that whales will swim near enough to risk it, and that people have, in the past, flouted these commands. Whales within arm's reach: the prospect zings between us. Eden's longtime tour operators trade on speculation that humpbacks know their boats by color and shape; rumor has it the whales not only *remember* specific watercraft, but express interest in, even affection for, some vessels over others. "Mugging" is the word the crew use for when whales approach of their own volition.

Together we loiter at the dock, shifting in our frigid shoes, until the last member of the tour jogs up, apologetic. In his fist he holds an antiquated brass telescope: the accoutrement of a pirate. But an heirloom, apparently. We're ushered across a gangplank. People find a seat, or hold on to a railing. The boat powers past the breakwater, a bank of concrete blocks on which pelicans bask untucked; origami, dropped in distraction. Ahead, the sea is the beige green of the inside of an apple. A singular hope beats its wings within me: Where are the whales?

Humpbacks are the cetaceans most Australians would draw, if asked, to attach to the label WHALE. The whale's scientific name, *Megaptera novaeangliae*, means "giant-winged New Englander" and alludes to their long, notched pectoral fins, which have the heft of hardwood planks. Adult humpbacks are the length of a limousine. Their heads have the dimensions of anvils. Studded with a type of barnacle that adheres only to whales, the humpbacks exhibit the countershading common to many cetaceans: a mottled white underside and darker, zinc-blue skin where the whale's body is orientated toward the sun. Dorsal-ventral coloration moderates sunburn and helps to camouflage the humpback from any animal eyes, looking up into the light-filtered sea from below.

A cosmopolitan, adaptable species, humpbacks inhabit the waters off New England and a range of marine environments beyond. These whales can also be seen near Hawaii and Brazil, in the Sea of Japan and in the Mediterranean Basin, in British and Norwegian waters, in the Bering Sea, off Costa Rica and Bermuda, and along the east coast of Taiwan. Almost all humpbacks travel with the seasons, with the exception of one population found in the Arabian Sea. During warmer months, humpbacks tend to converge nearer to the poles. As it gets cooler, the whales move in toward the equator, following coastlines and passing seamounts to enter luminously clear, yet nutrient-poor waters. Some cetacean species stay bonded as lifelong, familial pods, but at either end of the Earth the humpbacks are inclined to come together in social assemblages that are periodic and circumstantial, before migrating alone, as pairs or, rarely, in triads.

Those humpback whales that are seen traveling past Australian isthmuses and headlands reside for much of their lives at elemental latitudes in the south. There, the humpbacks forage for krill: matchstick-length, shrimplike crustaceans, translucent and spotted with fire orange, like drips of heated glass. The humpbacks' throats are furrowed into wide folds that expand, as the bellows of an accordion do, when the whales lunge to seize upon their squirming prey—they then strain the water out, catching the krill on the inward fibers of their baleens. Pack hunters, the humpbacks will sometimes send a hoop of bubbles up from below to cluster the krill together. Then they rise, open-jawed, into mouthfuls impounded by the aerified nets. Infrequently, the whales have been observed stalking krill by skating along near the seafloor in parallel, as though on dolly tracks, concentrating the crustaceans into a channel between them. Electronic tagging of humpbacks has revealed that during feeding the animals roll preferentially toward the left or right (their dextral or sinistral side); like us, humpbacks are left- or right-"handed."

Whether or not the origin is apocryphal, the word "krill" is said to be onomatopoetic. The krill are individually voiceless, but *krill* is, likely, the sound made by millions of tiny legs ("swimmerets") pattering beneath the seawater's meniscus; a minuet of scurrying, up and down, along the verso of the ocean's surface. From the Weddell Sea to the Antarctic Peninsula, krill congregate in great plumes that glow (their eyestalks are luminescent) around the pack ice. Krill sometimes move into open water as round, roiling balls, thirty thousand thick. The aquatic analogue of a billowing sandstorm, in this configuration they turn tracts of the ocean maroon red. Krill can collect in formations so numerous, they are visible from the upper atmosphere. How krill coordinate with one another remains a puzzle, though biologists have observed what may amount to hierarchies in their schools.

Ecologically, krill matter most as biomass and the basis of a polar food web. There are many more krill on the planet than any other animal; some four hundred trillion individuals. Their daylong industry is to scoff clouds of phytoplankton, high in the water column. They also consume the vast vegetable that is algae an inverted, inch-thick pasture, thatching the ice, which the krill both graze on and huddle in with their larvae. Krill are a faunal tide within the sea. They clamber up to glut on plankton, and then the krill descend again, jerky as little cocktail umbrellas, while they digest.

Throughout the Antarctic summer and spring, the micro plants and plankton the krill consume flourish, and so, too, do the twiddling krill. Then, as the season turns and the planet's most erasing cold begins to sweep in, the ice cover expands across the ocean's topside and deepens into it. Sunlight hours dwindle, decreasing the opportunity for photosynthesis. Underwater, the ice is pocketed with caves, divots, and sharp terraces, into the vaulted ceilings of which the algae contracts, fostered by what meager illumination still passes, through the thickening freeze above. It is believed the krill then die. Or, they follow the smeared plant life into the architrave of ice. When krill starve (over the unstinting nighttime of the austral winter, most do starve) they shrink, though their eyes stay the same size. They end up bug-eyed before spring comes around again.

As the season sets to piecing together its all-white jigsaw across the nocturning sea, the humpbacks migrate northward, away from the privations of the Southern Ocean at the end of the globe and across the Antarctic Circumpolar Current. These animals will travel more than six thousand miles in the course of their annual migration: elsewhere, humpbacks go even further, taking round trips of nearly ten thousand miles (equivalent to double the length of Africa, north to south). Knowledge of a migratory path is passed from parent to offspring—how, exactly, isn't known—but a whale's solitary orientation in the open ocean still demands innermost feats of navigation. What few stable landmarks there are on the seafloor are cloaked by marine darkness. The Southern Ocean runs four miles deep at least, in different places: more people have visited the moon than have ever journeyed those abysses. In the whales' forebrains, or in cells in their eyes, particles are thought to draw to geomagnetic fields—a faculty not visual or tactile, but nonetheless sensory, and which likely aids in their migration. The humpbacks' route splits, like old rope, with the weakening winds rushed up from the pole, and then the whales cruise along both sides of the Australian continent. On the western seaboard, some will reach as far as Cape Leveque. In the east, humpbacks soar past the Ribbon Reefs and Cairns, before pivoting to return. The whales come to these shallower, coastal seas to breed and give birth at a time during winter when their main food source near Antarctica is depleted and tricky to obtain beneath the widening ice. Though it may have grown chillier to us where we, terrestrial vertebrates, live, the whales flee a more crushing cold.

A female humpback matures at between five and seven years old, after which time she is capable of bearing a single calf every third year or so. She gestates for eleven months. When humpbacks are born their fins and flukes are supple, though they are instantly able to surface to breathe, and to propel themselves through the water. A fact about whale rearing that stays with you: humpbacks feed their young on pink milk. Pink, as an indirect result of their rosy diet of krill. Around one hundred gallons per day, and ropy, on account of the milk being close to 50 percent fat. (Q: How does whale milk taste? A: More or less, it would taste of butter.) Humpback whales have no lips to speak of—their mouths have hard edges—so the calves are thought to curl their tongues into long, dexterous funnels, with which to suckle. On the return leg of their migration many calves sport roughened and blistered snouts from the friction of attempting to stay latched beneath their mothers. A humpback calf is weaned between six and ten months, and, at a year, it is independent.

Having gorged themselves on krill in the Southern Ocean and laid on thick fat reserves, adult humpbacks eat hardly anything during their migration. Very rarely, and opportunistically, the whales pursue offshore churnings of small baitfish to consume in Australian waters. In the northern hemisphere, thirteenth-century Norse shipmen held that the whales they observed fed only on the evanescence of darkness and rain falling on the Atlantic. Which it can seem the humpbacks do—eat night and rain alone—off this continent, at least. The whales grow lean, then skinny on their return journey. At the last, they're starving: their body size whittled down by close to a third.

As the tour boat enters open water, I notice jellyfish. So many, splattered into the sea all around, as if dropped from great heights. Primordial things. Too soft to be harmed by the hull, they are the color of tea. Theirs are some of the planet's earliest eyes: simple ocelli, capable of perceiving light, though little else. *The first things with faces were only faces*, I think, looking at these jellies, gelatinous like beauty masks, scrolling past the boat. Is this a so-called bloom, or is it a natural aggregation? The question drifts away with the jellyfish in our lee. As yet, there are no whales to be spotted.

Eden's humpbacks are supposed to recognize, in particular, the voice of Ros Butt, our tour leader. People in town tell me whales come to her as dogs do, when whistled for. Ros doesn't dispute this, though she hedges when I ask her whether it's true. "I guess we'll see," she says, ripping the top off a box of Bushells. With her husband, Gordon, she has been running Cat Balou Cruises for twenty-eight years. *Cat* for *catamaran. Balou* derives from the Old French for "good water."

Our boat, with around thirty people aboard, slows after the better part of an hour. My body feels like a birdcage from which a bird has fled, its perch left swinging. The noise of the ocean and the swishing winds overlap the boat's buzzing engines. There is nothing to exclaim at in the wide yonder yet, but the crew speculate on whales nearby, sighted by another vessel (tours and local fishing operators stay in touch via shortwave radio). A hydrophone is pegged into the water to eavesdrop on the babbling world below. A moment of static. Then, trilling is what we hear. Chirping, high-pitched, amplified through onboard speakers. We smile and nod, expectation rising. But it's prawns, this sound, we're told. Everyone falls silent. *e e e e e* go the prawns—animals that fit snugly into the printed vowel of their sound. We concentrate hard for the tonic note of a whale. Nothing. Only the gritty pestle of the currents, rasping the seafloor.

The hydrophone is returned to deck. Next we're told to look for "footprints"—unwrinkled patches on the topside of the sea, indicating where the water has been displaced by the upward stroke of a whale's flukes beneath the surface. Ros reminds us the migratory whale species can also be identified by their spouting (not jets of water but geysers of steam speckled with nasal moisture: the condensation of breath heated in the whale's lungs). Southern right whales issue V-shaped blows because their blowhole bifurcates two distinct cylinders of vapor, vented at right angles. These exhalations tend to hang about, like unspoken thought bubbles. Gray whales (not found in these waters) are described as "sending out love hearts": a twee, breathy *amore*. We, the watchers, are most likely to encounter humpbacks, which emit spikes of misted air, some ten feet high, that are prone to being bowdlerized by the wind. The humpbacks' exhalations conjure the action of some bellicose deity reaching down to puncture the Earth with a pin.

A few people stare through their binoculars, agog. Minutes earlier, the man with the telescope could be seen turning it in his hands like a pepper grinder, but now he drums his index fingers on the rail; bored, maybe. An older man zips up the hood of his granddaughter's jacket, as well as his own, and kneels down to align their faces in a cottony tunnel of talk I cannot overhear. Other discussions taking place around me trail off. The ocean shines like marble.

What environment was ever more shielded from our collective imagination than the underside of the sea surrounding Antarctica? Unlit omnisphere, far-fetched. White noise; ice shifting, krill krilling. Trundled by see-through salps, orbital sponges, and other questionably animate organisms, the seabed shilly-shallies into murk, lacking all tactility and aspect. No writer, in good conscience, could reach for a word like "terrain" to detail it. A void. The Southern Ocean is galactically dark. A mirror for the Vantablack of the cosmos.

Yet, there is an allure to this space. An allure connected to the whales' spellbinding charge, it has something to do with why many of us have been drawn out on the tour boat. We have come to be put in touch with a vaster world, one that exceeds human dwelling—if not human influence. We long to make contact with a wildness without.

Writing in *The Baffler*, Barbara Ehrenreich has observed that contemporary wildlife tourism sells something "people have more commonly sought through meditation, fasting, and prayer. . . . Whale watching companies from Sydney to Baja promise whale-mediated 'spiritual experiences.'" As environmental degradation grows to be a source of widespread anxiety, seeking encounters with animals from remote habitats dramatizes a distinctly modern yearning; the urge to feel oneself minified and mystified by wilds where nature remains demonstrably sovereign, abundant, and vigorous. If this is not quite the sort of spiritual hunger Ehrenreich is referring to, then it seems, to me, a keenly psychological one. We ache to meet the limit of the human world, and to look past it.

Gaze out to sea and back in history. People once feared there was a terrible, existential emptiness in the ocean, an unpeopled and unending *openness*. In antiquity, cartographers populated the seaward frontiers of their maps with drolleries. Hand-drawn pictograms, drolleries are whales hybridized with sea serpents: monsters adorned with antlers and tusks, scales and sprigs of feathers. Inheriting an Aristotelian fear of the void (the Ancient Greeks suspected the sea might go on forever, like the sky), medieval mapmakers used drolleries both to enclose the known world and to indicate where nautical knowledge gave way before a front of myth. It made sense to find whales cavorting in this habitat, crisscrossing the line between rumor and fact. The conviction that every terrestrial creature would be revealed to have an inverted, underwater twin meant that mariners' descriptions were routinely twisted to affirm comparisons with land animals—a pod of narwhals, for example, each sporting a single, pointed horn, might be recalled as a gigantic, swimming porcupine. Whether whales were natural animals or fantastical beasts, few people of this era could claim with authority. On the fringes of the ancient void, whales shape-shifted, thrashed, and fought. Cetaceans long resisted surveillance, to such a degree that even as late as the 1950s, Rachel Carson could be found writing that whales arrived in coastal currents "having come from no one knows where, by no one knows what route."

At this moment in time, we fear the opposite of those early cartographers: we fear that there is *no void*, no part of the planet untouched. This fear intensifies on hearing of, among other obscenities, debris fields in the deeps, where tresses of plastic shopping bags are picked over by pallid, flea-sized amphipods (the only ghouls so versatile). A powerful, creeping horror. Our hunch is that what lurks beyond the end of the mapped territory now is neither a host of exotic drolleries, nor large, enchanting wildlife—there, instead, lies the dross of our everyday lives. A haunted house never inhabited but nonetheless built by our hands. This loss of containment destabilizes not just how we see nature but also our sense of self. Were *we* the monsters of trespass, all along, closing down the dominion of the wild?

Spotting humpbacks migrating up from the polar sea has constituted a reassurance, ever since Antarctic whaling ceased. Nature *can* return from frailty. In the course of these reveries we have tended to view the Southern Ocean as an inscrutable *mare incognitum*, too removed from the dailiness of human life to be spoiled. This is not that other sea we have read of, agglutinating into rubbishy gyres. Withheld pristinely, neither is it that ocean concealing the sleeping injuries of oil slicks. No symptomatic fronts of urchins march there. Only in recent history has the Southern Ocean come into focus, in the public mind, as the mainstage for a new, more encompassing disaster: climate change. That white, crystalline-edge world, disappearing into black salt water; the retreating maxima of Antarctica's sea ice makes global warming strikingly visible.

If what we hope to experience, encountering whales, is the validation of a bigger nature—a world that is separate and autonomous—then how will whale watching change, in a time when humpbacks force us to confront our involvement in remaking even the most isolated environments, the whales' half-year habitats? Can whales help us to conceive the extent of our impact on places so stark and elemental, they rebuff imagination; places few can envision, and fewer yet, ever visit?

From the back of the *Cat Balou*, I can hear Ros giving an informal talk about whaling to pass the time. At the peak of twentieth-century whaling, she says, the number of humpbacks migrating to eastern Australian waters dropped to 250—seeing even *one* off this coastline would have amounted to a miraculous visitation. In recent decades the international community has united to preserve humpbacks for future generations. More than twenty *thousand* humpbacks visit this coastline today. Ros radiantly reminds her audience that people all over the planet know whales, even if they live in apartment blocks far from the coast. These are the world's largest migrants: "W" in the child's first alphabet book. Whales are no nation's exclusive claim. Whales belong to a part of the living Earth that every individual has a stake in preserving—a commons. Crossing our maritime boundaries, whales have brought us together.

In Australia (not irrelevantly, the nation with the highest rate of mammalian extinctions worldwide and the country with the greatest number of unique, endemic animals), the story of the restoration of humpback populations from the brink of their extirpation is at the center of what it means to go whale watching. That the whales themselves were, very nearly, relics, is key to their spectacle, and the breadth of their appeal. As Ros continues her lecture, I walk around the boat to watch Eden's retreating shoreline, its trees shrunk to asparagus tips. This remains tourism in proximity to extinction, I think, an attraction formed by the thwarted specter of death. We've come to see what we've saved.

Ecotourism operators package encounters with whales as a double marvel: the spectacle of witnessing immense animals, unhindered in their marine habitats, offered hand in glove with a chance to contemplate our own species' capacity for temperance. Each cetacean is an evidentiary exhibit both of human care and the resilience of wilderness. To spot whales is to animate a redemption narrative; to reflect on the moral turnabout of our kind. We wonder at our ability to command nature to such decimating ends. We wonder, too, at our capacity to control ourselves.

But how did people come to *love* whales, after so many centuries of inflicting violence upon them? And how did the defining quality of that love become its supposed universality?

As far back as 1972, the United Nations Conference on the Human Environment had pushed for a moratorium on whaling, seeking to "declare whales the common heritage of mankind"—but it would be close to a decade before the IWC was able to mobilize such a ban. Over the intervening years, anti-whaling advocacy groups (along with former whaling nations) helped to reconfigure the membership of the IWC by providing financial backing for non-whaling nations, some of them landlocked, to become members of the commission. Oman and Switzerland joined in 1980, followed by Jamaica, St. Lucia, Dominica, Costa Rica, Uruguay, India, and the Philippines; the next year, Senegal, Kenya, Egypt, Belize, Antigua, and Monaco were recruited. China was purportedly induced into the IWC by an offer, from the World Wildlife Fund (WWF), to establish a million-dollar panda reserve (as the world was wrapped into the whaling debate then, so, too, were the fates of a handful of other endangered animals).

Since the IWC had become less dependent on whaling nations for its operating budget, biologists on the commission's scientific committees felt emboldened to adopt a more vocal, conservationist stance. But it would take recording artists, filmmakers, and authors to marry scientific concern to public action. Greenpeace UK called their approach "more an imagology than an ideology": footage of mettlesome orange inflatables, forced between harpooners and whales in the lawless arenas of the high seas, drew the sympathies of many. Photos of a rally in London, fifteen thousand strong, were broadcast around the world, while books like Farley Mowat's *A Whale for the Killing* (1972) began to appear on syllabuses. "A modern Moloch," Farley called Norwegian whaling. Letter-writing campaigns recruited children. From New York to Buenos Aires, Canberra to Athens, young people and teenagers signed off, vehemently: "a voice from the future."

It helped that whales had, as a consequence of underwater sound recordings, morphed into icons of the electronic age. First recorded in the 1950s, whale song testified both to the expansion of technology into the far reaches of the sea (cetaceans were initially recorded during acts of military snooping) and to new kinds of compelling, cosmological foreignness *within*. Field recordings from the frontier of whale endangerment unveiled a covert communications network: whale to whale. What had been secret about them might also be used to uncover what was hidden in us—whales became a natural soundtrack to shamanic meditation, channeling, Jungian analysis, past-life regression, opening the third eye, and treating the aura with quartzes. They were co-opted as the spirit animals of an awkwardly pantheistic suburbia. *Songs of the Humpback Whale* (1970), produced by the biologist Roger Payne, went multiplatinum. The album remains the largest single pressing of any LP, to date: ten million copies were made.

Under civic pressure, Australian whaling ceased in 1977, and though Australia was the last English-language nation to persist with commercial whaling, it was also the first country to shift to an official anti-whaling rhetoric rooted in a notion of whales as "special," "intelligent" animals, rather than as stocks for future reaping. The Australian Whale Protection Act 1980 became the earliest piece of domestic legislation, worldwide, to address cetaceans exclusively. Close on its heels, the global moratorium on commercial whaling stuck in 1982, with a three-year window for its implementation by 1985. Having lodged a reservation to the moratorium, Iceland continues to whale today, and, perhaps more contentiously, Norway does not consider itself to be bound, having objected at the point of ratification. Russia objected to the agreement but ceased whaling. Japanese whaling scaled down but continued, as sanctioned by the agreement, for scientific research up until 2019—when Japan left the IWC to whale, openly once more, for a supply of meat protein.

The rights of traditional whaling cultures to continue to hunt a set number of whales were also established by the moratorium, a dispensation that had been earned, most notably, by the efforts of an Inupiaq delegation. Whales are still hunted by the Inuit in Alaska and Canada, the Chukchi people in the Russian Federation, in the Atlantic islands of Saint Vincent and the Grenadines, and by the Lamalerans and others off Indonesian islands including Lembata and Solor. The Makah tribe seem set to commence hunting gray whales again, off the northwest tip of the continental United States in Washington State— having sought a waiver to return to their traditional haul, taking one-to-three symbolic whales a year.

The Sydney-based international relations academic Charlotte Epstein has called the 1982 whaling moratorium "the first global, precautionary suspension of a natural resource's commercial exploitation"—a statement that places the ban firmly within the context of extractive and cash-crop industries, rather than considering it a part of broader efforts to preserve animals because animals, of themselves, have a right to life. Zooming out along these lines, it is possible to see that the anti-whaling movement pulled its momentum from other efforts to curtail pollution and protect wilderness—it was not, at heart, an endangered species campaign.

At the turn of the 1970s and into the early 1980s, the transnational nature of environmental crisis had come into focus across several quarters, as the result of a series of prominent calamities. In Europe's black triangle (land thirded between the Czech Republic, Germany, and Poland), acid rain clouds wandered over borders to debark forests and sulfurate waterways, sending fish belly-up. Ozone depletion, caused by the long-term mixing of solvents, propellants, and refrigerants into the air, was shown to have created a hole in the atmosphere through which UV radiation spilled. The nuclear explosion at Chernobyl, near Pripyat in Ukraine, elicited fears of fallout wafting across continents. A growing consensus maintained that remedial and preventative action addressing these sorts of threats could not be limited to petitioning national governments in isolation. Solutions needed to be enacted across regions, if not on a world level.

Against this backdrop, whaling was the most striking example of a globalized, *zoological* disaster. Its history not only spanned hemispheres but converged on a segment of the Earth that major treaties had declared to be held, by all nations, "in condominium"—a polar reserve set off for nature (secondarily, for science). Beyond the carnage the whales had experienced, whaling represented something more: an industrial incursion on Antarctica and its southern sea, and the debasement of a universal park—nature's cleanroom. Whaling was a threat not just to real animals, but to *an idea* of wilderness namely, that it should be insulated from commerce. Saving "the environment" and saving the whale from whaling were yoked together by virtue of scale and of polar regionality; these goals became, implicitly, the same endeavor.

The logic parlayed one step more, then. To be anti-whaling was not simply to be pro-nature, pro-wilderness: it was to be pro-*worldliness*. The most important legacy of the movement was to establish a cooperative, planetary remit for "green" values and the sentiments of a global environmental citizenry; to demonstrate that compassion, as well as capital, could be an oceanic force that joined people together across borders. As we love those things that elevate us to the position of savior, so whales became loved not because they were innately lovable, but because, in being made subject to human mercy, they both revealed the extent of our power to change, and put us in touch with some loftier part of ourselves. And this is true, too, I think: we didn't just save wilderness *for* nature. Not for nature alone. Some self-regarding componentry, in us, craves the knowledge that wild places persist.

It is in the slipstream of this history that we define what we go looking for, when we go out to meet the whales in Australia. What we seek to locate, within ourselves, might be a capacity for awe and humility; we may want to join with others in being a part of this story of worldliness, celebrating life writ on its biggest dimensions and the collaborative effort that restored it. We wish to connect with an idea of the wilderness that maintains, in mystery, without us.

The beachfront narrows to an ocherous ribbon, belted by blue, above and below. After a while, a handful of shearwaters appear in the air above the *Cat Balou*. The birds flash around us; like knife-thrower tricks at a circus. Diving through the water, each is crowned in a diadem of bubbles. The shearwaters come from Antarctica, like the humpbacks, and also Siberia, South America, and Japan; they arrive in Australia, where they often die in large numbers from exhaustion. Such bird deaths, en masse, are known as "wreck events." A single wreck event used to happen every ten years or so—the result of irregular, rough weather overtaxing the birds' reserves—but flock-wide collapses occur almost biennially now, the feathered bodies washing up on the tideline, emaciated with hunger. Their prey are vanishing from the migration route as oceans warm. These shearwaters fly away, to where? I wish I could write *These birds leave in the direction of exquisite fortune, to live long, robust lives.*

From the cutwater, and either side of the boat, the watchers squint in all directions, and still there are no whales. A few people have given up, withdrawing to the lower deck for a rallying snack. I nurse a cup of instant soup, freckled with rehydrated corn and carrot, warming my hands. The light changes, the ocean turns waxy. Little slaps, notches and gloss.

"The whale who will come soon," a boy begins, in the rising pitch of hope.

"The whale *that* will come soon," a man—I suspect, his father—corrects, not unkindly.

As we wait and the tour meanders on, I ponder whether the pronoun choice isn't, in fact, more than a semantic niggle. Designating a whale as "that" is distancing and turns the animal, grammatically, into an object. "Who" is personifying and brings the whale into a closer nexus with the speaker by implying it possesses human qualities. Australian newspapers, I think, use "that" for animals. (On returning to shore, I will go to *The Associated Press Stylebook* to clarify: the stylebook stipulates "who" is appropriate only where an animal has been given a personal name, and it offers this example: "The dog, *which* was lost, howled," . . . "Adelaide, *who* was lost, howled.") No doubt other languages have more middle-ground pronouns for animals—something between "thing" and "kin."

I go back up on deck to find the swell has lifted, shortening the depth of our perspective: the *Cat Balou* is dropped and raised. Water hammocks heap into hummocks, then fall back again. Clouds overhead loose bright patches to blow across the expanse, as if someone had unleashed a stash of foil Mylar blankets, the kind used for hypothermia. The effect is cyclically sedating and awakening.

You believe you could step off the boat and walk for days. Things go on like this for a long while.

Then Ros gets on the loudspeaker. "*Whaleeey*," she calls out, across the sun-blown sea. "*Whaleeey whale! Here, whaleeey whale!*"

We come to whales to be drawn out into the world, and yet the unpeopled sea is also a state of mind—an inner space. Our smallness, set against this sublime sea, is connected to the notion that not all the thoughts within us are accessible, that there are thoughts, in some sense, *exterior to the self.* Vast feelings that slide, beyond command, beneath the wakeful tending of our days. The pleasure is that we are mysterious, even to ourselves. The fabulist ocean, this inscrutable outer space whales return from, in a metaphorical sense, enchants us with our own inward enigmas. Many throughout history have figured the sea analogous to the human unconscious.

The Austrian psychoanalyst Sigmund Freud most notably made popular the expression "oceanic feeling" to account for convictions of limitlessness and eternality that might lead to an inclination for religious observance. Freud connected this belief to the irrecoverable sensations of being a baby—before an infant is aware of its separateness, when the contours of the self are blurred. A bygone time when, Freud claimed, each of us felt at one, with every surrounding object and phenomenon.

It's often forgotten that Freud began his professional life as an aspiring naturalist with a passion for aquatic environments. Freud's college drawings are of the nerves in crayfish and lamprey. He studied under the Darwinist Karl Claus, and undertook fieldwork at a zoological research station in Trieste, on the Adriatic Sea, attempting to identify testicles in eels. Later, the concept of oceanic feeling came to Freud through correspondence with the dramatist Romain Rolland, a mystic who described himself as a member of an "antique species," and whose work with eastern religions in the early twentieth century sought to explain the appeal of pious temperaments. After Freud, the ocean came to denote unbounded inner space, whether that sense of limitlessness formed around a spiritual consciousness or otherwise.

What ocean is this, that I have begun to call a *mare incognitum*? *Mare incognitum*—undisclosed sea. An imaginary place. Also, a mentality cast upon the globe. The purpose of this creation? The *mare incognitum* stands in for the unconscious, the unexamined interior. Déjà vu, inner visions. Misreadings, mishearings, the ineffable. What fruits in dreams. Slips of the tongue and free associations. A place where subcurrents of pain are metabolized, and from which erotic desire and creativity supposedly spring. Intuitions: alien, animal, sordid, and stray. All of this has a sealike texture. Doesn't it?

The idea that the unconscious can be envisaged as a sea has its own history. It owes its existence to narratives about what the ocean represents in the Western tradition and the twentieth century, narratives addressing how human mentality functions, and whether an individual's sense of self, separate from her or his community, is even interesting to plumb (or decorous to expose). It is a trope composed in art, worship, literature. How does a space as boundless and ungraspable as "the sea" come to seem personal and intimate? It took a long time before I came to realize that the processes by which the sea was turned into a metaphor for a person's unconscious were neither natural nor private. I belong to these narratives—maybe you do too. I pick up whatever thing a dream washes onto my shoreline and turn it over to look, hoping for the shimmery incandescence of pearl shell and fearing, instead, the exposure of a writhing, unspeakable underside—an impulse unmentionable.

The question I am left with is: What will it mean for our inner lives—those of us who cannot disavow the ocean as a psychological motif—if the twenty-first-century sea turns out to be not full of mystery, not inexplicable in its depths, but peppered with the uncannily familiar detritus of human life? How will we delight in a feeling of unboundedness, in the apparent limitlessness, the unknowable space of our unconscious, if the real ocean demolishes its symbolic history?

An emotional landscape, it turns out, can also begin to acidify. Something hard to quantify, I suspect, disappears from the palette of human experience, from how we articulate our selfhood to one another, and from our relationship to our own private, inner depths, when we encounter evidence of the ocean's despoliation. Much fascinating scientific work is being done on how animals are learning to navigate the changing world's new behavioral demands, but less is dedicated to how human cultures are adapting; how our language and imagination is corroded, and in what deficit of narrative and nature we seek for fresh expression.

Without notice, we're off. The engine strains, and the prow of the boat tilts back, hit by slab after slab of spray. It becomes necessary to un-lock your knees to take the impact and bounce of each wave, as when surfing—and it's excellent fun. No one can juggle a camera or binocu-lars now: we're all gripping to any available railing with both hands. A woman near me pukes a spectacular mustardy arc over the side. The woman's daughter looks on alarmed, from within the clutched tent of the waterproof jacket she has borrowed from her mother. She can't de-tach herself from clinging to a fire extinguisher, buckled to the exterior of the cabin (the only anchor point she's been able to locate). I give her a grin, and my lip catches on a dry eyetooth, studding in the cold. All knuckles whiten. Seawater griddles the windows, glistens the boat. Whole-head smiles sweep through the group, like the beams from bea-cons atop dozens of crazed lighthouses. "Whoop!" yells a man in a Sea Shepherd sweater, his hair whipped up on end. "*Whooop*," retort people on the portside, louder. There are whales, there are whales! If this is being announced on the speaker, Ros's voice is lost to the drag of air.

Humpbacks? Yes, humpbacks. Now visible, white like little teeth. Are the whales leaping? Leaping! How many? Three? Four whales? We are gaining on them! Three. They are white, and also black, chrome blue, and mineral gray. Will the skipper slow the boat? At that very thought, the engine modes down and the reek of exhaust pulls over us, thick as a quilt. The fumes dissipate, and the whale-watching tour motors forward at a more moderate speed.

People are putting on their sunglasses and taking off sweaters and lens caps. We're jostling around the bow to watch the whales in the distance, salt air high in our nostrils. Homunculus whales at this inter-val, erupting from a patch of ocean without pattern. Gradually the boat homes in on them, and we can see the humpbacks sending up great crescents of salt water, slamming their bulk, over and over, against the sea's surface. I want to describe it as beautiful, and sudden, like catching dandelions pinging open in cracks in the pavement, though it's more akin to witnessing a demolition from afar. You think of cathedrals fall-ing into their basements after a brutal reformation, the dynamite con-taining, yes, diatoms—the silicate reliquiae of ocean microorganisms.

Alleluia. Then, sometimes, a whale's flukes come down with the force of plunging iron—a move known as "lobtailing." *Boom.*

In the air, the humpbacks are majestic. They spear far higher than seems probable, pulling themselves almost wholly up out of the sea. At the apex of each leap, the whales barrel-roll, or capsize backward. The humpbacks look built for flying, as much as for their undersea peregrinations. They launch. The humpbacks only decide, at a point, to fall. These whales are wonderful. People with binoculars stumble, gasp, or cry out. The humpbacks stay below for long minutes, reappearing to the north or the east. In the time the whales are underwater, no one speaks. Cameras pop and whirr.

Ros lets us know that the whales' airborne stunts are believed to dislodge any "barnacles, critters, or algae" that grip to them on their journeying. What is happening here, she says, is as grizzly bears scratch their backs on tree trunks. However much the cetacean acrobatics look like mirth, we would be wrong to put that idea onto what we're seeing. The whales are dislodging pockets of lice (whale lice!) that irritate them.

Though humpbacks are a balletic species above water, the energy required to boost a whale out of the sea comes at an immense cost, physically—particularly as these whales are in the end phase of their migration and likely haven't eaten for months. A breach happens, as the cetologist Hal Whitehead puts it, "at close to an animal's maximum power output." Viewed from an evolutionary standpoint, therefore, breaching should confer on the animal some significant advantage. After returning to shore, I will send an email to Robert Harcourt, a scientist working on southern right whales, to ask his theories as to why humpbacks perform these vigorous leaps. In reply, he will pass on a recently published paper suggesting that, counter to Ros's idea of whales ridding themselves of parasites, the breaches might effect a kind of sonic communication; a splashing dispatch that thunders between humpbacks, miles apart, indicating something different to whale song.

It's also plausible, and I insist on it in the minutes of seeing them, that the humpbacks are playing. Unproductive frivolity—joy, pleasure—might need no evolutionary explanation.

The afternoon that I first arrived in Eden I passed two teenagers who sat talking at a picnic table in the courtyard of my hostel, truant from a street parade, both dressed as whale mascots. The pair had removed their whale heads like motorcycle helmets and pushed their hands through their felted flippers to smoke. Crowned with bluish tobacco, these young men, mer-whales, made me think again of the humpback stranded in Perth—the insuperable heat flaring within it as it died, and the apprehension it might have contained something human. Yet for the lively animals in the sea off Eden, the overall recovery of their species in recent decades has been remarkable. The narrative of these whales, I thought, could only be one of hope. What *is* the future for humpbacks?

To look at the top-line statistics is to be buoyed by optimism. The picture the data paints is this: the Antarctic-Australian humpback subpopulation has rebounded to 90 percent of pre-whaling estimates on the west coast, and to 63 percent in east Australian waters. These whales are no longer at risk in the way they were after the crucifying efforts of factory ships brought their forebears down to mere hundreds: a near miss with extinction. Whale numbers continue to climb. Humpbacks off Australia are experiencing a baby boom.

In order to gain a more accurate, quantitative snapshot of the species' stability, in 2016 the US National Oceanic and Atmospheric Administration (NOAA) pushed to segment the worldwide humpback population, allowing for independent groups of whales, which do not interact, to be treated as discrete data sets. A designation as "vulnerable" or "threatened" indicates a species is subject to ongoing, hazardous conditions that, uncorrected, stand to push it toward disappearance in the medium-term future. "Endangerment," a more dire category, represents a higher likelihood of extinction ("critically endangered" denotes the probability of a species vanishing in the short term). The terms are officiated by the International Union for Conservation of Nature (IUCN), in the maintenance of a biodiversity "red list." Separate populations of humpbacks—those passing by Central America in the winter, and humpbacks feeding off California and the Pacific Northwest in the summer—remain vulnerable or endangered. Modifying how the data was analyzed and represented, however, clarified what had long been true for the Antarctic-Australian humpbacks. Their category is "of least concern."

Yet, while present-day whale numbers have been reappraised to nominally elevate some humpback populations out of harm, a new generation of bioinformatic, laboratory technologies have simultaneously forced a reconsideration of pre-whaling estimates. The yaw between the baseline of the past and "recovered" humpback populations may yet be revealed to be wider than first calculated. Plunging down into the nuclei of whales' cells, these studies examine gene diversity and rates of mutation in existing whale populations to inform mathematical models of former abundance. Prior to the nineteenth-century impacts of commercial whaling, the genetics show there may have been as many as six times the numbers of humpback whales, worldwide, than had been previously estimated. It may even be possible to make downstream inferences from whale numbers as to prey quantities, and so imply ocean fertility and other conditions that might have sustained those prey. That so much of the sea's history might be minutely encircled inside the single *cell* of a whale: this seems, to me, a sublime feat—even as it could demonstrate that the extent of whaling was far greater than once believed.

Setting aside how many humpback whales there might once have been in the world's oceans, recent studies of the population structure of humpbacks on the east coast of Australia suggest their numbers may be approaching overload, followed by a crash projected to fall sometime between 2021 and 2026. Michael Noad, a scientist who studies whale population dynamics, has warned reporters: "We might find the whales becoming very thin, and we might find a lot of strandings along the coast, as a lot of whales die." Whale watching could become, as a result, quite gory: bony whales in the sea, and perished ones washing up on the shore.

Yet if only it were so easy, to view each whale death as the result of more whales being born than could be sustained by the abundant food shared between them—then future strandings, though they may wrench our hearts, would confront us with the *success* of the anti-whaling movement; the balance of nature correcting for conservation's overachievements. (How antique that expression sounds now: "the balance of nature.") The technologies used to quantify animal populations in the past, and to accurately survey animal numbers in the present, continue to improve, yet however refined that data becomes, it belongs to seas lost to history: to oceans that have since been chemically, meteorologically, sonically, pathologically, and ecologically altered. The future is less certain: new pressures will play out in the lives of humpback whales. Today, we indirectly set the conditions in which life elsewhere makes its migrations, its abodes, its nests and niches. Predictions grow shaky, as the variances introduced into marine ecosystems, including the effects of climate change, introduce a wider range of extremes to underwater habitats.

More than 90 percent of the warming that has taken place on Earth over the past fifty years has occurred in the oceans. The seas' absorption of thermal energy has buffered the rate at which temperatures have climbed overall, but as the climate changes, and Antarctic summers extend their duration, the ice cover shrinks. More, and larger, icebergs fall from the continent's edge, dragging their roots along the seafloor and scribbling there a truer script of the atmosphere's accelerating machine.

The added warmth in the system might result in a longer growing phase for the photosynthesizing plankton and algae (overshadowed for a shorter winter), increasing numbers of keystone krill. Populations of humpback whales, nourished by the krill, would likewise expand. Thereafter, more whales might remain in the Southern Ocean for longer periods, and, as the tropics migrate toward the poles, the whales' northward migration may be truncated. Or perhaps humpbacks will eventually never leave the seas surrounding Antarctica. From Australia, we would cease to see them, but not, as before, because the whales were in danger. Instead, it would be because the humpbacks had located a different Eden. This scenario matters because these large, long-distance travelers of the animal kingdom relay nutrients not just through their vertical movement in the water column, but by moving between the polar sea and ecosystems other than where they have fed—there to excrete, to lactate, to die, and otherwise input energies into coastal waters. Some scientists call this circulation of matter "the whale conveyor belt." How the absence of whales would change the ecology of biomes closer to the equator is unknown (though presumably mining companies would welcome the disappearance of whale nurseries, which have, in the past, placed limits on the expansion of coastal infrastructure). One scenario.

Another scenario: the erosion of the ice poses serious problems for Antarctic krill and humpback whales. Krill populations have already declined by 80 percent since the 1970s—suggesting, to scientists, that krill are significantly dependent on ice-fixed algae over free-floating plankton and other bits of pelagic detritus they consume. As ice surface area contracts, their coziest habitat decreases, and there is less for the krill to eat. Ice melt is a symptom of warming driven by increasing carbon dioxide (CO_2), which has other, molecular effects on the sea. The decrease in the ocean's pH (a result of the sea absorbing more CO_2 out of the air, which it does especially where it is coldest, at the poles) has the potential to harm krill spawn in much the same way as pesticides can cause the shells of songbirds' eggs to thin and crack in their nests. Even though CO_2 is taken in at the sea/air boundary, its effects intensify down the water column: krill lay their eggs in a high-risk zone, deep down, away from predators. Antarctic researchers have shown krill eggs fail to hatch in waters that are more acidic (lower pH). Krill, in time, dwindle. Whales relocate, or die.

On the microscopic scale, ice melt poses other risks elsewhere, simply by removing physical barriers between whales and distant, waterside settlements. In the Beaufort Sea, west of Canada's Arctic islands, wild beluga now contract a minor disease from house cats; toxoplasma, a hardy parasite transmitted into the ocean with feline feces in wastewater from coastal townships. Where ice might have impeded the spread of this infection before, beluga acquire the minute parasites in the unimpeded flow of ocean currents.

As the environment is changing, so, too, is our understanding of how extinctions happen. In past vanishings, animal populations tended to "spiral" and noticeably decrease at a set rate, perhaps becoming rafted into environments that were geographically separated, or preserved from the introduction of a feral predator. But today even very abundant species can be susceptible to abrupt mass die-offs; the likelihood of populous organisms crashing to irrecoverable levels increases in a time of fast-paced ecological change. Migratory animals—reliant on the sequential timing of conditions across more than one habitat—are particularly vulnerable, because migration itself is energetically costly. And endangerment is also no longer the only, or indeed the main, drawcard of conservation. Ecologists are keen to impress that preserving animals *at specific levels of abundance* is key to the function of an ecosystem. Sometimes many more creatures of a species are needed to adequately cycle nutrients through a biome, for example—well above the threshold of "vulnerability" that might be set, looking only at that single species in isolation. Numerousness matters. Which is one reason that researchers looking at the positive impact large whales have on the atmosphere have argued that the preservation of whales ought to be included in the objectives of the 2015 Paris Agreement on Climate Change; whales count, from this standpoint, as remediators of carbon—not as a result of their endangerment.

Mass strandings of whales might be naturally occurring events, but an uptick in their frequency—or the quantity of animals involved—can point to an unnatural warping of the weather. Closer to the equator, balmier oceans curate other biohazards. In 2015, researchers discovered 343 dead sei whales in fjords along the rugged coast of Chilean Patagonia. Seis turn bright orange when they begin to rot, and so biologists enlisted satellite imagery to survey the surrounding region; the whales appeared up and down the coast, like marigolds peeping from roadside grasses. Some carcasses in the fjords were bone heaps, shrouded in tattered skin; other bodies were fresh. On closer examination, scientists were able to delimit the deaths to a six-month time frame. Likely more seis died than were discovered—between the fjords are rocky, high-energy shorelines, grated by storms, where any whale body would soon disintegrate. Sei whales are endangered: losing more than three hundred in one region was a species-wide disaster. They are also not a "gregarious" species (seis do not collect in large pods but tend to live alone, or in groups that consist of, at most, six individuals). That hundreds of these whales had died along one stretch of shoreline was made doubly alarming by the fact that no one had ever seen so many sei whales together, alive.

Necropsies revealed the most likely cause: the whales had been exposed to a toxic algae bloom. Roughly every third or fourth year, El Niño, a cyclical weather event, weakens trade winds and brings warmer water to the eastern and central equatorial Pacific Ocean. Prior to the mass die-off of seis, an El Niño current had entered the Golfo de Penas, where it met nutrient outflows from croplands and fostered the growth of a poisonous, golden-brown species of phytoplankton. The sei whales had been drawn in by the same climatic conditions. El Niño typically disperses little schooling fish like anchovies into colder, lower parts of the sea and creates coastal upwellings where the whales' prey is concentrated. The sei whales, in a massive aggregation, had come to feed but met, instead, a wash of biotoxins. Such blooms are set to become more frequent in heated waters—and though El Niño is a natural phenomenon, it is likely to exhibit more pronounced effects as the climate rachets. In their report, the scientists called the seis "the first oceanic megafauna victims of global warming."

The humpback whales' future is hard to predict, for it is far harder yet to try to preserve the whales' world than it ever was to stop whaling. Wilderness does not go on without us, if we withdraw from going to see it: keeping our distance no longer amounts to care. And yet, to tend to whales now is to acknowledge a greater worldliness than ever before— one that doesn't just bond people to people, against governments and companies operating from whaling nations. The worldliness garnered by looking at whales today knits us together with the planet's weather patterns, with what is happening to the sea on its most granular level, and to many, many more organisms than those we love.

How people will *feel* about seeing more humpbacks in poor condition, that is much easier to imagine. I picture distressed tour groups tossing blocks of burley chum to the waiting whales to keep the animals from starving. Not such a preposterous idea, as it turns out. In April 2019, the people of the Lummi Nation, on the edges of the Salish Sea, began to feed Chinook salmon to wild killer whales suffering from a paucity of prey and the effects of pollution. "Those are our relations under the waves," said a spokesperson.

The whale-watching boat comes to a halt, though we are too far away from the breaching humpbacks for this to be legally necessary. The watchers are disappointed. We cry out to be puttered closer. Who would know? What official is here to enforce the restrictions? Break the regulations—they're bad regulations! No one wants to rely on the whales' curiosity. Nothing about the catamaran seems likely, to us, to pique a humpback's interest. Begrudging glances are aimed at the bridge. A fractious moment passes in mentally urging the engines to restart, no anchor to drop.

Then "Look behind *yooooou*," Ros calls over the loudspeaker. People rush to the stern and lean over the transom. What we see is a huge, adult humpback with her calf. So close! Theatrical in their punctuality! There can't be more than the length of a tennis court between us. Ros announces that while we were oohing at whales on the horizon, this maternal pair crept in underneath the *Cat Balou*—and this surprises us; that something so huge as a whale can act with such stealth. The adult female stays between the boat and the calf, but the juvenile seems eager to get a closer look—it keeps waggling forward, then retreating, bold in increments. Mugging. The calf is the size of a sedan and charcoal-streaked: like something dragged from a chimney. As it presses nearer, the female pulls alongside the boat. She's as long, if not longer, than the hull. More than fifty feet, is my guess.

The ocean is pellucid. Thick weeds detail the seafloor—olive, sage, jade, onion green. Water thinner than water: methylated spirits or gin. The female humpback bulges below the surface; the same doming of vision as if looking into a glass paperweight. The barest undulation of her flukes moves her rapidly toward us. The whale, in her element, I see, is muscle all the way through. Tamped power resonates within her, a reserve of ferocity, a fastness in both senses: something internally secured, something quick; spring-loaded. Standing only yards above this mature humpback, my impression of a whale as blubbery is lost. The sheer *isness* of the humpback whale. I am awake. I am slammed into a state of readiness. At that moment, however alive I am, how much *more* alive is she! Blood thudding into every corner of her titanic body. The flex of her peduncle, the base of her tail, is a twitch of the largest muscle on the planet—in a moony voice, an older woman behind me repeats this factoid.

The whale steers below us, full of intent.

"She's introducing her calf to the boat," Ros says cheerfully.

No, I think. *She's about to flip us.* My skin raises into goose bumps. This dreadful apprehension; biologists call it the *"heiliger Schauer"*: the holy shiver of prey sensing a predator's gaze. There was no iota of comfort in knowing that humpbacks feed only on tiddling bits of life. Her aim, I felt, could not have been more direct. Here she came again, sussing out the boat and its occupants. And we were loose in the whale's world, skidding around, all incautious enthusiasm, all eyes. Our pursuit of spectacle had made us oblivious to instinct: I saw myself as tiny as the head of a pin, pressed into a classroom globe.

A fool would try to touch this animal, the whale. No wonder that's the first prohibition laid down on a whale-watching tour. People must really try to put their hands on the whales, or to dive in. But seeing the colossal humpback below, I thought: *Only the most witless individual would believe in a benevolent connection with real whales, with any affinity that runs outside of metaphor.* This is a one-sided intimacy. To wear masks in their image, to make whales into lucky charms: what a ludicrous notion. Ludicrous and inoculating. The adrenaline in me was the kick of imminent danger and the absolute futility of clearing it.

Later, I'd learn some scientists believe whales approach watercraft with the motivation of scratching their bodies, irritated by microscopic creatures, along the underside. ("They'll come right up to boats, let people touch their faces, give them massages, rub their mouths and tongues," the researcher Toni Frohoff said of gray whales in the Pacific, to the *New York Times*.) But this humpback whale didn't graze by the boat to dislodge any parasites. She made a pass. Then she turned, and did another. The light fallen through the sea came rising to us, a bright and whirring net, slid across the whale's back. Air huffed out of her as she broke the surface.

How to describe the sight of the immense whale, so near? Her back, lined with knots. Not sheeny like a shark, or glinting like a fish, but reptilian, almost. A dinosaur. A nasty thought presented itself: *Killing one must be as it is to kill a dragon.*

Was this humpback guided by fear, affinity, curiosity, or aggression? Call it emotion? Call it instinct? Or could it be that the whale was *aware* of making herself an exhibit for our enjoyment. A whale viewed annually by whale watchers, was she playing it up for the cameras? The associations evoked in the whale's mind on hearing Ros's voice on the loudhailer, and what any humpback might think, perceiving the hovering shape of the tour boat; all this was impossible to imagine. But on the deck we did; we did imagine. We asked ourselves what she wanted from us, and what oblations we had to give up.

Meanwhile, the whale slipped back under the water and turned onto her side once more. I could see her skin was stippled with tiny points, hatchings and crosses. Kiss-kiss. Strike. Foreign signatures, astral data. I felt she was something prehistoric, chthonic. Almost *geologic*—even as she moved so quickly, in the way of an avalanche or an eclipse: that unstoppable energy. She was, for me, a force I recognized as being connected to those that come down-rushing from peaks and shattering out of fault lines; an event to make senseless your patience, your ethic, your voice, your limbs, the age.

The whale turned toward the boat again, directly. I was, I'll say now, very frightened. Hot, sour saliva in my mouth, cold-skinned. The human body, that animal you own unpacified, speaks in these flashes— it throws itself against the bars of the cage again and again. Piercing through the fear, fluttery in my ears, came what? What was that? Awe? Mortality?

I have looked out to the sea from the beach so many times, thinking: *Give me a whale. If I see a whale, that'll be the sign.* Who am I, a secular woman, appealing to? The sign of the whale is always changing its consequences: I'll know; I'll stop; I'll start; I'll change. I'll leave. I'll set aside whatever the shades of ambiguity, submit, and act. What the whale promises is always the conclusion of a hesitancy, or the end of a story. The Swedes use the same word, *val*, for "whale" as for "election." A decision, a deciding.

In his influential essay "Why Look at Animals" (1977), the art critic John Berger writes: "The animal has secrets, which, unlike the secrets of caves, mountains, seas, are specifically addressed to man." I've thought about that sentence for a long time, but I can't name the secrets Berger contemplates. He doesn't define them. Except that, at a studied guess, Berger might be offering the provocation that what goes unspoken in the supposed recognition of humans by animals, are the bestial compulsions of humans. We were once a wildlife. We call ourselves *Homo sapiens*—a genus, *Homo*, we occupy alone, as though we stood in categorical exclusion from our nearest biological relatives, even though on the genetic level (as the evolutionary scientist Morris Goodman once pointed out) bonobos and chimpanzees are so similar to humans, they should rightly also be grouped together into our genus. I think Berger is suggesting that there comes a moment when looking at animals triggers, in people, a recognition of all the familiar ways humans persist in being fauna. Our shared zoology gets released from the padlocked storehouse of the unconscious—and, in that instant, it's scary. Being, in essence, all flesh and raw instinct, stands to undermine human rationality, exceptionalism, and the social and political lives of our communities. We are not drawn out into the world as we might have expected; we dive back into ourselves, trembling.

Following on from what Berger describes as the watcher's initial revelation (humans are only animals), he observes that people, confronted by other species, typically reactivate their self-awareness and superiority. They remember that they are, on a fundamental level, a different kind of animal from the animal they are looking at. Humans alone are creatures that analyze themselves; animals cognizant of the formation of their ideas and their psychological interior ("the only animal that cries / that takes off its clothes / and reports to the mirror," the poet Franz Wright reminds us). Humans, unlike whales, have the capacity for symbolic thought and expression, for conscious education. Maybe this ego brings us into a tighter kinship with others of our own species, even as it distances animals.

And yet humans are also the only animals *that love other animals* for no greater reason than they nourish our emotions and bolster what we think of as our better inclinations. The love of nature is unique, or so we think, to our species. The whale does not love the ocean, the seabird, the jellyfish; it does not love us, does it?

A marine biologist once showed me a cartoon by Paul Noth he'd cut out of the *New Yorker* and pinned by his desk. Two dolphins are paddling alongside each other. The first says to the second: *If I could only do one thing before I died, it would be to swim with a middle-aged couple from Connecticut.*

Whether whales include people in their category of "who-s"—as opposed to thingified "that-s"—if and when they turn their minds to us; this gets to the heart of it. Did whales once call us "who"? Do they still? When whales speak (because their sounds contain *information*, so we're led to understand), what do the whales "say" about themselves in connection with other species, and in relation to . . . the human fauna? Is it their prerogative, too, to tenderly pare an "us" from the "other animals," a "we" from the world? Do they whalify, where we personify? And if they separate us out from other forms of life, is it perhaps *not* because they love us but because they remember the violence with which we incurred upon their past?

The triumph of the animal-welfare movement has been to widen, in the public's imagination, our definition of what types of *bodies* can suffer. But what scientist, what empirical person, would ever claim true access to what goes on inside animal heads; their subjective pain, their pleasure, their intentions, the extent of their memory? We may know humpbacks "think" in the rudimentary, neurological sense. The mystery is how, and what those thoughts are about. What is their understanding of time, for instance? What is a whale's understanding of place and how places change? Do whales have selfhood; do they know themselves as individuals, or are they in "oceanic feeling"? How might the answers to those questions change what we understand of their suffering in this revamped biosphere?

Who is to say that whales *don't* also have an unconscious, after all? What philosopher or scientist can offer proof that some kind of cultural iconography for dreaming states is not transferred across whale generations? A whale's unconscious—I can envision it. It is a hallucination of endless, undulating dunes. The geographic apotheosis of the cetacean psyche is, surely, a desert, just as ours is a sea. And as deserts expand today to claim a greater proportion of the globe, isn't it possible that many more human populations will come to inhabit—come to build their homes in—the dried-out dreams of whales?

I have extricated my photocopy of the "Why Look at Animals" essay again from where it has long stayed pressed between books—the top third of the paper stained yellow by summer's heat blasting an indelible cube of light through my bedroom window. "Animals first entered the imagination as messengers and promises." This sentence is twice underlined. Animals, then, were once taken to be ambassadors from a godly plane. How an animal moved, what it did, when it died, where it ate, and when it gave birth—all this had the force of law. An "augury" or prediction originally referenced birds' flights, through a sky quartered into prophetic segments. In ancient Rome, haruspices picked through the entrails of animals to discern the future. The future arrived piecemeal. But the first parts of the future arrived inside animals. Whether as oracles, harbingers, or sacrificial offerings, the world's wildlife initially represented human spiritual imperatives, according to Berger. The sacred was fleshy, and that flesh was beastly. It moved people to humility, it moved them to change.

The tourists on the *Cat Balou* are all squeezed in against the edge now, and breathless. The boat rocks laterally, steady as a metronome. Cameras are thrust out over the water, snapping at arm's length with neck-straps swinging neckless. "Take only pictures," someone says crisply, delighted at the felicity of this shopworn saying.

About fifty feet off, the adult whale lifts her tail out of the water and holds it up in the air. The patchy white side is turned to us, like a semaphore, bordered in black. The woman who has been seasick clutches her daughter's shoulders and exclaims: "Her tail looks like Australia! She's got the shape of Australia on her tail, you see?"

And I didn't; I didn't see the sign as I'd hoped to, like a map being slammed down on the surface of the sea, but other people did: they saw the semblance of this continent from above, imprinted white on the underside of the whale's black tail. The whale's flukes slapped down, and, at that very instant, the humpback calf sprang out of the sea, its entire body airborne. "Who got a photo of *that*?!" Ros screamed. "Get your cameras ready, folks—this baby's about to jump!" But that was the only time the calf breached out of the water.

The mother humpback made one more sweep, up alongside the boat. I hung over the railing. As she drew level, I saw her eye revolve to look straight at me; the only expressive part of her face. The Germans have a word, "*Augenblick*," which, if I understand it correctly, refers to the feeling of a second as measured by a slow eye blink. We don't have an exact equivalent in English to express the sensation of an instant, as it registers inside the body, using only the body's equipment, though we have other ways to talk of the impression of glutinous time, as in the stretched moments of a car crash or the bad news that arrives in slow motion.

Meeting the whale's gaze I felt held in that *Augenblick*, the density and anguish of a second that stiffens, and waits. Which is something I did not expect to get from whale watching. The time-freezing power of the whale, watching.

03

Blue Museum

Whales Inside Out — Western Australian Museum — The Blue — Touch Pool — Posing for Shark Attacks — Nocturnal House — Animals as Artifacts — Meteorites — Zen Explanations — Spirit Collection — Carl Linnaeus — Mouse or Muscle — Breathing Rainbows — Population Bottleneck — Not Inconsiderable Confusion about Theology — *Hierozoika* — Where Was the Whale in Time — Prehistoric Whale Spotting — Whales with Knees — Bigness — "Secreted the Bones of a Hand" — Forewarning: Scantling — Our Beastly Nature — Nin and the Blood Cell — How Many Whales in Museums — The Poor Replicas — Plastic Heart — Picking the Eyes Out — Consummation — Hope — And Pain — Inadvertent Ark — The Japanese Fishermen Who Helped Prepare the Specimen — *Wunderkammern* — A Body without a Voice and a Voice without a Body — The Tendency to Destroy — A Stromatolite — Self-Importance of a Whale Louse — Past Outside of Memory — Sand

I BEGAN IMAGINING WHALES BY SEEING THEM INSIDE OUT. DOES this go some way toward explaining why I am so consumed by whales, so taken in by them: why I can't stop finding out more about whales? If you were to ask me *What is the first real whale you ever saw?*, I have to answer that by the time I got to it, that whale was already gone. The animal had slumped off its ribs and rotted into the fast-flowing currents of the past. My cardinal whale, the one it thrills me most to remember, was a skeleton on permanent display at the Western Australian Museum.

What does it do to your ideas of an animal to first conceive of it off its innermost bones? What will you build upon it? How, then, will you go on to configure that animal's relationship to history and science? Maybe you, too, are familiar with the atmosphere of whale bones; their magnetism to a child drawn to guess flesh onto the fleshless. Before I knew how to begin with a whale, I had seen how a whale ended. I started with a whale's afterlife, the huge casement of its brain, the stony pickets of its chest.

Lucy, my sister, and I, seven and eight years old: in the museum, our minds flung upward. Soaring unseen, five flights of stairs overhead, were the white bones of the blue whale. We felt it jerk on our attention. If we dawdled then, scuffing our heels, our only aim was to distill the adrenalized trickle of dread and urgency, to extend our suspense. We were a duo with theatrical tendencies. All jest and hiss, good liars, sulkers in cahoots and picky eaters, we toppled over everywhere into handstands.

Like so many suburban, working-class kids whose families cared to make nature a part of their world, our enthusiasm for animals stemmed chiefly from well-thumbed copies of *Australian Geographic*, documentaries, and romping around the taxidermies in the museum's "Age of Mammals." Our parents did not neglect outdoor pursuits—family holidays featured camping trips and hikes through fly-hiving bushland—but any knowledge gleaned in the eucalypt forest and the scrub had less to do with nature in its own right and more to do with safety. Recognizing the sash of a snake, say, or symptoms of sunstroke. We were clued up on telling poisonous from venomous (it came down to "who mouthed what," versus "what mouthed who"). Ant stings were agonizing, but worse, we learned, was a bite you didn't feel—from a dangerous spider, or, if at the beach, a blue-ringed octopus no bigger than a drink coaster. This way to flip a rock: reach over and pull the farthest edge up toward you. Now anything beneath escapes in an *away* direction. Excursions to the museum, the zoo, and the aquarium allowed us to dwell on the habits of animals under less vigilant supervision. There, incautiously, we could play with nature.

At Underwater World Lucy and I were quick to shoulder through meandering groups of tourists and dash past the smaller tanks, aglow with fingerling corals, silky nudibranchs, and bric-a-brac seahorses: these displays like the bountiful jewelry boxes of silver-screen starlets. We wanted, instead, the touch pool: the sops of faceless things, the globule, knuckle-sucking anemones, and bicep-sized sea slugs. Creatures you could get your hands all over. After coffee, our mother and an aunt, or the chaperoning babysitter, would find us both flattened against the aquarium glass in postures of distress: lures for sharks. The sharks wended above, blinking their weird, eggshell eyelids without appetite. So on we strutted; our pumped-up, hipless waggle, a parody of two mermaids on land.

In Perth Zoo the boredom of the animals was more pronounced. We glowered at the exhibits as they pawed enrichment objects and paced parallelograms into climate-controlled simulations of their habitats. We rapped on burrows that terminated in observation windows, determined that no boa, numbat, or tree frog should merely *sleep*, while we were there to entertain it. Best was the Nocturnal House, where big-eyed things flapped and scampered behind smudged glass, under an unremitting red light.

The boundary between cultural artifact and wild thing grew blurry in these contexts. Animals were globally muddled: birds from several sites might share an aviary; "the tropics" and "the desert" were genres without geographic distinction. The captivations of the BBC's *Kingdom of the Ice Bear* (1985) meant we wore through the videotape, stranding the polar bears on snowbanks rastered by cathode-ray color (a deterioration eerily clairvoyant of the great dissolving that would befall their habitats in decades to come). The nature that most fascinated us—which we could inquire upon freely—was found inside, behind glass or wire, or on the blazing square of the television. That this was a nature personified, protected, cropped, narrated, synthetic, preserved, and institutional, was part of its appeal. We could get nearer to this nature, we could *manipulate* it, more so than anything in the wild.

The WA Museum—on Francis Street, near the old East Perth lockup—was a favorite day trip. Inside, the building brimmed with sedimentary air, as cold as vapors sealed in a cave. Two meteorites, fixed on plinths in the foyer, hinted at what could be discovered on higher floors. Brassy and stippled with holes, the meteorites had been rubbed smooth by the impatient fondling of queued school groups—their metal smell lingered on your palms for hours, the tang of outer space. The museum was a hushed, low-lit place. We tiptoed on carpets the dark green of cinematic banks. There were few visiting exhibitions then, no activity stations or audio tours: the education of children did not seem to be a prime concern. The museum's stuffiness hinted at the protracted archival labors of natural history: pinning butterflies, verifying sightings, and sorting seed husks. The docents scowled and shooed, aggressively *sotto voce*.

We loved it all the more because the museum wasn't meant for us. The many binomial names on the wall plaques proved unpronounceable. (Binomials, or scientific names, are two Latin words that indicate the genus and species of an animal, and place it within a greater order of relatedness: hence *Eubalaena australis*, the southern right whale; *Eubalaena glacialis*, the North Atlantic right whale.) Whoever had composed the museum's signage possessed a vaguely Zen sensibility. "A Fish's Shape Suggests Its Habits," read one. "The Plates Are Greatly Reduced," "Irregular Urchins," "This Body Has No Brain." Creatures, and parts thereof, stored under preserving liquids in the spirit collection, appeared to sample the sepia environments of old photographs (I long believed the past *was* this color). Objects that ought not to be, were mounted tantalizingly within grasp, unalarmed. To our great delight we could see up close what we would never get near to in nature: the fierce and fangy predators, taipans spiraled in jars, animals known to trample and gore. Over time, some of the taxidermies had grown shiny on their hindquarters and up their shins—a high tide of mange spread by affectionate patting.

Any footage or picture we'd seen of these beasts paled in comparison to their three-dimensional presence and magnetism in the museum. Forever frozen in a bardo of human fabrication, they loomed with intricacy—replete with fabric eyelashes, wooden incisors, varnished hooves, and faint trails of silvery glue snaking through their hair. We could get a fix on their dimensions, calibrated to our own. Our hunch: at least one living animal had snuck in to masquerade as fake. When the docents turned away we touched, with darting hands, the dry pink paint in the animals' mouths.

The blue whale circled out of sight in an annex on the top floor. We sensed it up there, as if shafts of starlight, slatted through the whale's rib cage, bisected the gloom around us. We talked of it as "the blue," a color with no noun. Its singularity was notable. Our museum had only the one whale, but it was the biggest one. It filled an entire room.

The whale in the museum belonged to a subspecies known as the Antarctic blues, a variety inhabiting the Southern Ocean. Blue whales are also found elsewhere, in the North Pacific and North Atlantic, and the pygmy blue whale (which still grows to seventy-nine feet) lives in the northern Indian Ocean. The blue whale's Latinate binomial is *Balaenoptera musculus*, a name that is often construed as a joke made by Carl Linnaeus, an eighteenth-century Swedish naturalist. "Winged whale, little mouse," is a common translation. The pun is that the world's biggest animal should exist in comparison to the smallest.

Linnaeus is widely held responsible for the withdrawal of whales from the category "fish." In classifying whales for his masterwork, the *Systema Naturæ*, published in 1735, Linnaeus pored over descriptions and drawings of cetaceans—details of dead whales' chambered stomachs, their lungs and ears—and declared that nothing within the animals looked, at all, piscine (fishy). As the progenitor of modern taxonomy, it was not unheard of for Linnaeus to name an organism in jest or spite: he christened weeds after people he hated. But *"musculi"* was also a term once used to refer to muscles, precisely because they were observed moving, like mice, under the skin. Probably Linnaeus only meant that a blue whale is long and sinewy, in the way of a single taut tendon: *Balaenoptera musculus,* an opera of muscle.

Blue whales are creatures so large that when they exhale, rainbows can form, so high are the canopies of vapor they emit. When the whales "spy hop," prodding their gigantic heads up out of the sea with their tails hanging down, the variance in pressure from the nose tip of the animal to its flukes can be as great as three atmospheres. They exist in drawn-out time signatures: blue whales' hearts beat incredibly slow, eight times a minute. (Our own hearts go ten times that fast.) Under the right conditions, it is possible to hear the reverberating temple-drum of the whale's underwater heartbeat two miles away.

Though my sister and I remembered the whale skeleton from previous visits, each time we pushed through the turnstile a current of uncertainty ran through us. Would it still be there? Would it wait? How long? The blue whale paddled the air, stiffened only by our anticipation. Death notwithstanding, it seemed charged with capricious wattage—an ultramarine temperament. We encountered it not as the museum intended, as osteology, an empirical construction of science. To us the whale was make-believe. Had there been no ceiling to enclose the blue, we supposed it would glide away into the sky, its skeleton fluttering like a paper chain.

The skeleton's installation predated our existence, so we had no proof of the whale's journey to the museum, no way of knowing how it got to be up there, or where it had come from. This part of its story preoccupied us: how had the whale alighted on the top floor? Weighing it in our minds, we couldn't reconcile the implausible process of pulling the whale up the stairs, or cramming it in the elevator. If the museum's front gates were taken off their hinges the whale *might* have been dragged through the entrance, but no further than that. To the pinnacle of the building? Never. In the days before the building's roof went on in 1971, a crane had, in fact, lowered the whale's bones, wrapped in tarpaulins, into the highest gallery—but my sister and I could never have conceived of the whale skeleton as demountable. The blue whale, to us, was a perdurable thing of impossible collapse.

What we didn't know then, of course, was that outside the museum's walls, populations of blue whales *had* collapsed. At around the start of the twentieth century, there are estimated to have been close to 239,000 Antarctic blue whales. As the whale's bones were first hoisted into the Western Australian Museum, the number of live whales in the Southern Ocean had decreased to a perilous 360 individuals, the result of decades of legal and illegal commercial whaling. Trevor Branch, a marine biologist at the University of Washington in Seattle, estimates that numbers of Antarctic blue whales declined by 99.85 percent between 1905 and 1973. In a single year, more individual animals were hauled and flensed than are left alive in the world now This is what biologists deem "a bottleneck": a drastic reduction in a species' gene pool, evident in the DNA of future generations (though, surprisingly, Antarctic blue whales remain genetically diverse compared to other whale species—likely they were even more so before extensive whaling).

As we drifted around the exhibitions as children in the early 1990s, Antarctic blue whale numbers were not yet up to two thousand. Which is roughly where they are today, as an estimate. More than two *hundred thousand*, narrowed down to fewer than four hundred, and bounced back to just two thousand. If there was an apprehension of this reality in the museum, it was implied only by the apparent rareness of the whale: the specimen's preciousness was a factor of how few such animals existed. Lucy and I didn't twig then, I don't think, that humans were to blame. Our blue whale swam out of the *mare incognitum*, that sea without history. In the museum, the more-recent industrial history of whaling stayed hidden, overcast by questions of the whale's place amid an array of other animals in the so-called tree of life.

After some debate, we decided together that the most probable explanation for the skeleton's situation in the museum was that the whale had first been stranded on the rooftop by a flood. Then we ventured, *the* flood? Bible stories, evolution, and the meteorological and geological past were all still muddled together in our understanding of the Earth before we arrived in it. Our parents were atheists, but despite inculcated godlessness, during a holiday camp staged by a local community group we'd been furtively inducted into scripture via the mesmeric animal tales—Mary's donkey, the benevolent beasts of burden in the manger, Jonah swallowed by the Leviathan, and the serpent that smote Eve. Aiding in our confusion, long troops of taxidermized specimens in the museum were set two by two, from titchy reptilian creatures to big furred ones—frozen in a ceaseless, antediluvian trudge toward no credible deliverance, the worst having already happened to each and every one. Though these animals were clearly creations—with visible seams and waxed scales, their fur gently brushed into cowlicks—we found their artificiality no barrier to envisioning them within creation myths.

Long, long ago, the only living things had been the swimmers: that much we knew was true—a conflation of biblical sagas and school science lessons. The whale, we believed, belonged to that ancient batch of wet-world animals; a huge, thrumming pre-mammal (not slimy as a fish), left behind when the other animals pulled themselves out of the sea and grew waterproof pelts. That the whale in the museum was a skeleton, not—as with the parade of creatures downstairs—a taxidermy, made it seem much more ancient than the rest. Theologically ancient. Birth-of-the-Earth ancient. I didn't know then that museums weren't the only scientifically important storehouses of whale skeletons worldwide. Such bones have also been preserved in the vaults of churches; real evidence of the Leviathan, amid selections of other *hierozoika*—natural objects made sacred by virtue of featuring in the Gospels.

Our obsession with the whale had to do with its size and with what its position at the zenith of the museum implied. We were always working our way up to it. But our inability to locate the blue whale in time amplified our fascination. Like a dinosaur, for us the blue whale belonged in a category of fossils that relied on our capacity for imagination to be real at all. The skeleton balanced on the threshold between living things and purely mental constructions—an order of beasts that, as well as extinct species, also included Falkor the Luckdragon from the 1984 West German movie *The NeverEnding Story*. Was the whale a real-life animal or a supernatural friend, like Falkor? How far back did it go? How far would it travel into the future? As I was to come to know much later, the evolutionary origins of whales prove even more fantastical than I ever saw fit to conjure up, at the age of eight.

There were marine reptiles that preceded early cetaceans—*Plesiosaurs* and *Ichthyosaurs*—but these aquatic beasts of the Mesozoic died out around the turn of the Cretaceous era and are unrelated to modern whales. Whales surface to breathe air; they give birth to live young and nourish their infants by lactation; they maintain a core body temperature, and, like all placental mammals, their evolutionary origins track back to a land-based existence. In the Eocene epoch, the animals that would go on to become cetaceans emerged as small, terrestrial quadrupeds: they walked on land, on all four limbs. Experts in the field sometimes compare the ancestors of whales to coyotes, dogs, or lowslung, trim hippopotamuses. One primitive proto-cetacean, *Pakicetus*, is thought to have been a canine-sized, shore-living creature with a tail like a Labrador and clawed paws. Probably, it had fur (hair typically fails to fossilize, so on this point there is debate). With its tiny wide-set eyes, *Pakicetus* displays a sheepish expression in many artists' depictions, as if ashamed at having gone extinct.

These ancestral whales, fossils exhumed in recent decades show, went on to acquire what we recognize today as whale-like eardrums and thicker braincases, as they hunted for food farther out into open water. Their nostrils moved to the tops of their heads. Their front legs broadened to flippers. Once at sea, they became an exceedingly diverse clade of animals—occupying a variety of different ecological niches and positions in ancient food webs. Forty million years ago lived whales that looked a bit like today's iguanas, albeit larger. Others appeared more fishlike. One possessed a barrel-fat torso that tapered into something like the head of an oversize ferret. *Basilosaurus*: the largest. A serpentine dweller of seagrass seas, like an eel, but sixty-six feet long with serrated teeth. *Basilosaurus* lived off other early whales that it killed by crushing their skulls. It had four legs, but its feeble hind-limbs could not have supported its weight out of the water.

By the time of *Odobenocetops*, the walrus-faced cetacean of the Miocene epoch, the course of evolution had streamlined whales' bodies and dispensed with the rear limbs. *Odobenocetops* had two asymmetrical tusks protruding downward from its squashy muzzle. The right tusk grew twice as long as the left, for reasons unknown (perhaps it had to do with its diet of mollusks, or with unknown courtship displays). To the twenty-first-century viewer, these tusks give *Odobenocetops* the lopsided charm of an oracular character in a Hayao Miyazaki film.

While I wandered the WA Museum with my sister, the first complete skeletons of ancient cetaceans with knees were discovered in a red bed of rock in a dry Egyptian valley. At least one *Basilosaurus* skeleton found contained the remains of a second archaic whale inside it (a kind of ossified cetacean turducken). The earliest fossils of creatures from the whale family are 53.5 million years old—remains entombed not on any modern seashore but in the foothills of the mountains in Himachal Pradesh, India. These whales swim in the same timelines as those that thrust ancient marine limestone up into mountain peaks, thousands of feet high.

Today blue whales have grown to be the largest animals ever to have existed, and many whale species are more gigantic even than their ancestors. Their size is thought to have been obtained due, in part, to the selection pressures of the Pleistocene ocean—a frozen epoch, during which ice sheets ground down earthy nutrients and carried them into the sea, creating fragmented regions of high productivity. Whales needed to become enormous to migrate between distant feeding sites—gigantism enabled them to store sufficient fat, and to traverse, by muscle power, great distances. The bigness of whales is of interest to cancer-drug researchers today, because despite consisting of so many cells (and therefore, theoretically, an increased chance of cell aberrance), blue whales suffer surprisingly low rates of malignant tumors. No one knows why, though recent studies have begun to identify what are thought to be suppressor genes.

Beneath the lower spine of some modern whales remain the vestigial knobbles of two back legs, folded up like an airplane's landing gear and enclosed in the flesh of the whale's tail—the bony legacy of hindquarters that might still have been used, even after terrestrial walking ceased in whales' ancestors, for gripping another animal during intercourse or propelling any proto-whale along the seafloor to rummage in shallow reefs. These defunct limbs are fleetingly visible on the exterior of the animal in the womb (when, for a time, some whale fetuses look a lot like huddled piglets). This is part of the magic of whales: they contain forms both familiar and stupefying.

Anatomically, modern humans began to emerge in the Middle Paleolithic about two hundred thousand years ago, but cetaceans go back fifty million years. Inside the whale: the hippo, the dog, the swampy chordate leaping out. Inside the whale: us. Historian of science D. Graham Burnett has noted that when the taxonomists of the eighteenth century first began to reclassify animals according to their internal structures—as opposed to their exterior appearance—anatomists were alarmed to discover that whale fins "secreted the bones of a human hand," so familiar were the long, knuckled "fingers" of whales. If this weren't already science and anatomy, it would be folktale. Which, of course, it was and is. The Inuit, skilled whale hunters, told such stories. Whales are often depicted in their oral histories as the fingers of Sedna, the Goddess of Adlivun, the Inuit underworld. Sedna's fingers were chopped off by her father as she clung to the side of a boat, trying not to drown—a bloodied beginning in familial violence. Her fingers turned into whales, her fingernails into whale bones, her thumbs became walruses. Through other species we think our way into ourselves, our origins as animals, and the uncertainty of the boundary between.

Anticipation squandered, when at last my sister and I entered into the blue whale's low-ceilinged annex, the spectacle we confronted there proved perplexing, and deflating. The curators had designed the display to lead museumgoers in a tight ring around the blue, which meant you were always too close to register the whole animal in a single instant. No one who entered the marine gallery ever shouted, *Look, the whale!* People pointed out the jaw, the eye sockets, the thin canopy of its palate: they noted the knobble and craquelure in its bones. The whale was many colors, all of them antique—shades of sweetened condensed milk, and sun-cracked things. *Gris clair*, calico, onion brown, treacle and tobacco yellow, soot, blushing to a dirty rose on its shoulders (if a whale can be said to have shoulders). That it was neither snow white nor a bright, eponymous blue disappointed us: it seemed sullied with age. The blue's skeleton tapered to vertebrae the size of knuckles (our own knuckles, we compared) and ended in a point. Being boneless, the great spanning tail flukes were absent. In place of the whale's fins were spindly, fine-tipped bones, witch fingers, that stretched out from perforated palms. The whale's skull—a brainpan and muzzle clamped above its now baleen-less mandible—was narrow and elongated.

The internal workings of the whale remained a puzzle. Nothing but air was chambered within it. In what heart's dark cubicles, what oceanic secrets? I heaved the whale's ghostly brain up into its head. A thought experiment: What was the size of its mind? Where was its throat for singing? Where was its stomach? Didn't it have ears, and lungs? Once there must have been a tongue, strung like a theater rope through the whale's midmost region, but now no tongue could be seen. We peeked into its pelvis, scandalized by imagined genitals. In other parts of the whale's hollow lodgings, we conjured fleets of its bobbing, beady prey.

The longer we looked, the many laps of it we walked, the more the creature lost its buoyancy and became, simply, a kind of masonry clamped together. The whale, wherever it was, slipped into the background. The ocean once held the animal—you could see, I reflect now, how it might shatter itself on land. The bottoms of its ribs were thin and its heaviest bones bore down on them. Up close, the iron fixtures and spacers supporting the whale's spine made clear that this was not, really, a blue whale, but a kind of sculpture clamped together by people. This was something close to what was once called a scientific "scantling," the barest possible equipment of an animal or plant able to be experimented on. So the blue whale morphed before our eyes, from an animal into an object.

Yet, a hesitation. I can't say why, but the desire to go *inside* the blue whale—as any person might walk through a room, kicking the skirting, rapping on the walls—was very strong. Lucy and I dared each other to dash up and down within its seventy-eight-foot length. When no one else was in the gallery each of us would, in turn, swing beneath the railing and dart a sweaty hand to tap a bone or stroke it, then rush out again. Perhaps we only wanted to flaunt the rules of the museum. Visitors weren't supposed to touch the animals, though they had been so endlessly touched before—by hunters, perhaps, and then by those who prepared the specimens, by the collectors and curators who installed and displayed them. But grubbier yet than strolling through the whale's palings or caressing the skeleton, we made a point of bringing snippets of ourselves to deposit onto the blue whale. A flicked-off scab, a loosed tooth trailing red filament, the translucent patch of skin sloughed from a healing burn. As though it were seasoning, I'd sprinkled hair trimmed with craft scissors onto the whale's jawbone.

Why did we do this? Why let go of what was letting go of us in this way, and place shucked-off bits of our child bodies atop the whale's great bones? I can only guess it had something to do with an urge to ritualize the shorter history of our own beastly nature. Putting ourselves into the whale's time frame—that might have been our intention.

In 1941, Anaïs Nin enters a gallery at the American Museum of Natural History in New York City and is confronted by mounted blood cells, replicas of them, many times magnified. Before deducing exactly what these objects are, she is struck by an instinctive, unmatched aesthetic intensity; she feels that these blood cells may be the most beautiful, most transfixing, and ideal structures she has ever seen in her lifetime. In her diaries Nin wonders if a sense of pleasing form in art and literature might be gleaned from an intuited appreciation of shapes that are inti-mate in us, but impossible to observe—the somatic design of the plate-lets in our blood, the uprooted fan corals in our lungs, the squat urn of a womb. Those things we know to be true but which have qualities of the unreal, being unseen and sometimes barely sensed. She feels ensorcelled by the model cells, as though she herself had shrunk to dimensions tiny enough to wander inside her own veins.

It's too loose an idea for me to get behind, this notion of an inward clairvoyance to our organic, anatomical design, but reading her diaries I nonetheless love to envision Nin approaching these huge, swollen cells, surging with aliveness and mystery. Nin knew that writing consciously from within a body arouses anxieties like: Where does that body start, and where does it end? How does this body feel what it knows; where are those faculties located? Can this body be made to feel *for* that body, a body unlike it—how much of a likeness is needed to identify or empa-thize with any creaturely other?

I imagine that Nin derived, in that moment at the Museum of Nat-ural History, an unruly but vital creative energy from perceiving herself to be tumbled in the ectoplasm of some bigger planetary entity; the flow structures of the world. And isn't this mentality—of projecting ourselves into places we cannot physically observe—also essential for thinking *ecologically*? Considering the deep sea and the outer atmos-phere, too, and how our livelihoods might connect to the damage done there. Contemplating, as well, the inside of a whale's stomach, and whether it contains krill and squid beaks, or a greenhouse.

At the launch of a poetry collection in Woolloomooloo one evening

I get talking to an inner-city choreographer who pulls me up on this point: "It's the colonial impulse, isn't it? Colonize the interior. Go into the whale. Drag what's inside out. All those white men, shooting and stuffing animals, giving them new names, and glass eyes, to send back to where? London. Amsterdam. Chicago. They got to empty lots of things, back at the beginning of this Australia."

When she said *this Australia*, she flicked out the fingers on both her hands, as if to shake some intolerable substance off them. I said I agreed. I agreed and later on, I thought about how, inside animals, and in animals' names, and museums, there were residues of power—as well as pieces of the past.

Many international collecting institutions count a blue whale among their centerpieces. There are blue whale specimens, and reproductions, of differing ages and conditions in museums in Vancouver, Ontario, Christchurch, Ottawa, Melbourne, and Reykjavík. In Pretoria their blue whale skeleton is set up outside, in a window-lined tunnel near the entrance to the museum's main building.

Where a museum cannot obtain a blue whale skeleton, replicas are popular. Suspended from the ceiling of the Milstein Family Hall in the American Museum of Natural History in New York is a ninety-four-foot replica of a female blue, found dead off South America in 1923. The replica—made of polyurethane plastic, a fiberglass coating, and six hundred pounds of paint—was constructed going off photographs of the dead whale's body. Later, because the model proved to be anatomically incorrect, it had to be renovated and reinstalled. In the early 2000s, the blue whale's bulging eyes were flattened against its head, its blowhole was modified, and its tail was narrowed. On the whale's pale underside staff daubed a belly button.

In Ueno Park, Tokyo, the Japanese National Museum of Nature and Science has positioned its blue whale—a true-to-life, synthetic reproduction, spray-painted and hollow—outside and open to the elements. For a few days each year snowdrifts heap up on the whale's back, evoking the real lives of the Antarctic blues, spent roaming beneath the sound of snow, cat-footing on the roof of the sea.

A blue whale's heart, salvaged from a blue whale killed in a heavy sea-ice season near Newfoundland in 2014, was sent by the Royal Ontario Museum to Germany to be plastinated. The fat and water in the heart was removed and replaced with plastic. This technology is a recent innovation: the huge organ was first soaked in acetone (the fluid used to take off nail polish), and then a plastic solution. Next it was placed in a vacuum chamber, under conditions very near to being thrown into outer space. The hardened heart—its organic tissues saturated with polymer and silicone—has toured various international museums. "The Frankenheart," technicians call it. It is the size of a wrecking ball and very pink. Because it is mostly plastic, I think it will last thousands and thousands of years. Perhaps it has become as impervious as a meteorite.

There are few whale taxidermies because a whale's body, and whale skin in particular, tends to putrefy quickly when removed from the water. Gothenburg's *Naturhistoriska* Museum in Sweden holds the only stuffed blue whale. In 1865, a calf beached alive outside of Näset. The first act of its discoverers was to poke the whale's eyes out. It should "not be able to see us," they are recorded as saying, meaning: the whale should not be able to observe the men who had set about the task of killing it. This took two days. One of its eyes was later retrieved from the sea and put in glycerin and alcohol—the second eye was lost, left wandering, bumping along the seabed. After it died its terrible, durational death, the whale calf was taken apart and its skin was preserved with arsenic and mercury chloride, before being pieced back together using copper furniture studs (this chemical whale recalls the whales hunted off Russia today that smell so strongly, and strangely, of iodine). The Gothenburg whale's mouth was remade to allow the jaw to hinge all the way open, in the style of a grand piano. Its interior was decked out as a lounge, with wooden benches, red carpet, and a ceiling lined in blue muslin decorated with little gold stars. On special occasions, museum attendants sometimes served dinner and coffee inside it. American tourists sought to have their photos taken there, praying in the fashion of Jonah confined to the Leviathan's belly. In the 1930s, two lovers were caught *in flagrante delicto* inside the whale, having consummated their passions in its esophagus—and then the whale's mouth was closed.

In the WA Museum, the blue whale skeleton was haunted by sound. A recording of a blue whale's underwater calls played mournfully on a loop. Low moans; notes that tolled as if they were being dripped from some distant nebula. The recording technology needed to capture the sound of any blue whale is much younger than the whale skeleton itself—the soundtrack of the WA Museum's Antarctic blue was not, therefore, a recording of that specific animal but of another whale. So there were, in fact, two whales on the fifth floor, not one. A voice without a body, a body without a voice. I can find no information about this other blue—the incorporeal, audible creature. Its history has fallen into the quicksand of the skeleton's story.

Whether visitors see whale skeletons, fiberglass replicas, plastic organs from a whale, chemically treated whale taxidermies, or if they listen to whale songs, museums encourage their patrons to view whales as larger than life: as sources of wonder. In mid-2017, the British Natural History Museum in London replaced the *Diplodocus* in their main hall with an aerially wired blue whale skeleton, lit with pale-blue spotlights. The museum named the dead whale Hope.

As a child, I looked at the WA Museum's blue whale as the apotheosis of all blue whales. I saw nothing of the individual animal's suffering, and I learned nothing of what the people who first encountered the whale felt for it, or what they thought they *should* feel for it—what a blue whale might portend to those people. The history encapsulated by the whale in the museum was global and ancient, evolutionary and natural. Only later did it occur to me that the blue whale emerged not only out of the ocean but also from a past, and a social context, that could be localized and examined. I had to go to the archives, to the microfiches, to dig out old news on the whale.

The reason I started tugging on this line of inquiry had to do with a different dead whale I was then reading about; a bowhead, the longest lived of all the whales. There are bowheads on record living to 211 years. Bicentenary mammals. In 1992, a bowhead killed off Utqiaġvik, Alaska, was stripped of its exterior blubber to reveal a deep tract of scar tissue, in the terminus of which was concealed a stone harpoon tip. The whale's wound had healed over the shrapnel long ago—evidence of a weapon used by indigenous whalers up until the 1880s, when metal tips introduced by Soviet whalers replaced hand-chipped blades of flint and slate. Other bowheads have also been found with stone points in their blubber. One had a fragment of a lance from a shoulder-mounted gun, shot last in around 1890. Another bowhead carried a piece of walrus ivory from a traditional spear. Yet another contained a whaling iron that could be tracked back to a specific Yankee whaling vessel. These objects have ended up in museums too: as artifacts of bygone people. They were valuable discoveries in their own right, because of what they revealed about the technology of whale hunts.

Scientists used the shards as evidence to substantiate, or corroborate, the stupendous age of the bowheads. But you need a chip of ice in your own heart to learn all this and not go straight to how it must have felt for these whales to live for so long with inanimate objects grating inside of them. Are there bowheads out there still, storing more precipitate of human history? How much is any whale, or any animal really, an inadvertent ark or cultural archive of things we have lost touch with?

Then I started wondering about how the blue whale in the museum had died—and if people had gossiped and conspired, hoping to profit from the whale, or were their feelings more melancholic and culpable; was the whale received with foreboding? Scanning copies of the *Bunbury Herald* and the *Sunday Times,* I learned from several sources that the blue whale was first sighted back in 1898, at the mouth of the Vasse River, near Busselton, in the southwest of Western Australia. This isn't far from my mother's hometown, and from where all those pilot whales have lately come to strand. The blue whale out of the Southern Ocean died in the shallows of the Indian Ocean, likely drawn to feed in a canyon off Perth. Daisy Locke, a local woman, discovered the whale. Locke rode a horse to her uncle's property to call on him to harpoon the whale. It is unclear whether this spiking killed the ailing whale, or if the animal was already dead and the action was meant only to lay claim to it—"the monster," "the fish," "the finback"—for the family's financial benefit. That the whale is called monstrous and "a fish" leads me to believe that while it might have been considered a marvel, and perhaps frightening, it was not mysterious or ominous. Locke sought to have portions of the whale boiled down for oil, though in the end the "best of it went to waste."

From the explanatory panels that accompanied the blue whale specimen comes this thin account: over three years starting from 1898, "a taxidermist from the Western Australian Museum cleaned the flesh from the bones of this massive animal with the help of two Japanese fishermen." The taxidermist's name was Otto Lipfert. The men who assisted him were recorded in the newspapers of the day as *"two Japs."* Who were they? What were their names? As the flesh of the whale was removed, this story of the collaboration between Japanese fishermen and Australian scientists seemed also to have been mostly pared away— only the slightest details remain.

Having garnered the blue whale skeleton as a centerpiece, the establishment of the WA Museum was set in motion—a death project and an imagination project, coeval; set upon making Australian nature part of global natural history, and putting Perth on the map by way of a whale. Naturalist clubs embarked on collecting flocks of native birds to stuff, as well as geological specimens to encrust the beauteous chambers that were planned to surround the whale. They envisioned *wunderkammern*, many cabinets of exotic wonders, housed in plush rooms.

As I write this, the whale skeleton, as Lipfert prepared it, is officially 122 years old—older than a blue whale's natural life span (this species are thought to live to 110). The exact age of the whale when it died is unverified. 1898: a time before the country *was* a country. The slow migration of the blue whale's disaggregated bones, inland and up the coast by rail, toward its initial temporary mounting in a shed on Perth's Beaufort Street, spanned the period of federation. The blue whale stranded and decomposed in the colony, exposing the red of its flesh, dark decay, then the white of its bones. It had its afterlife in this Australia.

What I am describing, my own childhood experiences in the museum, took place some twenty-nine years ago, but it feels to me like it might have been a hundred. I phoned the museum's helpful records officer, a man named Brett Zimmer, and he sent me an electronic folder full of photographs of the exhibitions as I saw them back then, as a girl. *My god*, I thought, *remember that! The hideous stromatolites! The head of a baby dugong in a Perspex box!* There were things in that folder I had been convinced I'd made up. And there, too, were the photographs of the whale being lifted by a crane into an open corner of the museum building. Flicking through the images, I thought: *Where is the museum of museums, to showcase how museums once were?*

Such meticulous labor was undertaken to conserve this one whale in the museum—painstakingly mounted piece by piece, and dusted daily—all the while so many Antarctic blues were destroyed. How can this obscenity be true: that all the ingenuity it takes to preserve a whale can overlap with the industrial slaughter of them? Looking at taxidermies in museums reminds me now that the tendency to destroy individual animals might arise not from indifference, hubris, or greed, but from a very deep attachment. Specifying any organism for preservation and inclusion in an official collection suggests that, at the point of its death in the past, that animal was thought to ripen with a message for the future. People felt something for it. They expected their descendants would, too—or that they *should*. The desire to maintain animals in a state of suspended animation is a desire to sustain a relationship with them and to advocate for that connection into the decades that come. It is the pleading of disappeared people for our continued enchantment with this marquee of other species.

Museums are places for thinking about the past; the ancestral past of human cultures, the surge and violence of acquisitive imperialisms, and, too, those taproots of time that extend far further, beyond the reach of traceable generations and family trees. Some scholars call this "deep time" or "geologic time," because it is a past that inheres in the strata of rocks and fossils, shells, peats, tars, oils, and chalks, rather than in the artifacts of humans and their societies. In museums, the time spans of animal evolutions are made visible, and geological durations outside our daily perception become graspable. Animals morph into plants, and plants into rocks; everything a bit waxy and crumbled on its edge. Eerie, transitional states in the style of a surrealist landscape. What looks like an ugly carbuncle of marine stone turns out to be the compacted accretion of billion-year-old mats of algae: a stromatolite. The small, seized spiral of an ammonite reminds you the world's topside is always invisibly turning over, that you stand on a continuously renewing surface (and a vast, undetectable artwork). But in the museum, such telluric processes—the foisting of rock from the seafloor into alpine elevations, the molten yawns of volcanic cones—are also abridged in displays you can turn cartwheels through, if no one is looking.

You might come to see yourself as more than a unit of biological life, of biological time: a unit of evolution. You might grow up in the

Devonian, amid organisms of uncertain affinities, spore-shedding and jawless. You might get to hold a foot-long core of rock and feel a thousand cold years in your palm, or measure it mentally against the length of the ponytail you're hoping to grow by the start of term. So many types of time collapse together at the museum.

"Historic time overlaps with geologic time the way a whale louse overlaps with the blue whale it infests"—this is by the American biographer and nature writer Verlyn Klinkenborg in Yale's *E360* magazine. Klinkenborg goes on: "The scale of that comparison is too small by several orders of magnitude, yet it's all too easy to believe, with the self-importance of a whale louse, that we exist apart." If lice are indeed egotists, we inhabit time as they do: with mystical regard for the materiality of the deep time we imagine is beyond our influence. The whale's moment, meanwhile, is inestimably, insensately long.

Maybe what whales convey is the endlessness of the distant past: a tide of time that subducts outside of the realms of human life, and unfurls toward the bluesome distance of life's far origins. That stream of time—the slow-turning engine of paleontology, evolution, geology—has long been thought by us to be *glacial* and *unchanging*. Two synonyms that slide away from each other today, as the great ice packs of the world disappear at a rapid clip, and glaciers retreat to glass houses, glass fists, and then wink out. In the museum, it may be startling to discover that whales are older than humans, and then that sharks are older than trees, and jellyfish are older than leaves. But what concept does a shark have of a tree? A jellyfish can't think of a leaf: or if a jellyfish does, it has no word to describe it. We alone have concepts of the past occupied by whales and their ancestors. We are the animals able to envision the time to come and the nature that will abide it.

———————

The old Western Australian Museum building was demolished in 2003, and many of its specimens were moved to the museum's interim window-fronted home, in the square just up from the State Library. The blue whale waits elsewhere, in a temperature-controlled warehouse, pulled apart and packed in sheeting. When the museum's new building is completed, soon, the skeleton will be installed. For now, the whale's 1,763-pound skull rests on one pallet, its spine on another, like pieces of flat-pack furniture. The ribs bracket one another, wrapped in transparent plastic, still greasy. Oil trickles through the pores of the bones; an inky perspiration that's been going on for over a century.

For the longest time when I walked down Francis Street, I would look up to the place in the sky where the whale once waited for us. There's nothing there now, save for atmosphere. After the rubble was carted away the site below lay bare, a rectangle of smooth sand, yellow and uniform. Do you also recall hearing this?—that sand isn't what's recognized as a substance. It's a scale. Any matter can become a sand if it's ground right down. Glass, stone, bone, silicon. When every object is forced to self-same size, nothing retains the capacity to be divisibly miraculous. The numen loosens from the particles, particular on a fingertip, then identical in a dune, and what magic persists drifts into our perception of supranatural forces. Which is to say, that magic resides in a feeling of duration; the haul of a future that's already set to work decomposing us, scattering our knuckles, our ankles, our littler nodes of cartilage, out to the wind.

04

Charisma

Dolphin Selfies — Why Didn't They Stop? — E. O. Wilson's Biophilia — A Magic Well — Machine Division — Nature Goes Viral — A Nebulous New Pangaea — Dreams, Being Screen-Colored — The Big Data on Mini Mammals — Faking Solitude — Cold Hard Numbers — Windshield Phenomenon — Stacking Stones — Out of a Canon, the Endangered Species Act — Virtual Lions — The Agony of Loving the Disappearing — Children: Their Hostility and Hospitality — Caring More Than We Can Stand — Tableaus of Devotion — Cute Aggression — What Is Done as a Crowd — The Animal Light Bulb — Berger, Once More — Charisma Is Hierarchy — Migaloo — Endlings Are Babied — Whales: From Where Does Their Charisma Derive? — Tim Watters — Super-Whale — Ecosystem Sovereignty — The Narrowing of Killer Whale Culture — The Less We See, the More Aura It Has — A Pocket of Pure Nature — Atlantis — E. E. Cummings Defines the "Zoo" — Mosquitoes Kill Whales — Eating Stones — Whales' Eyes: Do They Mean a Double Self? — Does It Blink? — Sheer Scale — Bryant Austin's Enigma — *Nature Morte* — When a Miniaturist Meets the Biggest Animal on Earth — "For Whales the Ocean Is Not Blue" — A Half-Head Dream — Smiling Eyes — Biophobia

Santa Teresita, Buenos Aires: The pictures show a jostling crowd of adults and young people, shirtless and in swimsuits, standing on a subtropical beachfront. A thin band of staticky ocean behind them. Most are men and boys. People wince in the hard light. One or two toddlers, lofted onto shoulders, clutch fistfuls of sweat-wetted hair. Within the interlinked shove of forearms and frantic, outstretched palms, all effort lunges toward a center point. There, amid the crush, a sunburned, barrel-chested man holds up a dolphin. He hovers it in one hand. The dolphin is small, pudgy, and pin-eyed; its mouth hangs open. Just a few feet long, with little flippers: a baby. No one looks directly at the camera. Not out of shame; their focus is elsewhere. Many wave smartphones. These are photos of people filling their phones with yet more photos: a panoply of unseen images, retained in private, or since deleted. Something darker than glee steals across their faces. A hunger. The bulky man presses his thumb, possessively, into the underside of the dolphin's head where its flesh rucks into a fold. People grab for its tail.

The animal encircled by the crowd is an immature La Plata dolphin, one of the smallest cetaceans and known locally as a Franciscana dolphin because the species' biscuity color recalls robes worn by mendicant Franciscan monks. The IUCN lists the La Plata dolphin as vulnerable and declining: there are thought to be only thirty thousand alive. It is unclear what has brought this dolphin to the tourist beach in Argentina. The force of a tropical depression, perhaps. Misguided, maybe abandoned by its pod, did it wander within reach of the shoreline, or was it, in some way, seeking help? I scroll down; maximize a different image. Most shots are low resolution and slightly too reddened—still frames of newsreels, captions fixed on the chyron. What seems evident in the pictures is damned by the headlines: "*Tourists Kill Dolphin for Selfies.*"

I've found myself returning to these images online when what I am thinking about is the trouble of loving the natural world and its animals today. I try to look past my own disgust—a knee-jerk response. The reasons anyone would be drawn to pet a dolphin—these, I can understand. I am searching, instead, for a dispassionate answer to the question: *Why didn't they stop?* How does the urge to exhibit care, grief, and attachment overwrite the imperative to help the creature that triggered those feelings? What I see, there on the Santa Teresita beach is, I think, a tormented love. Antithesis to John Cage's maxim "Love equals making space around the loved one"—a need to connect, so dire, that it smothers the beloved. How to speak of this violent tenderness? Where, on our side of the human-animal divide, has it come from?

In the early 1980s—a time when the internet was scarcely a set of communication protocols, and a plaything for computer scientists—the American sociobiologist Edward O. (E. O.) Wilson coined the term "biophilia" for the innate affinity that compelled all humans or so he argued, to attach importance to other life-forms, living systems, and natural environments. Now two times a winner of the Pulitzer Prize, Wilson initially experienced biophilia on a research trip to the coastal forests of Suriname in South America—it was a feeling (more accurately put, an instinct) he devised a term for over two decades later. He wrote that biophilia inclined people to value nature by igniting wonder, a sense of mystery and curiosity. In infancy, the scientist observed, humans move toward animals and plants in preference to inanimate objects. The magnetism of shared liveliness functioned in his view, as prologue for the many cultures that lauded the natural world, populating it with myth and story. By adulthood, nature grew to be "the refuge of the spirit," a reserve "richer even than the human imagination."

Wilson saw biophilia as, very likely, a "part of the program of the brain," and cause for optimism. In effect, biophilia amounted to a psychological factory setting, predisposing people to care for the environment. Biophilia put a thumb on the scale for nature. Such instincts could be cultivated throughout one's lifetime through exposure both to nature's enchantments and to findings from the biological sciences—particularly vis-à-vis the animal kingdom. Per Wilson's explanation, the more a person learned about evolution, or the physiology and interactions of different species, the stronger their sense of nature's absorbing mystery would become. With the mesmerism of his vocational calling, Wilson wrote: "Every species is a magic well. The more you draw from it, the more there is to draw." Each new discovery about an organism—a mouse, a whale—suggested further levels of inscrutability to be inquired upon, down to the marrow, the molecule, and further, into the gene.

It was this apprehension, that nature would remain unplumbable (because each new discovery implied the existence of things not yet known), that provided—to Wilson's thinking—the foundation on which people were induced to want to protect the environment. Preserving nature safeguarded more than the wild: it shored up an emotional taproot of awe, humility, and wonder; and it provided a space for human knowledge to expand into. People sought to defend nature not because it made them feel at home in the world, but for the reason that nature promised to always exceed imagination; to be eternally strange and surprising.

Antipodal to nature, according to Wilson, were machines, which intervened between person and environment to "tear apart the paradise" and alienate humankind. Only in some instances, he conceded, might a love of machines be considered a covert expression of biophilia—particularly where technological design replicated elements of the natural world in an anxious eagerness to transfer people's inborn emotional affiliation with nature to their devices.

Over thirty years after Harvard University's publication of his book *Biophilia: The Human Bond with Other Species* (1984), people's enmeshment in the digital sphere—and their attachment to the handheld computers that capture and tabulate it—has not partitioned humankind from nature as Wilson once feared, though the machines we have since devised have generated new sorts of fixations on wild places and creatures. If technology was once, in Wilson's formulation, the intercessor that divided people from nature, by the late twenty-teens technology had become the driver, pushing people toward nature too much. Nature was going viral. On photo-sharing platforms, most notably. A great deal has been written about how digital systems underrepresent the reality of people's lives, but less so on how the compulsion to create idealized versions of nature *shapes* nature, where it is encountered, raw and real.

Maybe you, too, had the impression that the networked apparatuses of social media, many of them mobile, were busily assembling a nebulous new Pangaea—an idyllic supercontinent of pastel vistas, sunset monoliths, alpine lakes, powdery beaches, meadows, and waterfalls. Which were where? These places, scattered across both hemispheres of the planet (but concentrated more so in the north), conjoined in the ambience of various high-contrast filters. Online, the natural looked vividly beautiful. Nothing was threatened in any of these images, and nothing was threatening. You could tap into the supersaturated everywhere, from anywhere, with the barest gesture of thumbs. You flicked through it as though your presence were no more disturbing than wind, vapor, light.

A series of studies, began decades ago, showed that the preponderance of people surveyed remembered their dreams as being screen-colored. Adults who grew up watching black-and-white television tended to report dreaming in monochrome for the rest of their lives. By the 1960s, after Technicolor, 83 percent of dreams dreamed by the dreamers surveyed contained at least some color. Now I wondered if the gentling tones of the environments I scrolled through online were imprinting on my sleeping thoughts, touching up the nature I slumbered within. A hypnogogic nature cleaned, intensified, brightened. A nature that made the world beyond look, in truth, a little dull.

Millions of cute wild animals populated this digital world, their smallness and their appearance of tameness seeming to exist, in inverse proportion, to the unchecked vastness of the system that contained them. Furred things, big-eyed. Was someone—some agency?—collating big data on these mini mammals, manipulating their appearance for clicks and mentions? Would work like that amount to the hijacking of biophilia, or was it something else entirely? Where the animals were from, and what they were called, had little bearing on their popularity. Their new habitat was the internet.

Outside, the brunt was born by public lands, natural monuments, and national parks, which saw huge influxes of tourists bringing camera phones. By 2016, American parks were taking in 330.9 million visitors (which, as journalists for *The Guardian* noted, was a number near to the extant population of the entire United States). In Australia, eco-tourism rose too—by 30 percent between 2014 and 2016 in New South Wales, to isolate one statistic. The rise in tourism created traffic jams and incited petty acts of aggression at the outlooks: fistfights broke out in woodland car parks. Whale-watching ventures purchased faster boats to outrun one another, offering a greater return on the cost of a tour (more chances to see passing whales in less time)—around fifteen million people, worldwide, booked whale-watching tours each year. Sanitation workers swept through postcard landscapes on daily shifts, removing the squattage of human waste. In the United States, wildflower "super-blooms" got trampled by small-time celebrities sprawling, for portraits, in the pollen-clouded rainbows, while hundreds of citations were issued to recreational drone operators who hassled the wildlife and buzzed the serenity. Off New Zealand, a woman struck out into the sea to be filmed swimming freestyle alongside several killer whales.

Meanwhile, park management authorities reacted to the pulse of internet tourism with a series of contradictory initiatives. Signs were erected asking tourists to refrain from geotagging their photographs and thus drawing ever larger crowds to once-lonely sites of wonder. But, too, visitors were offered scanners, along with the frequencies of radio-tracker collars worn by wild animals within the grounds—they were told they would be able to drive right up to where the wild things were. Additional mobile-phone towers were installed, poorly disguised as very tall, very straight trees. Wi-Fi networks got threaded across the backcountry and through alpine ranges.

All this seemed a striking turnabout. More and more, people went to wilderness not to seek solitude but to connect with one another online. And when they got there, many found that it was increasingly difficult to take shots implying they were peaceably alone. One thing that united the digital crowd was their preference for beautiful places "off map." Places of autonomous leisure testified to a person's self-sufficiency and their resourcefulness—and though, in the past, this might have meant having the wherewithal and liberty to battle the elements, now it equally stood for an aspirational lifestyle that could be monetized via product placements and promotional deals (a different kind of self-sufficiency). So people pushed on. They crept farther out on the overhang, and trod dangerous parts of the atoll, battered by huge waves. They fed the local wildlife from their chip packets and yogurt cartons to draw the animals closer. Then they stunned them with the *tsk*-ing of their camera flashes.

There was a sloth craze, a sugar-glider phase. Dedicated fans of dolphins. Big baby elephants in baths too small. Frenzied hearts on fennec foxes, slow lorises, and tiger geckos. Real animals were a new category of kitsch, and kitsch was, again, compelling. People groomed the conversation pieces of their online collections—their animals in cubes.

A report commissioned by the World Wide Fund for Nature (WWF) declared 60 percent of vertebrate life—mammals, birds, fish, and reptiles—had disappeared off the face of the Earth since 1970. French biologists estimated 130,000 species (including invertebrates, excluding sea creatures) were already gone. The UN said marine pollution had increased tenfold since 1980, and a million species now inched toward extinction. The total earthly biomass of wild mammals dropped 82 percent. Comparatively, the biomass of agricultural species soared: 70 percent of all the birds on the planet were revealed to be poultry. Livestock (cows and pigs) now made up 60 percent of all Earth's mammals.

These are almost impossible numbers to wrap your head around, I know. When I hear them I feel as though someone has thrown a handful of dead batteries, cold and skittering, into a space inside my chest.

The truly wild things have today been redacted into ever wilder, ever more inaccessible hot spots. Delicate moths, caterpillars like Day-Glo litter from a rave, beetles, and bees are all vanishing, while swarms of more pernicious insects—snakeworms, ticks, and stink bugs—slide in beneath drying forests, or between the wall cavities of houses on urban fringes. One study calculated that three-quarters of the flying-insect population had disappeared from German nature reserves. In the rainforests of Puerto Rico, bug life declined sixtyfold.

Researchers talked of "the windshield phenomenon," a shorthand expression to capture how ordinary people were made aware that insects were disappearing when they remembered back to cleaning smeared bug life off their cars in previous years and decades. Road-trippers used to have to stop every few hours, to wipe away the obscuring streaks of so many dead grasshoppers, flies, thrips, and midges. Driving in agricultural country or alongside a forest, the windscreen became an increasingly virtuosic orchestral score of wings, legs, and antennae. This was within recent memory, but now the glass stayed unsmirched. Though our computer screens filled with animals, windscreens—another interface between us, nature, and an older technology—emptied of them.

It wasn't that all the insects had become roadkill but rather that inadvertently killing them with our vehicles had once made their sheer abundance visible. The insect eradications were the result of multiple interacting causes: herbicides and pesticides, habitat loss, shifting and intemperate seasons. Yet even as nature was breaking (maybe *because* nature was breaking), people's emotional connection to nature intensified. Hiking and mountaineering associations in Europe implored visitors to stop scattering the ashes of their loved ones on famous peaks, because the phosphorus and calcium of so many incinerated bodies had changed the soil chemistry on which fragile high-altitude plants depended. In the shallow oceans some fourteen thousand tons of sunscreen was thought to have rinsed off sightseeing snorkelers and divers, contributing to the collapse of reefs. (Common ingredients in sunblock had been discovered to cause coral bleaching at very low concentrations.) The rush to see reefs still neon and jumpy had inadvertently sped up their decline.

Back on the beachfront in Perth, the euthanizing "green dream" had, to me, typified a troubling kind of compassion—the fluorescent injection represented mercy to the whale and condolence to its mourners, but it would also sicken whale-meat scavengers, being toxic in the biome at large. Yet the green dream, I saw now, was not an isolated concern: in many other places around the world, people's urge, en masse, to express their love of nature was throttling subtler gradations of life. Mountain grandeur threatened tiny alpine flowers; the vibrancy of reefs jeopardized gloopy coral larvae. Being discreet and lacking splendor, some life was overlooked—though the problem was not that individual people couldn't be made to care about spawn or tundra, exactly, but that the aggregate harm was done collectively and across long swathes of time. Stood atop a ridgeline with an urn of cremains, it didn't behoove you to imagine all the people who had done this before, or would do so after.

In that moment you were not an organism in the ecosystem; you were a person in pain.

Just as the natural world had begun to appear more halcyon—lusher, and less trashed—on the web, so digital iconography also crossed over into the nature we beheld before our eyes. Piles of flat stones called cairns or "fairy stacks," for one thing—rocks artfully balanced atop one another to be photographed. "Cairn," a Gaelic word, was Scottish, but now you saw the pebble stacks everywhere: along streambanks, on shorelines, and on the waysides of trails. What was the point of these? In the twitchy tumult of nature pictures, it seemed that it was no longer enough to have witnessed tranquility. People sought to record how nature composed them; how it calmed their mental state. The quietude found in assembling a tower of little stones was visual proof of a meditation that otherwise couldn't be seen. The cairns, as it turned out, were found to disrupt bird nesting grounds, to dislocate populations of inching invertebrates, and cause soil erosion. In England, stone stacking resulted in the piecemeal dismantling of certain heritage-protected walls that had stood, undisturbed, since the Early Neolithic. Documenting a hike, in the information age, had the power to erode the landmarks that made it distinctive. An ancient culture, a minor nature, pillaged for the construction materials of a new photographic tradition.

Bill McKibben, author and pioneer environmental activist, wrote once that "without Kodak there'd be no Endangered Species Act." Wildlife photography and documentary remain powerful tools for generating the public's attachment to animals, but today these important communication projects nestle into a historical moment when photographing nature has the power to destabilize conservation. A time when safari operators in Namibia require tourists to wipe the metadata from their images before uploading them, fearing the poachers who are thought to rely on social media as a proxy to track rhinos (the rhinos are hunted for their horns, powdered for medicines, demand for which is generated and met in marketplaces online). A time, too, when the average French citizen is presented with more than four "virtual" lions, per day, in adverts and electronic images: thus seeing many times more of the animals, in one year, than exist in the whole of West Africa (and readily misestimating how threatened living lions are). At this juncture when, only recently, a group of tourists are said to have patted a dolphin to death for a close-up.

———————————

The images from Santa Teresita: with a toughened heart, I turn back to them. To be so near but unable to touch the dolphin, appears, from the expressions of those on the periphery of the crowd, to be the source of terrible suffering. You can almost see the blood-heat loud in their ears: the crisis of their unexpended loyalty. I think, *This must be the agony of loving the disappearing.* Tableaus of devotion are recalled by these pictures. The struggle toward worship. As in: the masses ford the holy river braced beneath their icons. The sick at Lourdes; Kumbh Mela pilgrims thronging the Ganges; insurrectionaries in old religious wars. Or else, a fresco of toil and piety by a Flemish master—the gloomy glory of some alpine church. How did Barbara Ehrenreich have it? Contact with wild animals today provides "*that which people have more commonly sought through fasting and prayer.*"

I look, again, at the large, maroon man holding up the flaccid dolphin: its specks for eyes. In the crook of his other arm, I notice, he is also supporting a small girl, perhaps three years old, hugged into his torso. The girl's hair is pulled into a fountain ponytail. She reaches for the dolphin with a fist, peering at it sidelong, her head tilted against the man's meaty neck. In other photographs the baby dolphin is lowered to allow people to stroke it, which they do, many of them all at once, but the kids do it shyly, snapped mid-motion, wiping an index finger along the dolphin's forehead or tapping it up and down with a cupped hand. Their gentleness is excruciating. On the verge of tears, a boy in a blue T-shirt desperately, disbelievingly, glances back to a man he must know—he has reached the dolphin! With his palm he very tenderly covers its blowhole.

The distance between hospitality and hostility is so short at first. If biophilia *is* hardwired from birth, we must still be taught how to restrain ourselves from stifling what we love. These children can't know their menace.

In the lee of twentieth-century psychology, the suffix *-philia* has come to mean not just affection, but an abnormal attraction. Attraction that takes on an unmerited ardor, defiling that which it seeks to cosset, or cosseting the wrong thing—that which it would undo us, debase us, to get close to. For this generation—my generation and those younger—living through the slow emergencies of mass extinction, biodiversity loss, and defaunation, is there not also something weirdly thanatophile (death-loving) in the biophile? We find ourselves possessed by a savage urgency in relation to the animals we adore: we care more than we can stand. An animal's rareness—fear of its imminent decline—draws us closer.

Performing our love for nature can, for some, seem more important than not causing harm. The austerity of restraint ("take only pictures") has, after all, failed to remedy the crisis. Neither does restraint *show* how hurt we are: only a demonstration of love does that. Outsized love; a terrible glamour. A love that disgusts, but from which we cannot desist. The grief is so immense, in the absence of any official, collective, mourning protocol; individualizing our connection to it demands a damaging proximity. As the Laguna Pueblo author Leslie Silko once wrote, attempts to get closer to nature through rendering its features iteratively, and in particular detail, may betray deep feelings of *disconnectedness* rather than intimacy. So perhaps the dewy, digital Pangaea is not a place to hide out, a place in which to pretend that what is happening to nature isn't. Rather, the lush proliferation of idealized environments—that nature of our making, and the flocks of cute animals found there—might describe the diverse melancholia of our lost connection. Our unprocessed, inchoate loss, fruits gorgeously online.

When I thought about the small screens through which this glossy nature was encountered, I also thought, again, about the "windshield phenomenon"—how the vanishing of the insects became evident when you recognized the legion of bugs you, yourself, *hadn't* dispatched with your car. What had been killed, indirectly by pollution and climate change, had ceased to be only within your immediate sphere of action—the kill space extended out in front of you, and behind you, for miles, and for years. Even after hours of driving you could still see the horizon clearly. There was no mess. The insectless future you approached lay, chillingly clear, up ahead.

It remined me that one other thing we pursue in ourselves, when we seek contact with wildlife now, is absolution. An amnesty for the harm we and our kind have caused, but have failed, until now, to see.

Writing of the baby dolphin's death in Argentina, the Polish-American philosopher Margret Grebowicz refers to "cute aggression"—a violent impulse toward pictures of adorable animals, described in a study undertaken by two Yale University psychologists in 2013. The words of one researcher summate the findings: "Some things are so cute that we just can't stand it." Participants in the survey admitted to wanting to squash, squeeze, and throttle lovable creatures. When the researchers gave the study's subjects Bubble Wrap to pop and then showed them a succession of endearing animals, the participants mashed the plastic in their fists.

Cuteness, as the cultural theorist Sianne Ngai has best detailed, is not merely a matter of smallness, softness, the cartoonish and the infantile. All cute things invite fondling, but nothing is cuter than when it's vulnerable, helpless, or pitiful. Sloths are dear, but sloth *orphanages* are dearer. Being hobbled or injured, engaged in pratfall or blunder: that's cute. A baby dolphin is sweet. A baby dolphin that has stranded is sweeter. It needs us. It *needs*. The little dolphin has had a little accident. A diminutive object with an "imposed-upon aspect"—this is the sweetest thing of all. But such creaturely objects (for cute animals are objectified) can cause us to grind our teeth. Ngai writes that cuteness "might provoke ugly or aggressive feelings, as well as the expected tender or maternal ones," inciting "desires for mastery and control as much as [a] desire to cuddle." Cute things should be soft and twistable, because they need to be capable of withstanding the impulse to violence they arouse (think of the aggression young children sometimes display toward their toys). When cuteness, a quality of products and pictures, is turned back onto the natural world, then the urge to squash animals—to touch, pinch, and squeal—is amplified.

Grebowicz attaches this feeling—cute aggression—to technology. The need to connect, she argues, extends in two directions: the desire to be closer to animals and the desire to make meaningful contact with other people. A selfie with a darling animal might be one of the few remaining digital forms in which a demonstration of heightened pure emotion, and enthusiasm, is freed from irony. Miniature intensities, these pictures make a show of relinquishing power to the animal's untroubled virtue, its goodness. The animal is artless: it can't pose. It doesn't know what a camera is *for*. That kind of authenticity is currency, online. Yet, the crowd on Santa Teresita beach, I keep coming back to the brutality of their longing: what looks like a loss of control, rather than a carefully staged exercise of it.

I would like to grant these people some reprieve, in fact. I can shut my eyes and imagine the crowd from the beachfront, dispersed later that night. Let's say it's a hot night, and that they're pacing through the dusky evening. The westering sun is prolonged in slats between buildings. Winged insects stir and scintillate through each tunnel of light, like threads of saffron in hot water. The mood of that evening is, in some sense, a formation of the insects, which make no audible noise, but dramatize, by way of their brightness, the colors pooling behind shopfronts and hotels, lending to the outdoors the closed intimacy of the indoors. I see the people after the Santa Teresita photos, strolling barefoot or in sandals, a slight swagger in their bodies, heading down the empurpling driveways of recently built holiday complexes planted with waist-high trees. Their skin tightens with goose bumps, forewarning of the sunburn that sharpens beneath their clothes. Maybe they have bundled up a little washing for the coin laundry, or they pop the top off a bottle, and then each sits, on the curb, to swipe through the photographs of the dolphin from Santa Teresita. Only this time, their faces take on a cast of alarm. They see themselves as they will be seen. They see that what was done as a crowd would never have been done alone.

Neurologists have discovered there is a specific part of the human brain dedicated to the identification and classification of other animals. This animal "light bulb" is in a part of the temporal lobe called the superior temporal sulcus, a ridge that runs along the side of the brain. Perceiving living things, and imagining living things when their names are spoken to us (or even when we read those names), sends flares of electrical activity through tiny tracts in this zone. So when I write— and you read—*whale*: there's a shared zap in that part of our brains. Again: *dolphin*. Then *jug*—a different part of the brain fires. People who have suffered neurological injuries to the animal-recognition section of their brain sometimes cease to be able to identify animals and begin confusing them with objects. This is of particular concern when it comes to mealtimes; such patients routinely fail to distinguish between the edible and the inedible, having lost the ability to tell organic matter from inorganic.

E. O. Wilson's biophilia hypothesis holds that humans automatically identify with the living world and lifelike processes in all their various permutations but that this simpatico feeling can be upgraded through venturing into nature and learning more about it. Biophilia is the "quiet passion [that] burns, not for total control, but for the sensation of constant advance," Wilson writes. The overtone of conquest here I've read with unease, recalling, too, John Berger's idea that "what we know about [animals] is an index of our power, and thus an index of what separates us from them. The more we know, the further away they are." Both Wilson and Berger see nature's mystery as compelling, though that mystery leads the two thinkers in opposite directions. For Wilson, education in science and natural history provokes humility: faced with nature's enduring mystery, humans feel a duty to preserve wild places. For Berger, our knowledge of nature proves human superiority, and calls attention to the distance between us and animals—so we are led to diminish and coddle nature, creating zoos, for instance, in which animals live unnatural lives.

Our attachments to other animal species are frequently contradictory, and are produced in the interplay between closeness and distance: physically, biologically, and metaphorically. But it is not true, ultimately—is it?—that all forms of animate life attract the same level of affection. *Biophilia* suffers from the biologist's overly generous belief that every organism is universally interesting, when, in fact, some organisms feature recurrently in human stories and a handful are persistently more cherished than others. What the onlookers felt, seeing the humpback suffering on the sand in Perth, was not just different *in degree* to witnessing, say, the death of a coral larvae: it was different *in quality*. I'm talking of magnetism, of charisma.

The extent to which people perform feelings of affection and protectiveness toward other animals is not evenly distributed. Environmental campaigners refer instead to "charisma"—a species' capacity to function as a mascot, to sustain a riveting narrative, and to motivate a crowd to action. Such charismatic animals are readily "anthropomorphized."

That is, they are personified: assigned human characteristics and assumed to demonstrate humanoid behaviors, or values. Charisma establishes a hierarchy. Charismatic species are those that especially arouse compassion, so charisma influences the types of animals thought to warrant protection. Frequently featured by collecting institutions—museums and galleries—as taxidermy and skeletons, in sculptures and paintings, charismatic animals also commonly appear as logos and playthings. They need not be endangered, but as ambassadors for their kin, and for the environments they inhabit, charismatic animals are often exhibited in captivity, and feature as *dramatis personae* on television (e.g., Flipper). To belong to a charismatic species is to be a pack animal for human imagination.

A handful of individual animals can also become independently charismatic—either because these animals are uniquely storied and identifiable in the wild, or because, being held in captivity, people develop a close relationship with them. Off the east coast of Australia, an albino humpback called Migaloo, first spotted in 1991, is considered an exceptional spectacle. During its migration each year, the famous whale is logged up and down the seaboard by social media accounts recording its appearances. Migaloo is one of a select number of animals to be identified, as an individual, under Australian law. Legislation ensconces the whale in a "do not approach" zone a third of a mile wide—an expansion of the standard vessel exclusions in place for all whales. Access to the sky above the white whale is also restricted ("it needs our protection, not our attention," said one scientist, affirming the laws). Prior to the exclusion laws being enacted, Migaloo hit (or was hit *by*) a trimaran in Hervey Bay, and a scuba diver also attempted to sit on the whale's back to make a video. Recently, biologists have become concerned that Migaloo may be developing skin cancer in the harsh winter UV, a vulnerability worsened by the animal's lack of pigmentation. The white whale's back (a "landscape of snows," as Herman Melville had it) is riven with pinkening marks, cysts, and scratches, some of which are almost certainly not lesions, caused by the sun, but are striations from synthetic debris that the animal has come into contact with in the past. Though tourists may be kept at a distance today, there remains an inscription of human trespass, an anonymous autograph, on the skin of Migaloo.

The symbolism of the white animal as omen, ghost, or otherworldly guide is well-trodden in literature, but if an animal's novelty—albinism as a recessive, and atypical, occurrence—is one attribute bestowing charisma, then the quality of rareness that is being extremely endangered forms a particularly drastic subcategory of charisma. The charisma that attaches to those creatures known as "endlings"—the last survivors of species otherwise thought to be extinct—has a desperate and unstable voltage. These animals are commonly given personal names and may be exhibited while living, or preserved after death. Ben, the last thylacine; Lonesome George, the last Pinta Island tortoise; Martha, the last passenger pigeon; Turgi the last *Partula turgida*, a Polynesian tree snail. (Turgi, who died far from home, in the London Zoo, was given a tiny tombstone that read "1.5 million years BC to January 1996"—its species' evolutionary life span, rather than the last snail's.)

The word "endling" was first proposed by members of the medical fraternity in the mid-1990s, to describe people whose deaths meant the end of their familial lineage. It wasn't until 2001 that the National Museum of Australia established the connotation of a last animal, or plant, and popularized the term. Other alternatives were debated—"lastoline," and the imposing "terminarch"—but "endling" has stuck, being very nearly a homonym of "earthling," and perhaps also because the diminutive suffix confers the implication of a petering out, a waning. From the many, to the singular; the left behind. The suffix *ling* is not just a diminishing syllable—it works to cute-ify and infantilize the noun (see: "princeling," "duckling," "changeling"). Aptly, most endlings end up in captivity, dependent on human care. Endlings are darkly charismatic animals. They give extinction a face, and are made to stand in, as a denominator, for their entire species—even if their innate behavior is curtailed by their solitariness and containment. Endlings offer a version of "last chance to see" tourism, often within grasp of ordinary people. An endling may achieve longevity—Lonesome George lived to be just over a hundred—but it will still be babied, in large part because endlings will never pair off and have offspring of their own.

Animal charisma can seem a spurious idea, more readily attributed to cultural history than to any inherent, consistent bond with one animal species over another. Yet there *is* research to show that a principle called the perception of "phylogenetic relatedness" determines how humans infer the inner sensitivities of other species—their pain, sentience, or intelligence. Studies have shown that people feel a strong attachment to beasts with "an erect bearing," and to large-eyed mammals. If an animal walks on the land or flies in the sky—both these correlate positively with our impression of that animal's charisma. Having fur and being rotund matters. Being capable of facial expressions suggests emotion. Fingered hands or paws: dexterity. Offspring suckled, and carried close to the mother's body foster a belief in shared childrearing styles.

Dolphins may be cute, but whales? The humpbacks, blue whales, right whales, sperm whales, and others—undeniably charismatic animals, about which people feel ardently—lack many classic charismatic features. True, they are mammals, but whales live where people do not, in the most remote seas. Whales are hairless, legless, handless, and, most of the time, horizontal. Their faces defy expression by being prolonged sidewise, as solid and static as the stone Moai heads on Easter Island. Their eyes are antipodal. You rarely see more than one eye at a time. The whale's extreme bigness is not just stupendous, but a bit eerie; to think of aliveness, and sensitivity, on such a scale. Being so huge, there is little chance of a whale being babied.

So what makes whales charismatic?

I wanted to understand more about what underlaid the charisma of whales, and so I arranged to speak with Tim Watters. Watters's work as a photojournalist and videographer has fronted a wide range of prominent activist campaigns worldwide. Off the Kimberley coast in Western Australia, he shot footage that helped to propel the successful campaign against a proposed onshore gas hub, adjacent to where humpbacks migrate on the James Price Point headland. Watters is in the business of making animal injury visible, an undertaking that walks the high wire between shock and sympathy. In 2014, his photographs of bloodied minke whales, hauled onto Japan's *MV Nisshin Maru* in the Southern Ocean, made global news (and, he suspects, led the whalers to paint the ship's slipway a darker color in following seasons, so as to obscure the volume of whale blood on display). Over the past decade, Watters has developed a detailed understanding of the kinds of animals people respond to and how their range of emotional responses can differ between species; why this might be so underlies his work. Ethics *are* optics in Watters's trade.

Watters described cetacean charisma as the yield of the animals' size, intelligence, sociality, their playfulness and vocalizations, and the remarkability, or uniqueness, of whales and dolphins. We talked about how, while all these descriptors are apt to apply to separate species within the cetacean clade, the qualities that make whales captivating have tended to merge in the public eye, either as a result of people's unfamiliarity with the differences between whales, or due to the efforts of anti-whaling campaigners to extend markers of charisma, unique to one species, to all cetaceans (a phenomenon the Norwegian anthropologist Arne Kalland has referred to as the mobilization of "a super-whale"). Humpbacks sing elaborate songs, for instance, and beluga (known in some subarctic communities as "sea canaries") are also chattily vocal, but other whales merely drone or click. Sperm whales have large and convoluted brains, and humpbacks have spindle neurons—the cells credited with causing human beings to suffer emotionally—but right whales likely possess neurological faculties comparative to cows (which is not to say that cows *aren't* intelligent, only that we consider their intelligence unexceptional).

Kalland has argued that aggregating all the varieties of cetacean specialness into one globally imagined, intelligent, songful "super-whale" has served an anti-whaling movement concerned with elevating whales to the status of animals distinctly "like us." The motivation, he contends, has to do with branding whalers barbaric, regardless of the types of whales they hunt. Watters and I get talking about what gets lost in this conflation—and, beyond that, what blind spots can persist even where whale *species* are treated separately.

In fact, whales are more particular than we could have guessed. The "killer whale," for example, has at least seven identifiable, discrete types of animal clustered inside it. Killer whales, biologists have discovered, occur in "ecotypes": groups bracketed off beneath the species level that won't naturally interbreed, and which have minor but distinct variations in their body shapes and coloration (some grayer and squat, others with white oblongs extending from the corners of their eyes, or faint "cape" markings on their backs). The ecotypes are sorted into separate seas and are genetically divided. Only very close scrutiny reveals the dissimilarities: in the open ocean, the distinctions between them are detectable to specialist scientific spotters—although it still takes a DNA sample to confirm whether a whale is within a known ecotype. To the whales themselves, though, the differences are believed to be striking. Behaviorally, in their social customs, their prey types, in their ways of making sound, and in their "languages," it is as if the ecotypes belong to long-partitioned and estranged nations. Killer whales don't just have a culture; they have cultural *diversity*.

To acknowledge that there are discrete, subspecies-communicative populations of whales—killer whales ecotypes—might change how we imagine endangerment and extinction, for even if a species persists worldwide, the extirpation of a group of animals can nonetheless mean a language, or something close to a language, is extinguished. And we can agree, can't we, that this erasure—of different kinds of expressiveness—is disquieting. That it should *count*, somehow, as a kind of de-animation, a limitation on the range of ways that an animal is able to *be* in the world.

In the course of talking with Watters, I wondered aloud if what we call extinction would ever be expanded to include the disappearance of animal *cultures*, as well as animal species. Perhaps the moment to debate this is now, when attention is beginning to shift away from the question of preserving biodiversity—minimum viable populations of the widest range of animals (a Noah's ark sample size)—toward preserving abundance, and proportionality between species; privileging the sovereignty and "health" of ecosystems over that of species. A focus on animal abundance means that endangerment may not be the threshold at which our concern for a species warrants interventions to preserve it: preserving numerousness might matter more in ensuring the ongoing resilience of the ecology, and all the organisms that compile it. Likewise, diversity within a species—the diversity of cultures, say, as it is represented by ecotypes—is one thing we miss, when we concentrate on numerousness alone and disregard the located histories of populations of animals. That there may be many individuals of a species held in one zoo can lead us to overlook the importance of preserving distinct, separate groups of one species of animal, in multiple different environments. The resilience of a species might be expanded by its cultural diversity, which increases the opportunity for a greater range of adaptations to novel threats. So it matters not just that we sustain populations of animals, but that we preserve the widest range of possible *contexts* in which they can continue their unique behaviors.

Watters is inclined to talk with his hands, revealing, from time to time, the blue outline of a sperm whale tattooed onto the underside of his left forearm. When I asked about it, he showed me the red krill—almost as large—inked onto his right arm; a counterweight to the whale's charismatic exceptionalism, and a nod to the importance of its ecosystem. Also, just body art: another animal etched into an animal, and a way to measure time on the body, in the gradual blurring of the line.

We moved on to the question of rareness: how a substantial part of the whales' charisma might derive from a history of endangerment, as well as how infrequently people inland encountered whales. "It's the same with gems and gold, isn't it?" Watters observed. "The less we see, the more aura around it; the more we care about, talk about it, write about it." He described the difficulty of photographing a whale underwater, dealing with the light-scattering effects of sediment in the water column. One of the most striking observations made by the biographer and author Philip Hoare in his book *Leviathan or, the Whale* (2009) is that people had looked back on Earth from space well before a picture of a free-swimming whale in the ocean existed. ("We knew what the world looked like before we knew what the whale looked like," writes Hoare.) In Watters's childhood, he said a deficit of good whale images paradoxically enchanted him with the animals.

The less we see, the more aura around it. I was still turning this statement around in my head when, toward the end of our discussion, Watters wondered aloud if a big part of what amounted to animal charisma was how readily any other species could be imagined to exist in a world apart from human beings. "I think with whales, why people care for them," Watters said, "is that there's a connection because of the distance we have from the ocean: a respect for an ecosystem that we're not so intrinsically connected to, a symbol of a pocket of pure nature that we haven't messed up." What he was talking about, I think, was that while animal charisma may feel like an affinity with other species—dependent on "phylogenetic relatedness" or personification—it flourishes best in the distance. The charisma of cetaceans hinges on whales and dolphins being far away from us, existing in worlds separate to our own. There, they can be continually mysterious: they are geographically removed, and their states of interconnectedness—their social worlds—are hidden from us. Whales' charisma inheres in mystery, and in the hope that there are, out there, minds perhaps akin to our own, that one day we could meet.

The trouble of loving animals now is that even those creatures that live very far away from us are almost constantly in a relationship of unwanted intimacy with the material trace of consumer culture. Each year, when the whales migrate out of Australia's coastal waters, they turn away from us. But not for a single second are we intangible to the whales. It can seem doubly tragic, then, when a whale washes up—perhaps having swallowed plastic, or because the krill it depends on for food are disappearing. Not only must we reckon with the death of the animal, but we are also confronted by the end of what had made it charismatic. The idea that it had come, to us, out of a "pocket of pure nature" and a void beyond our influence, collapses. At the bottom of the "magic well" of an animal shines something from industry and human culture, from a home, from a fishery. Not a compelling mystery, but a shameful familiarity.

In the 1980s, the charisma of whales moved people to save them, which meant preserving the animals from hunting and permitting them to live their lives, unaffected by humans, in the wild ocean. But as the concept of wildness has dissolved, cetacean charisma compels people to behave differently. In this context, it becomes possible to see our modern human behavior, including cute aggression, and a tourist's mollycoddling of endlings, as disorders of closeness and charisma, born of a misplaced need to personalize our relationship with animals, in a time of defaunation and mass extinction.

My own impressions of dolphins as playful and smart, as charismatic, were first formed in the leisure precincts of their captivity. At Perth's short-lived Atlantis Marine Park, children were invited to touch the dolphins; you could stroke their taut sides in the water, or pass them fat paintbrushes they held, cackling, in their mouths. The dolphin trainers wore costumes lifted off the covers of pulp science-fiction novels: the men in white jumpsuits open to the navel, or sometimes they dressed in gold loincloths with metallic shin guards, like gladiators. The women commanding the dolphins strode the poolside concourse in high heels, with white swimsuits or gold bikinis. Sometimes they removed their footwear to ride the dolphins, standing upright on their backs—an act for which the dolphins were leashed together with golden rope.

The dolphin yawping for fish out of a bucket, or responding to a whistle: nowhere is its charismatic cuteness more *acute* than when the animal is domesticated in captivity. Laws prohibiting the containment of larger whales were passed in Australia shortly after a Senate Select Committee report in 1985, but bottlenose dolphins, harbor porpoises, and finless porpoises, along with beluga and killer whales—animals that can grow to twenty-six feet long—have been kept circling in tanks or sea pens in Japan, Russia, the United States, Canada, China, and continental Europe. The anti-whaling sentiment of the 1980s may, indeed, have implied whales had a right to life, but cetaceans were never endowed with a global right to freedom. In oceanariums the animals are presented as idealized versions of their natural counterparts; responsive to verbal instruction, their vastness bottled and instincts neutered, their charisma diverted into lucrative corniness for the purposes of entertainment.

E. E. Cummings liked the word "zoo," which he described as having originated "in that most beautiful of all verbs zoo, 'I am alive.'" He went on to write, "Hence a zoo, by its derivation, is not a collection of animals but *a number of ways of being alive*." But Cummings also understood that zoos were mirrors, and that they revealed "not only our powers, but our weaknesses; not only our docility, but our cruelty, and our will to crush." Most cetaceans don't thrive in captivity—there, they do not exhibit "a number of ways of being alive." Their lives are shorter; their health, overall, worse. Whales kept in enclosures suffer from candida, dental infections, kidney problems, and potentially fatal bouts of pneumonia. They contract diseases transmitted by mosquitoes, such as West Nile virus and Saint Louis encephalitis.

If a whale's brain can be said to be the subject of psychological inquiry, and injury, how does captivity impact a whale's *mental* health? One relevant case study: a distressed female killer whale, held in a sea pen in Japan, consumed 179 pounds of rocks off the floor of her enclosure. It is one thing to be passively exposed to toxicants in the water and air, or to mistake a floating greenhouse for food, but consuming, over weeks, a weight equivalent to that of a washing machine, in rocks—this seemed to require willful intentionality on the part of the whale. Why didn't the whale stop? Was this evidence of some sort of cetacean pica, the disorder that leads people to want to eat soil, metal, flaking paint, hair, and other nonnutritive objects? Or was it suicidal, the way Virginia Woolf filled her pockets with rocks, before walking into the Ouse River to drown? In the presence of what mind did this whale eat the inedible?

Animals in captivity might warrant this sort of inquisition into their emotions: after all, creatures in zoos and aquariums, with their basic needs met, gain both the luxury of exceeding the urgings of their instincts, and the chance—the time—to suffer the chaotic rot of boredom, repetitive behaviors, and self-sabotage via disordered eating. To speak of the mental health of wild creatures, outside of enclosures, made less sense to me: free animals surely live under such lean conditions of wilderness that their habits are wholly responsive, not neutered or diverted, as they might be in captivity. Yet the idea that putting something in a cage, a pool, or a tank *expanded* the range of its moods struck me as both a repellant and untrue conclusion to arrive at. Perhaps the point was that the psychological well-being of an animal proved impossible to identify until it broke, and broke such that people could recognize the deterioration—in close proximity. That being so, the disparity was too reprehensible to transact: that something as buoyant as human fun should come at the cost of darkening whales' inner lives. If we go to nature to settle our minds, it should come as no surprise that taking animals out of nature unsettles theirs.

No threatened species of cetacean was ever saved from extinction by being brought into a captive-breeding program. Whales born in captivity are very seldom released, and even more rarely do those released animals survive. Enclosed animals are not meant to propagate their species in the wild; where they are bred, it is to produce offspring as future entertainment spectacles. In this way, whales are different from some terrestrial mammals—for example, cheetahs and pandas—numbers of which have been revived via zoo-born populations, and their species-wide genetic diversity bolstered by conservation biologists pairing unrelated animals to produce offspring. A whale in captivity is a creature that is, at once, personified *and* objectified. Trained behaviors, like leaping through a hoop, are often described to audiences as being expressions of "happiness," while, at the same time, the whale is expected to robotically respond to the trill of a whistle, or the clap of palms. "I don't know what belugas are like in the wild," said a former SeaWorld employee in an interview with *Smithsonian Magazine* in 2014, but in the tank, "it was almost as though they were staring at me through a veil. Like they weren't really home."

Berger wrote of the animals in zoos as "a living monument to their own disappearance," but for so long as we have believed whales were worth saving, they have had to be saved in the wild. Cetacean instincts are adapted for the dimensions of the sea, and—depending on the species—for a life in familial and cultural groups more numerous than can be contained in a built environment. Killer whales have been recorded moving in pods of close to fifty, and, during their migrations, clusters of beluga often join into flowing herds of thousands. It makes sense that these types of whales wouldn't flourish in tanks. There they are not just physically impoverished but socially. Whale sounds are echoic. Bounced back off the flat confinements of pool walls and viewing panes, it seems doubtful that their voices are still capable of conveying the same meanings as in the open ocean, or in coastal inlets. Over generations, captive-whale culture may develop such cloistered protocols, whispered dialects, and refined diets that it is no longer meaningful to talk of these whales as interchangeable with wild ones. They belong, now, to the ecotype "amusement."

A sperm whale looked squarely at him, in the Azores, and the writer Philip Hoare said, "This was not the eye of a horse, or a cow. It absolutely was reading me." A male gray whale returned his stare off Baja, and the journalist Charles Siebert wrote in *New York* magazine, "I'd never felt so beheld in my life. . . . [I]t felt to me as if he were taking one impossibly long and quizzical look in the mirror." A killer whale trainer from Florida said to documentarians, "When you look into their eyes, you know someone is home. Somebody is looking back." A whale's stare, according to the marine biologist Ken Balcomb, is "much more powerful than a dog looking at you. A dog might want your attention. The whales, it's a different feeling. It's more like they're searching inside you."

A whale's eye is looking, unexpectedly. Its vigilance seems to resemble ours—thoughtful, inquiring, knowing. Some*body* is in there. Making eye contact with a whale runs a deeper taproot into the human psyche than catching the eye of another sort of animal. They don't want your attention. *They're searching inside you.* For what? What is it that migrates with the look passing from whale to human, human to whale?

I had tried to make eye contact with the stranded whale in Perth, but, behind the smokiness of its eye, whatever message it had brought with it slipped from my grasp. Off the coast of Eden, there I had seen the humpback's eye click from left to right, scanning the length of the *Cat Balou* whale-watching boat. The moment passed. The whale went under, and it looked away. The whale's eye remained within my mental space, and I was carrying it around with me every place I went. This was not the sheath eye of a fish; a thin eye you could see all the way through to the silvered backing behind. The whale's eye had no resemblance to the shiny black eye of a bird, or the jerky, conical eye of a shark. How much it had appeared *like a gigantic human eye*—though what emotion was funneled there, I had no words to put to. In the merest sense the humpback's eye seemed human in the way of having an iris and a pupil, and in the white that surrounded it when it widened (it *had* widened). Surfacing on a more cogent arrangement of features, the expression might have been taken for one of startled recognition. But the flare of attention in the whale's dilated eye fought any easy familiarity. Something else I later realized: the whale hadn't blinked.

When, in 1851, Melville's *Moby-Dick* was published, even whalers—those people most familiar with the visceral gristle of cetaceans—remained perplexed about how whales saw the world. Most confusing was how the whale's vision reconciled two entirely separate lateral fields—one on either side of the whale's body. "The position of the whale's eyes corresponds to that of a man's ears," the narrator Ishmael notes, "and you may fancy, for yourself, how it would fare with you, did you sideways survey objects through your ears." That the whale's eyes were divided "by many cubic feet of solid head, which towers between them like a great mountain, separating two lakes in valleys," led to speculation that the animal had, in sensory effect, two backs and two fronts. Its blind spots were directly behind and in front of it (the backs), and it saw two "distinct pictures" alongside its flanks (dual fronts). Ishmael suggests it may be possible, therefore, for the whale to be, literally, in two minds—"attentively examining two distinct prospects"—in opposite, sidewise directions, with either side of its brain. This would be like solving two geometry problems simultaneously, he declares, each mathematical calculation taking place in a different neurological hemisphere. Having two eyes, greatly divided, might result in the whale having a double self, and being sentient twice over.

The problem of the whale's eyes, I think, has something to do with consciousness, and with how humans desire to see a reflection of our mental processes in other "high order" mammals. We hope to recognize an animal's quickness of mind crossing its sightline—though, instead, what is often reflected is our own solipsism. In *Moby-Dick*, the white whale is almost too conscious: its eyes might permit it to be occupied by two mentalities *at once*. Thus, the whale is supremely crafty, and monstrous: it holds a kind of twinned whale mind inside it. Or else, sometimes the eyes are too absent, as in captivity ("[The belugas] were staring at me through a veil. Like they weren't really home"). Confined to a tank, it can seem that the mindful part of the whale wanders, like a ghost in an off-season hotel, passing the windows of its eyes only at odd hours. The whale is a stranger, even to itself.

Yet the importance of whales' eyes, for me, leads back to the fact of their size. Whales are *so big*, that, faced with a wall of animal, I look to connect with this smaller nexus of liveliness and communication, the eyeball. What proof of being beheld is otherwise possible to gain? So perhaps this is where most of the whales' charisma, in fact, resides: in sheer scale.

Based on my experience of whale watching, I now hold this to be a true statement: certain standout "big things"—sequoia trees, for instance, notable feats of ancient architecture like the Sphinx of Giza and, of course, the great whales—naturally distort the attention of the psyche. Being too large to be possessed by the human eye up close, in a glance, where we encounter these big things, they overawe us, and loosen something electrical, or maybe adrenal, to circulate through the nervous system, long before the intellect awakens to it.

Tim Watters and I had discussed the work of an American photo artist named Bryant Austin, whose images of whales were said to capture the animals' magnetism as a quality of being *mega*fauna. Setting aside the animal's distantness, and the rarity of seeing a whale, for some species—the blue whales, the sperm whales, the fins, and the sei whales—their sheer size was, surely, a crucial component of their allure. Watters had heard Austin took pictures of whales that were the size of whales, he said—reproduced at a 1:1 ratio, and in exhaustive detail. People found the detail not just aesthetically but ethically confronting. One attendee at Austin's show in Tokyo had been recounted saying, "I feel like the whales are talking to me with their eyes." A number of reviews described gallerygoers welling up with tears before the enormous images.

So, on a day made glary with cloud light, I headed to the National Maritime Museum, in Sydney, to see one photograph in particular in Bryant Austin's exhibition *Beautiful Whale*. Nearly twenty-three feet long, the image is titled "Enigma." The photograph took three hundred hours to develop, and there is only one original print, weighing more than four hundred pounds. "Enigma" shows a three-year-old sperm whale calf, entire, tremendous; motionless and in profile. The animal's body occupies almost the entire field of view.

Stood before it, alone in a downstairs gallery at that midday hour, I wondered whether the photo was a black-and-white render, or whether Enigma was natively shades of gunmetal, ash white, and silver. The animal was a part of an extensively studied Caribbean pod known as the "Group of Seven"; its members also included mature whales called Fingers, Tweak, and Pinchy. Fingers had two calves: Thumb and Digits—names like pickpockets in a Charles Dickens novel. These seemed diminutive, even consciously belittling, designations for whales, in comparison to the glamour of "Enigma."

The water that surrounded the whale in Austin's picture was a uniformly faint and rippled gray, a color the taste of tungsten steel. Though Enigma was young, its head peeled in patches, and its body was wrinkled and wizened. Seeing the whale, flaking and monotone, at first you might have thought, as I did, that the image was a *nature morte*—that genre of Dutch still life in which the fruit bowl decomposes, pushed past ripeness to a perversely organic afterlife, decaying before the painter's eyes. But near to the surface, apparently all sperm whales appear this way—shriveled, mortifying—regardless of their age. At depth, the whale's skin is smoother and its body becomes streamlined, as the organs inside it are compressed, and a huge volume of oxygen is pushed into the animal's exterior blood and tissues, allowing its lungs to fold in, at great pressure.

I looked more closely at the photograph. There were apparently no sensors or alarms to prevent me from bringing my face to within an inch of its surface, or nearer. At this range Enigma was less animal than mineral. A lunar topography, pocked, scuffed, and starburst, as if for eons it had been bestrewn by hot meteorites. I repressed the desire to spot a petite flag, the claim of some Lilliputian space-faring nation, on its skin. The photograph had both the focus of the microscope and the sweep of the satellite. A small pleat above the whale's pectoral fin suggested its ear.

I arrived then, almost too quickly, at the eye. Enigma's eye, low in its head and almost at the corner of its jaw, was weirdly opaque. Like the mother humpback off Eden, it also had the semblance of a large human eye—the white sclera, the iris, the round pupil. But the whale's gaze was lodged at the viewer's feet; a downcast expression. The impression it gave was not of despondency, but a powerful, prolonged indifference, edging up to fury. As if the whale were irately willing itself *not* to look at the photographer. The eye possessed the clouded pupil of a megalithic Buddha, stranded anciently in a dripping jungle. If Enigma were disturbed, or simply decided to turn, you felt the whale would open up a second, translucent eyelid, and flash with a sudden and pitiless rage, a great energy for destruction. What any person had coming was already right there, in the frosty set of that eye. It was not a dead eye, but a shrouded one.

For near to an hour I wandered the hallways of the exhibition, seeing parts of minke, and humpback whales, that I had heretofore only imagined. There were whale belly buttons. Whale nostrils. Blunt teeth, tongues, and stray bits of organic filament. Whale parasites. I saw what I took to be a scar inflicted by a cookie-cutter shark on one whale's side: a neat wound for the shark's signature, quite literally the shape of a biscuit mold stamped into dough. The faintest of wrinkles beneath a whale's tender armpit; an encrusting ecosystem, just visible, on the gossamer fin of a suckerfish.

The eyes of the humpbacks in the photographs appeared in startling resolution, shot by Austin side-on, and from an angle above. Some of the images were so clear I could make out the whales' corneas—the thick, gluey lens on the outside of the eyeball. I had hoped to catch the reflection of the photographer in one whale's pupil, a wind-up diver in his wetsuit, maybe even with the camera held out in front of him. But Austin was nowhere to be seen. Instead there was, at the center of a few whales' eyes, a surprising rustiness, a clutch of something that appeared mineral, gritty, and red. This wasn't the sequin-like "red-eye" effect of a camera's flash rebounding off an animal's smooth retina, but a feature that was somehow further back, and powdery. I wonder whether what I was seeing was the so-called magnetite, the bio-receptors whales are believed to use to detect magnetic fields to navigate by. But these would be subcellular particles and therefore, I think, impossible for any ordinary camera to pick up.

When an artist's project is to inflate small objects to huge stature, the impact, conventionally, is nostalgic. As an artistic strategy, enlargement creates a pleasantly dreamlike and infantilizing impression; such artworks reacquaint audiences with the perspective of childhood. Pursue this intention in the opposite direction, making miniatures and things diminished to dollhouse dimensions out of large subjects, and the effect is to charm. What, then, of an artwork like "Enigma," where absolute fidelity to the acreage of the animal is the artist's most intense fixation—where the image proves an exercise in extreme precision? Realistic reproduction, at life-size, is a potent gesture, in any medium. It suggests not just that the work is documentary, and truthful to a real experience, but that rigorous exactitude can transport specific qualities of that encounter directly to the artwork's viewer (which, I imagine, might capture something of the purpose of the Balls Head whale engraving).

And yet, an encounter with gigantism was not what these pictures were, at heart, about. Austin's driving preoccupation was the granular, not the grand. Making a very big image was not enough. He insisted on doing it in the tiniest increments, over a very long time. He had set himself a test of endurance. This was also a test of technical mastery—aimed at demonstrating the artist's control over, and orchestration of, his chosen technologies. Austin approached the behemoth whale, Enigma, with the micromania of an artisan found elsewhere in history, piecing

together intricate Fabergé eggs, lover's eye portraits, and finely decorated models of the solar system. Austin's ambition was to exhaustively
record Enigma's data-rich finitude; to know its size by replicating this
whale's every concrete feature, painstakingly, as it appeared through the
lens of his camera. His fascination was as much with pushing artistic
machinery to its edge, as with the whale itself. This was biophilic art,
wherein seeing each new level implied that there was more to know,
more visual information to collect.

What happens when a miniaturist meets one of the biggest animals on the planet? When everything is equally in focus, everything—
the photographer implies—is equally interesting. This is not the way
the human eye usually works, distinguishing close from far, and center
from periphery. The human gaze seeks to resolve faces out of detail,
and looks to locate the eyes within those faces, which we instinctively
take as expressive "windows" to the mind. Austin's images, however,
adopt the machine-eye's perspective. "Enigma" shows us how accurately cameras and computers are able to see (to *reconstruct*) whales—for
"Enigma" is not one photograph but several, imperceptibly stitched to
one another digitally. Equipped with scuba gear, Austin had gradually
flippered along the length of the juvenile sperm whale as it floated near
to the ocean's surface, taking multiple consecutive shots over the course
of around forty minutes, gathering what amounted to an external body
scan. Later, using a series of solid-state hard drives, daisy-chained together, Austin compiled the composite, a mosaic. These same technologies are sometimes used to create images of other planets. But if this
much data on Enigma had been taken in, then what was withheld in the
whale's misted eye?

A few years ago a neuroanatomist named Leo Peichl, from the Max
Planck Institute for Brain Research, wrote a paper with the doleful title
"For Whales and Seals the Ocean Is Not Blue." Despite the familiarity
of cetacean eyes (eyes that have corneas, retinas, and photoreceptors:
mammalian eyes that appear to mirror our own), it turns out that whales
see a world utterly unlike the one we look at and live in. Whales have
very poor color vision, for one thing. This is about the rods and cones:
most whales have only one or two of each, whereas we have around six

million cones and 120 million rods. A whale's focal point is barely a few feet in the distance. It is thought they do not, as Ishmael proposed in *Moby-Dick*, see separate images in their heads, just as we do not see distinct, slightly offset pictures broadcast out of each eyeball. In sperm whales at least, a portion of the front sweep of their field of vision has binocular qualities—meaning they can determine depth, to a greater degree, in an arc above their foreheads. This might mean they hunt upside down, scanning for squid below them (if they use vision in combination with sound to home in on their prey, which is debated).

Whales have thick eyelids, but they blink very rarely, only to clear their vision of irritants. Underwater, a layer of transparent mucus coats the whale's eyeballs, replenished infrequently by glands under the whale's eyelid. Sperm whales can retract each eye partway into their heads, a bit like an anemone curling into itself (an adaptation to life in the deep sea; the whale's bone socket helps protect the eyeball under pressure). Most cetaceans sleep with one eye open, or both. There seems something Delphic in this open-eyed slumber: how whales are permanently receiving the world. Dolphins in captivity have been recorded "talking" in their sleep. Scientists say cetaceans do dream but that their dreams are staged in one hemisphere of the brain at a time. As whales' breathing is not automatic but conscious, a deep sleep might suffocate a cetacean. REM (rapid eye movement) sleep is rare: pilot whales get six minutes of it in every twenty-four hours. In some species calves are born with insufficient blubber to float, and so they must always swim, even when napping. Because a whale's dream is housed in only half its brain, I think their dreams would move small and fast, like rolling dice. Are whales' dreams visual, at all?

Some types of whales have sharper eyes than others. Northern right whales and bowheads are thought to be nearly blind close to the surface of the sea: their cone cells lack any photoreceptor proteins. These whales probably see best in the dim and in the icy dark—inverse to our vision, as with other animals we call nocturnal. Generally speaking, all whales see their world in monochrome, and as if under a nightclub's black light. People who have set out to design whale-deterrent fishing tackle often reach for high-vis, fluorescent materials: but ropes that are fluorescent don't stand out well to whales. They might appear not very different from the way the dull color of kelp looks to us.

On a gray or black watery background, krill, fish, and other prey species zing—a bright white—though, even then, the whales' visual acuity is still very low. Those white shapes aren't particularly defined unless they're very close. Alexis Madrigal, writing for the technology section of *The Atlantic*, draws the comparison that if humans have 20/20 vision, whales have approximately 20/240 vision—they must be as close as twenty feet to be able to see what humans can see at 240 feet. That's a rough working—Madrigal notes that it doesn't account for what is most unusual about the cetacean pupil. The human pupil constricts uniformly in direct light to a smaller, tighter circle. But most whales' pupils close into a semicircle like a smile, in each corner of which a circular dot remains open. That is, whales looking straight into the sun have two pupils in each eye.

The female humpback that I felt looking at me in Eden would not have seen me at all, nor anyone else on the deck of the *Cat Balou* whale-watching catamaran. Against the luminous sky above, the whale watchers would have been as indistinguishable as clouds scudding into clouds. A whale's eye was a vanity, then. It reflected back what we hoped was true about our own significance to the animal world. I thought of the eye of the whale on the beach in Perth, and the eye in Austin's "Enigma" photograph. That eye was a bad tarot card, waiting to be flipped over. Enigma seemed to be *deliberately* turning a blind eye to Austin. The whale appeared to refuse connection. It was as though the huge, sentient animal had drawn a visual scrim over its eye to hide behind, even as the rest of its body was exposed to Austin's obsessive gaze. Enigma's name was a sort of key, I guess. Austin had seen so much of the whale, looked at it exhaustively; yet somehow the animal remained elusive. Which might be a very rational demonstration of intelligence, on the whale's part—evading human depiction. I thought perhaps that it was not the photographer who had made the whale's eye eerily out of focus; it could have been the whale itself.

A diary entry by Walter Benjamin begins, "In an aversion to animals the predominant feeling is fear of being recognized by them through contact. The horror that stirs deep in man is an obscure awareness that in him something lives so akin to the animal that it must be recognized." A kind of bio*phobia*, as opposed to the biophilia of the photographs from Santa Teresita. What we dread now, in that returning gaze, is recognition of the dozens of material ways we are already deeply connected, deeply intimate, with the whale—from influencing the whale's environment, all the way through to its guts, in the form of plastic pollution. Our inability to make space around beloved creatures today speaks not just of wonderment, but of our fear. What we stand to lose is more than mystery, more than cuteness or charisma: it is connection.

05

Sounding

A Whale Is Never Heard in a Vacuum — First, Signal Interference — Swan Song — Whales Played to Congress — And Harpooners — The SOFAR Layer — Biophony — An Aquatic *Silent Spring* — Ships Made Sirens — Did Their Voices Equate to Personhood? — "A Nation of Armless Buddhas" — *Feral* — Whale Song: Is It Futuristic? — Golden Record — The Pole of Inaccessibility — A Threshold Between Usefulness and Hopefulness — How to Split a Common Tongue — The Various Sounds Whales Make — Humpbacks Rhyme — What Is "Song"? — Holosonic — Ventriloquism — Hands as Humpbacks — Two Voices in One Bowhead — But Whales Have No Vocal Cords — "Water in Water" — Cutting Junk — What is "Noise"? — Anthrophony — Commodity Migration — Seismic Mapping — Having Internal Antlers — Soft Pollutant — Even Animals with No Ears Are Affected — Bounce Dive — Ship Strike — Terrorism as Recorded by Right Whales — What Is "Silence"? — Blue Whales Drop Three White Keys on a Piano — Voices with No Origin — The Beluga That Learned to Talk Like a Man — Spies in Sperm Whales — Standout Soloists — Humpback Pop Songs — Cultural Revolution — Magnetoception — Auroras — Soft Mountains — The Spirits That Move in Them — Sperm Whale Gods

Sound: *adjective and noun*

1. Entire; unbroken; not shaky, split or defective; *sound* body; *sound* health; a *sound* constitution. Undisrupted and profound, as a *sound* sleep; or

2. A narrow passage of water, or a strait between the main-land and an isle; or a strait connecting two seas (this name perhaps given to a narrow sea because wild beasts were accustomed to pass it by swimming); or

3. Noise; report; the object of hearing; that which strikes the ear; an impression of the effect of an impression, made on the organs of hearing, by an impulse or vibration of the air, caused by a collision of bodies or by other means. Sounds not audible to men may be audible to animals of more sensible organs.

———————————

Inside the nighttime house of itself, the humpback sings.

The tonnage of a whale (the imaginative force of its fleshly presence) is oddly defused by our fascination with that part of the animal that roams, untouchable and ethereal, far from its physical body, out into the ocean: whale song. The sounds whales make are multiple, and we, their on-land listeners, have likewise played them in a variety of contexts: military and political, pseudo-spiritual, and, occasionally, as pure entertainment. Whale song is never heard in a vacuum. Over any other trait, the voices of whales built their charisma—at a time when people conceived of such sounds as tragic and timeless, whale vocalizations would prove to be the fulcrum of a turning point in our culture. Only later did scientists come to learn that, like us, whales make sounds sensitive to context. Whale songs are neither purely natural, nor static; they have evolved in response to the shifting timelines of their cultures and the changing broadcast environment of the sea.

The calls of humpbacks were first audiotaped by chance, in the 1950s, when US naval engineers, sweeping distant seas for the signature whirr (a "teaspoon stirred in a teacup") of Soviet submarines, caught, instead, an acoustic backdrop of flurried *whumps*, arcing whoops, and bouncing, stair-fall blares: whales. The technicians, by and large, gauged these sounds frustrating signal interference—noise, not captivating sonance. Decades later, during the 1970s and '80s, activists sought to elevate the whale to an icon of environmental stewardship via the transcendental qualities of its singing. A growing awareness of whale endangerment portaged humpback voices into living rooms with paisley drapes and corduroy lounge suites, where listeners were primed to hear cetacean sounds, played off vinyl, as mournful. Record sales chart an enthusiasm for whale song that peaked with the apprehension that whales, and their presumably ancient oral traditions, were becoming harder to hear, because the animals were disappearing. Whale song's intrigue lay in harking to the communiqués of receding creatures—a "swan song," or disclosure on the threshold of exit. The voices of whales, far-reaching and recondite, instantiated what it meant to listen to a planetary extinction.

That whale vocalizations entered the public sphere framed as the sounds of a vanishing world was not mere happenstance. Throughout the congressional Marine Mammal Protection Act debates in America (a prelude to the 1972 UN Conference on the Human Environment, where the global ban on commercial whaling was first tabled), recordings of whales were presented as testimony. Christine Stevens—called "the mother of the animal protection movement" in the United States—played the sounds of whales to Congress in lieu of giving verbal evidence. (*The animals rest their case*, proclaimed placards outside the Capitol.) Elsewhere, off the coast of California, whale song boomed stereophonic from Greenpeace's small speedboats: intercessors in orange life vests laced wakes between Russian whale catchers, the immense flensing ship *Dalniy Vostok*, and surfacing sperm whales. The sounds did not deter the harpooners, but humpback calls went on to stir the compunctions of composers and musicians. Asked what she thought of whale song, one of the earliest performing artists to insert the cries of humpbacks into her tracks, the singer Judy Collins, replied she felt "angst for being a human being on a planet where they also live." The conservation movement had found its soundtrack.

Sound cemented whale charisma to scale and mystery. Not only did some species migrate staggering distances to feed at the poles and give birth in the equatorial tropics, acoustic research revealed that their voices were pan-oceanic. Whales lived in a sensory realm that humiliated human faculties in contrast: a few species were sensitive to sonic amphitheaters extending across entire sea plateaus. Songs sung by humpbacks off Puerto Rico, for instance, would be heard by whales in waters near Newfoundland some 1,615 miles away: the equivalent to a shout on the streets of Moscow being made out, whisper-quiet, by people in London (though sound waves travel farther and faster aquatically). Hydrophones likewise recorded the very loud, low frequency calls of blue whales rumbling for thousands of miles inside a channel known as the SOFAR (sound fixing and ranging) layer; a distinct belt of oceanwater—as vinegar underlines olive oil—where temperature, pressure, and, to a lesser extent, salinity ally to refract sound waves beyond their subsea range elsewhere. The clinks made by sperm whales last bare microseconds, but they proved to be among the loudest single-source noises on Earth. The 1967 launch of Saturn V, the heaviest space rocket ever fired, was quieter by comparison.

Because whale sounds crisscrossed hemispheres (transmitting far beyond the vicinity of their vocalists), even where no whales lived their calls could be configured as part of the ocean's natural ambience, or its "biophony"—a term later coined by the musicologist Bernie Krause for the environmental soundscape collectively broadcast by animals and weather-tossed plants. Mobilized by campaigners, whale song was rarefied beyond the status of an animal sound to exemplify the biophonic grandeur of the hydrosphere at large. The defaunation of whales signaled not just the diminuendo of individual species, then, but a critical change in what the whole sea sounded like. An aquatic "Silent Spring."

It didn't used to be this way. The oceans were once filled with whale song, for those who were fortunate enough to hear it there. Of the oldest documented occasions wherein whales were overheard, many took place aboard whaling ships, roving whale-rich waters. Before the mechanization of seafaring vessels began to churn an occluding, on-board atmosphere of white noise (the chuff of diesel generators and propellers), ships were quieter, and, being made of timber, their hulls acted as amplifiers. Shipmen claimed to have detected all manner of mythical creatures caroling in the ocean. These wood-chambered whale sounds were mistaken for Odyssean sirens, most of all, and mermaids: women–slash–sea things. So what was built for the purpose of violently de-animating whales—the harpoon ship—became the means by which whales were pluralized into all sorts of enchanting life-forms, as wildly imagined as drolleries.

After the invention of recording technologies capable of relaying humpback voices into suburban households, whale vocalizations heard indoors sparked listeners' minds to similarly bewitching ends. The turn-table's stylus spiraling across Roger Payne's compiled, multiplatinum *Songs of the Humpback Whale* (1970) elicited an aura of enigma around whale song. To the ears of the 1970s and '80s, the analogue whales on LPs sounded paranormal and ghostly, like missives from "the other side." Why whales sang, and what they sang *about*, were indecipher-able questions, but they nonetheless induced reverie. Activists rushed to capitalize on this fascination, arguing that the haunting voices of whales corroborated the animals' intelligence and, tacitly, their quasi-personhood; beyond conveying pain or pleasure, whale song was cred-ited as evidence of a cultural dimension in whalekind.

Anthropologists, linguists, and philosophers furthered these views, weighing in on whale song to contend that because cetaceans evolved their communicative abilities in the absence of hands, facial expres-sions, head-on eye contact, and within an underwater environment unfurnished by graspable objects, they had evolved a "language" cen-tered exclusively on their interrelationships. Whale song was thought to be an exchange of affinities—a psycho-social conversation. ("A nation of armless Buddhas" was one Greenpeace cofounder's description of whalekind.) Killing whales, therefore, meant killing another strain of communication, and, so too, a species of *consciousness*. Within these var-ied understandings of what whale song represented, it made sense that humans should feel obliged to avert their cruel deaths by ceasing to hunt them. From the viewpoint of a swelling number of people, the charisma of whale voices provided the impetus for a right to life.

All this took place during a period when technology's transformative potential seemed limitless, and the era of terrestrial exploration was coming to a close. Those few remaining "precontact" human cultures preserved themselves, defensively, in pockets of isolation, while the increasingly interconnected globe beyond shrank. In *Feral: Searching for Enchantment on the Frontiers of Rewilding* (2013), George Monbiot observes that a preoccupation with aliens, UFOs, and galactic travel figured in the zeitgeist. Whale call recordings offered a similarly tantalizing prospect to that of outer space; the colonialist prospect of "encounters with unknown cultures could continue," in Monbiot's words—if not between human communities, then in the animal kingdom. Though whale song seemed to unveil an older, immutable world, its "discovery" also supplied utopic, pioneering fantasies of technic mastery; voyaging outside the human family with microphones and submersibles to commune with other beings.

After all, whale song as people heard it on land, over airwaves, was never a purely whale-made sound. Funneled through subsea recording equipment and pressed into so many flat, shiny plastic discs, whale song came, instead, from somewhere *in between* them and us. It was, innately, a biotechnology; a marvel of mechanical creation, as much as a disclosure of animal culture.

What is it about whale song that feels, still, so futuristic? Is it that whale song holds the postapocalyptic bona fides of a language that has survived the eradication of so many of its speakers, worldwide? Or do we imagine that the songs of whales are tantamount to the voice of the environment without us? A misanthropic fantasia; reversing stories of the starry interior of the whale, in which Jonah crouched, and turned, again, to Eden.

Vinyl is not the only substance into which whale voices have been inscribed. The third track grooved onto the surface of two duplicate "golden records" installed aboard both unpiloted Voyager spacecraft in 1977 is: a humpback whale. The astrophysicist Carl Sagan curated the records' contents before the Voyager probes were launched as "bottle[s], into the cosmic ocean," adrift in anticipation of visiting with extraterrestrial life. "A love song"—Sagan called the records' contents—"a love song, cast upon the vastness of the deep." He meant deep space, of course, but also a deep duration: an as-yet-unquantified spool of time. Having completed their assigned investigations of Jupiter's trajectories and Saturn's braided rings, the probes now drift, unnavigated, beyond the solar system's heliosheath, into the mysterious regions of the gas giants and fields of galactic plasma. They are semi-dormant capsules, like the scintillating larvae of some minuscule, marine detritivore. Voyager I is the farthest-flung artificial object humans have ever ejected off the face of the planet. You might call it space junk, if it weren't still transmitting; if it weren't—in almost the most passive sense of the word—functional. Both probes send back weakening signals as radio waves.

In the South Pacific Ocean, between New Zealand and South America, lies the world's spacecraft cemetery; a site where at least 263 orbital vehicles and rockets, once controlled by the United States, Japan, Russia, and Europe, have been directed to fall, to scatter their mechanical remnants across a section of seafloor. The cemetery is centered on Point Nemo (Point "no-one"), also known as the Pole of Inaccessibility: the farthest seamark from a peopled landmass, not traversed by shipping. Strange noises are heard there: ultra-low-frequency "bloops" originally thought to be huge whales, and later confirmed as the telegraphing pulses of cryoseisms: far-off icequakes. But the Voyager probes will never be returned to wreck at the Pole of Inaccessibility. They will stay, forever, in outer space, until they are pulverized, soundlessly, by cosmic forces.

NASA's mission control will soon begin powering down the instrumentation on the first probe, Voyager I, as the spacecraft migrates toward being scientifically defunct by around 2025. At an indefinite point before its obliterating rendezvous with a star known by several names, some forty thousand years in the future, Voyager I will surely pass over the threshold between usefulness and hopefulness, between tool and junk; from fact to fiction. Departing its categorization as an object of science, and an optimistic symbol of our species' potential, the spacecraft will, finally, be transfigured into something like a myth. For the time being, though, the distance the Voyager I probe unravels behind it—more than thirteen billion miles—is the longest length scientists have ever measured (as opposed to theorized). Animal voices from the depths of the oceans exist today, at the extremity of space as we have been able to plumb it. The scale of the universe is, tick by tick, unwound in sound.

On the golden records, the whale's voice is listed in a section of audio dedicated to greetings in dozens of human languages. Placing the whale alongside these recordings treats it as a speaker—whereas other sounds of life, of nature, get a billing after our multilingual introductions. Which makes you wonder about how a common tongue is defined; and by what criteria the harmonious background (biophony) is cleaved from sounds sufficient to rise to the level of a language (and, out of language, *salutation*). From the standpoint of the whales we are the extraterrestrials nearby; and, like the aliens the Voyager probes anticipate, we, too, examine recordings of whale songs, filled with expectation, assuming their sounds are meaningful but encoded; the dialogues of another knowledgeable species.

During the making of the golden records, Sagan insisted that there should be no sounds or images of war scored into them. "A love song," he said. Yet the biophony inscribed onto those metal plates—not only a humpback's call, but, further on, the polytonal music of insects, amphibians, birds, and an elephant's trumpeting—all this feels, today, like evidence of an ongoing conflict. That which was meant to manifest the axiom of astral travel, "We come in peace," now transmits so many creatures under siege. It is a sonic archive of what stands to be lost, and, in some instances, that which is already diminished, muffled, or stilled. What stays with me most of all is a line President Jimmy Carter wrote for the golden records, which reads: *We are attempting to survive our time, so we may live into yours.*

Sperm whales click and crackle; reduced in volume or heard at a distance, their sounds recall the snicking of an old-fashioned cinema reel, the celluloid strip running through its gate. Beluga purr and chirp; killer whales whistle, ping, and pip. Minke make a noise like an old-fashioned dial tone, the ringing of a dormant signal exchange. Beaked whales buzz. Bowheads wail up and down; glissando. Right whales *crack*; their sounds are known as "gunshots." At lower latitudes in the Southern Ocean humpback whales make numerous sociable noises. The whales grunt, rasp, thwop, and moan; they shriek, whine, bubble, gurgle, and fin-slap on the sea's topside, as well as generate what are called "pulse trains": subsonic resonances, only recently discovered, that thump across the lower thresholds of human hearing like rain drubbing a tarp. These short, communal noises are less renowned than the longer compositions recognized as whale songs—of which humpbacks are the most elaborate vocalists.

The songs humpbacks sing are myriad and cannot be sufficiently characterized using only this printed alphabet, our human language. Though, an attempt? The songs are ululating bowwows that wind up and unwind, interspersed by clanks and spackling chitters, as if flinging open a cutlery drawer. There are coughs, squeaks, and tightly embouchured mewls. A licked fingertip dragged over rubber. A metal train bridge rattling. Sometimes a whale will sound like glassware, tinkling in a flight attendant's trolley during turbulence—and then it is surprising, to identify that degree of airy daintiness in the voice of one of the largest animals on the planet. And yet, despite the possible range of sounds a humpback is capable of making, their songs can also prove tedious and unvaried. A whale might seem to grow obsessed with a two- or three-note composition and cycle it insistently, like an automaton, for weeks.

Repetition is a key feature of what we call "song," defined, in mathematical terms, as patterned sound. Specific sets of notes, phrases, and refrains reoccur throughout the songs of individual whales, giving their noises structure, a type of scansion, and musicality. From time to time, humpbacks rhyme. This is perhaps (as is the most rudimentary function of rhyme for our species) a mnemonic tactic, a memory aid for whales. Do their voices substantiate the extent of their memory, then—how much it might contain? How much does a whale remember? How much does a whale forget? We cannot tell. Inside the nighttime house of itself, the humpback sings.

I have been thinking about why the utterances of humpback whales are called "songs," beyond meeting the stochastic threshold of being "patterned noise." Why are they not called whale *groans* or *bellows*? Why are whales not considered to be speaking? Scientists have long surmised humpback whale songs are meant to attract a mate and scare off rivals, though it's never been proved that the singing plays a role in courtship. Either way, whales have reasons for making sounds—and we have reasons for calling those sounds "songs."

A song is not unilateral communication, as talk often is; a message aimed directly to another, or a small group of others. Song is, instead, exhibitionist sound, a broadcast—to, who knows? An audience undisclosed. Even people who don't speak the language a song originates in can intuit the emotion with which it is freighted (or so it seems, dancing in a foreign country). What makes any sound a song might be that *how* it sounds is inflected by what it *means*. Too, its meaning exceeds the transmission of a fact, or directive—song overruns whatever basic information is needed for survival. There is an art part, an adverb, added to the communication of a song. A song speaks grandly, forlornly, exuberantly, excitedly: it expresses a deeper significance than its surface meaning. A song *moves*. So calling a sound a song assumes a certain amount of crosscultural common ground; it suggests that within the singer there subsists an emotional inner life, recognized by, and akin to, that of the listener.

Beyond possessing the physical means to make a sound, to vocalize, being capable of song requires having a distinctive voice; taking enjoyment, even, in one's own voice. To put a finer point on it, being capable of song requires having some consciousness of having a voice. It requires the competency of recognizing the uses to which a voice might be put. Song infers the ability to—thoughtfully, deliberately—modulate.

The English writer Heathcote Williams believed whales were capable of conveying more than information to others of their species; he argued the animals could articulate "a sense of the unknown." I feel justified consulting a poet like Williams because of something else: the work of a handful of naturalists and bioacousticians that suggests cetaceans might deploy a kind of "picture language" and that their missives are "holosonic." The implication is that what a cetacean communicates arises from where its sounds are projected *to*—where those sounds exist in three-dimensional, aquatic space—instead of how sounds, syllables, enchain together sequentially, across time, as with most human sentences. A whale sound might mean something different if it is projected down along the seabed, instead of near to the rumpled underside of a wave. Its loudness or quietness (the next-to-ness of its projection) may also carry a connotation, making the matter of whether "phrases" occur in proximity close to, or far from, the singer's body, an additional source of nuance.

Imagine if what a word denoted depended not only on how that word was spelled, how it sounded, and whether it was whispered or whether it was yelled, but also where on the page it was placed.

Don't we know that form already, reader? Don't we call it a poem?

As a girl I tried hard to throw my voice off—*away*. I wanted desperately to learn ventriloquism. I believed being a ventriloquist was something that, with sufficient training and dedication, I should be able to do, and so I practiced often. I sat a row of stuffed animals up on the mattress. Then I lay under the bed and tried to jump my voice, convincingly, into the felted mouths of each of them.

When I look back, I think that what this was about, this impulse, was the thing that happens when you first become aware that you will die and you are afraid to. Being flesh and bone: that dawning horror. Your life, my god; it is the life of just one physical entity. I found it creepy that a person's voice—to my mind, the real essence of their identity and character—disappeared with them when they died. It seemed perverse that nothing of the voice remained in the body after death; no imprint, no organ called "voice" to autopsy. Sure, a person might leave a trace of how they sounded on a cassette, but from then on they could never say something *new*. Their organic voice, and every sensitivity it imparted, just dissolved.

The other thing that happened to fuel my ventriloquist aspirations was that I saw the movie *The Little Mermaid* (1989). How the witch-cephalopod, Ursula, traps Ariel's singing voice in a nautilus, threaded onto a chain around her neck: I wanted to get out ahead of *that* fearful prospect. Throwing my voice was how I intended to preserve it against death or the confiscation of envious harridans that, needless to say, were a threat as real as any other in the hot, quiet suburbs of Perth. That a living person might be struck dumb by having their voice thrown into an inanimate object taught me that just as things could be personified, people could very easily be thingified.

How did I come to think it was possible to cast off my voice and vest it somewhere else, to hide it in a safer place? After a period of travel, one of my uncles brought over a conch, glazed lustrously pink in its cleft. Inside the shell, he professed, surged the tides of the far-off continent he'd returned from: the hammering of an exotic, timeless sea. Nature's one-way telephone. The shell was the first device that proved it could be done—this embedding of a voice within an object. A vortex in the shell, its spiral a constant curl inward and outward. It transmitted to . . . where? Simple logic to infer, as I did, that anything shouted into the conch would bleat from its twin on an antipodal shoreline, startling beachcombers. The conch collected all my early cussing, my secrets, doubts, and confessions.

It would be easy to point out, now, the obvious resemblances of seashells to books. I want to say that the second discovery proving it was possible to throw my voice without seeming to speak was writing—that out-of-body experience, the exhibitionism of introspection. A delusion of warding off mortality. But what you hear in a conch is not breakers belaboring a blushing beach. The sound in the shell is the murmuration of the blood in your own ears, the hothouse hum of your interior environment, a circling in the column of your body. What once lived, and died, inside that shell, I never knew. Only much later, when it came to writing this down, did it occur to me that something very like ventriloquism was its effect: the conch was the means to move from being a listener, inside the shell of a body, to being a listener listening to the body, inside of a shell.

Pocketed into the world's natural objects, small, observable cacophonies of the self take place. What we find to listen or look inside ourselves with may have its own life, its own story—though it's a narrative in a frequency that proves tricky to separate out from the tales we tell about who we are and where we've come from.

Humpbacks tend to stay stationary when they sing; they hang on a diagonal slant. Here's how: hold your hand out flat, palm down, then point the fingertips toward the ground. The whales' heads are canted downward, and their tails bend a little, as does, now, your wrist.

I have been told that a singing whale can seem to contain dozens of separate voices, coming out all at once; and that hearing it, underwater, compares to plunging your head into an aviary. From many perches on its inward branches, the sound ricochets off coral and is dampened through sand. The whale's voice becomes omnidirectional—seeming not to emanate from the whale, but from the world. Some types of whales, the bowheads in the Arctic, for instance, do sing with at least two voices simultaneously—making dual frequency, FM and AM sounds, at the same time. People—the traditional vocalists of Mongolia and Tuva, that is, trained as throat singers—are able do this, too, albeit with less range between the notes.

Harmonizing with yourself: I think it must be like being doubly alive.

A whale does not exhale as a human does when speaking; its jaw does not swing open. Neither are its syllables shaped by the muscles and soft tissues of its mouth—however loud and reverberant a whale is, it has no vocal cords, and its voice is not projected out between its lips, shaped by its tongue. Instead, air cycles from within the red sacs of its lungs, up toward the whale's dark head, then back and forth, vibrating across a U-shaped ridge of cartilage, which also forms an aperture into a laryngeal pouch that the whale contracts or expands, changing its resonance.

Picture it this way: the whale throws its voice from room to room, and the effect is how it is in some old mansions: conversations telegraph from the lower floors, up lately bricked-in chimneys and embrittled staircases, through wall cavities, and along hidden beams. An eerie warp. *Who was that?* The voice cartwheels around, stirring scuds of dust off the floor. The whale's cranial sinuses, its jawbones, and maybe also the integument in the ribbing of its throat, are engaged in making sound in ways that are unobservable and so remain, to us, inexplicable. We only know that the animal, entire, is sonorous. A whale's whole body is its speaker.

A fetal whale in the womb is washed with the sluicing sound waves of its mothers' vocalizations, though in its airless state it cannot yet reply. As an acoustic object inside the maternal whale, the growing calf changes her sounds, the way attempting to talk with a mouth full garbles our speech. A symptom of late pregnancy, then, is a shift in the timbre of the maternal whale's voice.

It seems apposite, here, to put down that which the *poète maudit* Georges Bataille knew to be true: "Every animal is in the world like water in water." Every animal, including us. When we listen to recordings of whale song, we're eavesdropping on more than the utterances of another mammal. Our ears take in the ambience of the whale's body vis-à-vis its ocean setting; we apprehend, however subtly, the architecture of the animal's interior, the depth of the water, and the broader contours of the seafloor—all the grottos and barrens from which the whale's sounds have rebounded. Hearing whale song, we're hearing the shape of the ocean.

Most sounds whales make rely on the aqueous medium of the sea to be conveyed, which is why baleen whales are largely silent when they strand; the atmosphere is too thin to sustain the breadth and frequency of their calls. Toothed whales can click, chatter, and whistle above water, but they, too, need the sea to channel their full range of sounds. Killer whales, beluga, dolphins, porpoises, and sperm whales rely on what is best described as a fatty amplifier, or transducer, in their forehead—called the melon in some species and, during the peak of commercial exploitation in the nineteenth century, known as "the junk" in the case of sperm whales. ("Junk" not because this part of the whale was thrown away, but because the organ is faintly divided by cartilage, into segments, and *to junk*, as a verb, once meant "to cut up into pieces"—the whale's junk, then, being a reference to the design of the organ, and, too, what the whalers did to it.) Behind the melon, and beneath the blowhole in most toothed whales, is an internal organ that appears to be a pair of pursed, ink-dark lips, encased within the whale's head. These lips are made to fizz, generating a sound that enters the melon to be redirected. The whale's internal parcel of wax and triglycerides functions as does the pupil in the human eye: it focuses the whale's warblings, its pings of echolocation (used to identify congregations of prey, and to navigate sea ice), providing the animal with an acoustic "field of vision" in a lightless environment.

Being animals that variously inhabit the pitch-black and low-light seas, animals that sometimes hunt using the biosonic tings of echolocation, and which are thought to create, and maintain, their social attachments by vocalization, whales exist in a delicate envelope of biological sound.

So it surely surprises no one to learn that the clamor of marine traffic, seismic surveying, and underwater infrastructure creates channels of aversion, for whales, in the sea. As one Canadian scientist and Pew Fellow put it, "Noise is reducing acoustic habitat for whales in the same way that clear-cut logging is reducing habitat for grizzly bears." Yet the problem does not distress us the same way as, say, the deracination of a forest. These spaces are offshore and underwater, and as the harm is sonic and invisible, such blared-out regions are less routinely perceived by people. Like the deaths of the coral reefs—which would horrify if we could smell them (dying reefs would reek of rotting flesh, were they above water)—our own senses limit the suffering we can imagine. It proves necessary to project ourselves into the sensorium of whales, to sensitize ourselves to the extent of the damage done by artificial noise.

When Rachel Carson wrote the game-changing environmental book *Silent Spring* (1962), she envisioned a sinister season, absent the aubade of birdsong, because the birds had all died out from pesticide use. Conversely, noisiness has always represented a state of emergency underwater. Jacques Cousteau's renowned 1956 documentary series, based on his memoir of underwater exploration, was titled *The Silent World*. But by the time of its release, the oceans were already drenched with artificial sound—with what has come to be called "anthrophony" (*anthrōpos*: "human"). The seas have only continued to get louder.

After World War II, innovations in containerization—the transportation of cargo in "roll-on, roll-off" steel boxes of standard dimensions—facilitated a massive boom in the global shipping market. The accessibility of coastal nodes of transport, growing demand for energy drawn from offshore infrastructure, and increased freight entering expanding cities, have all called for larger ports and the intensifying traffic of heavier vessels (called "Ultra Large Container Vessels" and "supertankers"). The largest of these ships are propelled by engines that can approach five stories high, weighing 2,300 tons. The fuel these vessels run on is, essentially, the dregs of crude-oil refinement—it looks like asphalt, and is so thick that, when it is cold, a person can walk across it. In 2009, data released by maritime industry insiders showed that fifteen of the world's biggest ships were emitting as much pollution as all the cars—760 million cars—then in existence. As of 2016, 80 percent of the world's merchandise trade relied on shipping. Minerals, chemicals, lumber, fodder, and other bulk commodities are now shuttled worldwide, in immense quantities, over the sea, to be concentrated around sites of development, manufacture, and agriculture. This migration is inordinately noise-generating.

Marine "roads" bisect the sea, creating laneways of spoil: oily surface waters, akin to road run-off, that get tugged to and fro and dispersed by ocean circulations. We don't often consider these roads part of the built environment, but while the waterways may be impermanent and permeable, their long-term effects are anything but. Connecting the "here" to the "faraway," each freight-loaded vessel emits a mobile territory of commotion. A carrier ship can have an acoustic report equivalent to an airplane (age, and even minor maintenance issues make the ships louder yet). Ship noise volleys off the crenulations of the coastline and seafloor, off islands and the fretted hardscape of underwater pipelines, wellheads, manifolds, and risers. It can be insulated behind natural atolls but then blast out on either side. The movement of sound underwater is hard to predict, and harder yet to limit.

Additionally, seismic mapping—done in the course of oil and gas exploration, and for scientific purposes—entails the use of low-frequency, high-volume air guns, fired down toward the seabed to reveal a profile of its shape and mineral density. Early in 2019, a US congressman let off a blast of a 120-decibel air horn before a hearing of the natural resources subcommittee, in a stunt meant to demonstrate the effects of seismic mapping on the North Atlantic right whales the committee was debating (411 such whales remain alive as of this writing: each individual is known to science). The caper was ineffective, and, at any length, it did not represent the true extent of the noise created by these surveys. At sea, the bangs of the air guns can go for months, sometimes occurring at intervals of less than a minute.

The effect of all this sonic disruption is to shrink the whales' world. The whale's spatial perception, and their communication with other whales, decreases in range. Their horizons are drawn in, because the noise of the ships and the surveyors set limits: transparent walls the whales can't sense their way through. The International Fund for Animal Welfare (IFAW) reports that, per the calculations of their experts, for a whale born seventy years ago, "the distance over which her vocalizations can travel has decreased from 1,600 km [994 miles] at the time of her birth, to 160 km [99 miles] at present." What would this feel like? In childhood, your voice could be heard at the end of the street; approaching adulthood, it carried inside the house. Today, your voice is only audible in a hand you hold cupped over your mouth.

The trade the ships enable has contracted geography for consumers in all the prosaic ways we demand, making it possible to get things out of season and items not locally available (berries in summer, *kawaii* Japanese cosmetics, toys from Europe), but the cost of this daily traction, our pulling the faraway close, is borne by the animals below. They see less of their world, because we see more of ours.

Even though whales are among the world's loudest animals, anthrophony disrupts their lives to a far greater extent than for animals that make quieter, more localized, sounds. Many cetacean species have been observed going out of their way to avoid mechanical noise, or else they fall silent, and wait for intermittent interruptions to pass, as though abiding the loud interval of a storm. Some whales can adapt to a limited range of noises. But ultimately, noise pollution is more than auditory disturbance: it is also social disturbance in the case of whale species who sing to find a mate or whales that make sounds to communicate with their infants and rivals. Noise is foraging disturbance to whales relying on echolocation to identify and pursue squid, krill, or small fish. For a small number of cetaceans, noise likely impedes their ability to "see," for some animals are thought to use rebounding sound to form a mental picture of the physical features of their surroundings, and one another.

Scientists believe, for example, that, as a group, beaked whales are particularly sensitive to changes in their acoustic environment. Beaked whales—rarely seen by us—are highly reliant on echolocation and sound to obtain their food and socialize in the deep, dark, pressurized zones they inhabit. The whales' morphology suggests that they may not only identify one another with sound, but perceive, through their tissues, the *insides* of their bodies. The skulls of both living and extinct species of beaked whale are characterized by strange lines of puckering and protuberances that appear to serve no useful purpose to the brain they contain, and which are not visible on the exterior, domed skin of their heads. Typically, the males of the species have more pronounced skull formations than the females. Some scientists believe these bone structures are weapons, and that male beaked whales fight one another in the gloom by headbutting, their lumpen skulls being reinforced for impact (running counter to this explanation are the whales' porpoise-like beaks, which would seem to be too fragile to withstand such collisions). Other researchers have suggested that the bizarre bony knots and crests aid, perhaps, in the amplification or convergence of beaked whales' calls. But a third theory holds that beaked whales' skulls function as "internal antlers," not to tussle with, but to signal a male's strength, size, and reproductive fitness to other whales; a competitive display of dominance. The whales would apprehend these "antlers," not with their eyes, but with echoic sound, rebounded back to them. In which case, when beaked whales perceive what is inside one another, the "face" they respond to, in all its intricate grandeur, is like nothing we have ever seen alive.

WEBSTER'S REVISED UNABRIDGED DICTIONARY
[1828 & 1913]

Sound: *verb*

1. To ascertain (the depth of water in the sea, a lake, or a river), typically by means of a line, plummet or pole; or

2. Metaphorically, to try; to examine; to discover or endeavor to discover that which lies concealed in another's breast; to search out intention, opinion, will or desires; or

3. The use of any elongated instrument or probe, usually metallic, by which cavities of the body are sounded or explored: to *sound the lungs*; or

4. (Especially of a whale) dive down steeply to a great depth: *he sounded, arching his back steeply and raising his rubbery flukes in the air!*

But what *is* "noise," as opposed to song?

Noise interferes with perception—hearing, "seeing," sensing. It yields no useful information; noise is the ratcheting of machines, a disturbance of the peace. Once, whale song was considered noise, when it obscured the rat-a-tat of enemy watercraft. Now these sounds have flipped categories: the roar of ships is noise pollution, a whale emits a song. Pollution, in Mary Douglas's famed definition, amounts to "matter out of place." Noise is more ephemeral than other pollutants, like smog, trash, and toxicants, yet because it has the capacity to disrupt matter, noise is sometimes referred to as a "soft" pollutant. A misnomer: recent science has inferred that noise could have insidious impacts that endure, long after its cessation. Researchers across a range of different fields are discovering artificial noise has ecosystemic effects, and that it harms even organisms with no ears. Seismic blasting turns out to increase scallop mortality. Air-gun testing has been shown to kill all krill larvae within a three-quarter-mile range. Ship noise eradicates sea hares (a plentiful kind of underwater slug). Cuttlefish, which change color to communicate, are less able to interpret one another's pigmentation signals in the presence of acoustic interference, suggesting that noise can impede visual modalities as well as auditory ones. Distracted by the pattering of motorboats, young reef fish fail to recognize the warning signs of the predatory dusky dottybacks that hunt them—noise apparently derails their sense *of smell*. So anthrophony can disarrange entire food webs, and its effects can be permanent, even after the source of the noise is subdued, or removed.

For whales, noise pollution can be lethal. When dozens of Cuvier's beaked whales stranded in 1996 and 1997 along the Peloponnesian coast, and again in 2014 on Corfu and the east coast of Italy, the whales appeared, to the naked eye, to be in ruddy health. Nonetheless, they all soon died. Necropsies ("autopsy" refers only to the examination of a human corpse) revealed gas-bubble lesions in their internal organs, ruptured ear canals, and accumulations of nitrogen in the whales' blood and tissues. These were decompression injuries—barotraumas—thought to be caused by "bounce diving" (rapid movement up and down without surfacing) in the Hellenic Trench. During the weeks preceding the strandings in 2014, Greek and Israeli navies had been engaged in joint anti-submarine warfare training off Crete, and a NATO vessel had likewise been testing sonar in the region. The whales' ears bled: not from the direct force of the noise, but from hemorrhaging caused by frenetic zigzagging in the water column. Research undertaken in the aftermath of these events showed that between 1960 and 2004, 121 strandings took place around the Mediterranean, at least forty of which were linked in time and place with naval activities. There may be any number of Cuvier's beaked whales sunk invisibly to the seafloor with irreparable injuries gained in their disorientation caused by loud mechanical reverb in their habitat.

Because of these effects, there have been attempts to curb subsea noise pollution in some places. Multinational oil and gas companies bring biologists along on their surveying vessels to scout for cetaceans, setting paths to avoid them. The International Maritime Organization (IMO), a UN body, has instigated "traffic separation schemes" in the Gulf of Panama to shift shipping transportations away from whales during their visits to the region. Speed restrictions, and the use of mufflers or low-volume engines, have been instigated in a few marine roadways known to cross migratory routes. Around the Canary Islands, mid-frequency active sonar, used by the military, has been banned since 2004, while, near British Columbia, quiet zones have been implemented in seas where killer whales hunt (though the whales must first be sighted, for the ships to then steer clear of them). These changes are positive, but they remain small-scale, and address only the most cacophonous effects of noise.

The larger, faster, and louder the vessels get, of course, the harder it is for a shipping crew to adjust their course away from whales where they are glimpsed, or overheard via hydrophones. Those whales, like humpbacks, that have evolved to gargantuan body sizes, so as to be able to migrate between distant regions of periodically and seasonally productive ocean, are the species most likely to thrive in seas that are changing—their immensity allows them to travel farther, to be more adaptable. But these whale species are also, because of their dimensions, and their need to breathe regularly, highly vulnerable to ship strike. Today's ships are so very big that, when the hull collides with the body of a whale, it often goes unnoticed by anyone aboard.

Traces of human conflict implant in the crushed dust of geological strata and are fossilized: atomic weapons leave their tell-tale footprint in the radioactive isotopes of silt and sediment; deformations of the seafloor show where surplus munitions have detonated, after being dumped; rusted bullets, driven into killing fields, become flecks of crystalline and metallurgical matter. With the right toolkit and expertise, all this can be exhumed and identified, layer by layer. The realm of rocks—inert, accretive—is a repository of the historical past. Less well understood is how the wars of our kind might show up in the biorhythms of other species. Yet the flesh of whales, scientists are discovering, can also, indirectly, register trauma in human communities. Our emotional lives are connected to those of animals, in more material ways than we tend to imagine.

When the terrorist attacks of 9/11 occurred in 2001, a group of researchers from the New England Aquarium documented North Atlantic right whales exhibiting a significant drop in stress hormones, called glucocorticoids. Before the hijackings researchers had been monitoring the rare cetaceans some five hundred miles away from Manhattan, in the Bay of Fundy, retrieving their feces to test for the biochemical signature of environmental pressures. The low-frequency rumblings of large ships in the area were known to overlap with the acoustic signals of these baleen whales; so the scientists were looking for indications that the sounds affected the animals' interactions and distressed them. Then, suddenly, nonessential shipping and boating was suspended in the wake of the attacks, creating a unique experimental "control"—or as near to—in the wild. Many American and Canadian ports were temporarily halted, or stayed open to only restricted small-vessel traffic. Looking at traces of the whales' hormones, the scientists saw evidence of the new serenity.

The US National Park Service's Natural Sounds and Night Skies Division estimates noise pollution double, or triples, every thirty years. Today not only the seas, but our urban centers and industrial fringes are getting louder—the sirens of emergency vehicles keep having to add extra decibels to cut through. "Having emancipated itself from the human hand," writes Bianca Bosker in *The Atlantic*, "noise is becoming autonomous, and inexhaustible." There are fewer places of quietude in wilderness, too, intruded upon by snowmobiles, motorized riverboats, photographic drones, and, higher above flight paths. The hazards of noise are evident in the physiology of the human body, as they are in right whales: rising blood pressure, slowed digestion, and the pumping of adrenal glands are symptoms, as are despairing psychological states and difficulty absorbing new information.

In the rawness of sorrow and mourning, even distant anthrophony, to which we have grown habituated, can seem to be pushed to the sensory foreground to become unbearable. The *zizzz* of the highway, the *rrrrr* of a data storage tower in a room, the *bp bp bp* of a light bulb: all this, suddenly crucifying. People say such aural sensitivity is indicative of the sort of grief which yearns to turn away from the world, to close off; the griever would shut their ears, had evolution equipped them to do so. Yet what we call "silence" is, so often, the solace of nature. The minute's remembrance that marks a national tragedy fills with biophony—a breeze in the grass, snowfall's plump aerospace, the drowse of insects, or a fluttering in the hedge. Silence is the resource we most overlook, though its potential is inwardly replenishing, and its absence is often taxing.

Perhaps you are unsurprised to hear that truly quiet sanctuaries of natural sound are endangered in a global soundscape, riven with mechanical noise. This much, I think, we expect; it marries up with our own acoustic exposure. But what I did not anticipate was finding out that even in very remote places, where people do not go, and where machines make no noise at all, the sounds made by animals are changing as a result of human intervention in the environment; that the line between anthrophony and biophony is blurring.

Antarctic blue whales—inhabiting, as they do, latitudes where the sun scarcely rises for near to half the year—exist in an environment sharply defined by bioacoustics. Earth's largest animals, these whales use sound to identify schools of krill, their prey, and they also call to distant others of their kind, though the social sounds blue whales make are less songs, more a low, flat humming. If gravity had an audible noise, this would be it: a tectonic rumble, a black-hole fugue on the furthest perimeter of human hearing.

That the sounds of blue whales are monotonous and simple might lead you to believe that they are, as they have always been, unchanging across the generations. But blue whale calls are evolving. Since limited protection of the subspecies begun in the late 1960s, and was bolstered by the successes of the anti-whaling movement globally, their pitch has downshifted the equivalent of three white keys on a piano (keys that once might have been made from whale bone). Scientists have several theories as to why.

In the modern sea, this deepening of sound is not unique. As the voices of Antarctic blue whales have descended to a more baritone frequency over time, so, too, have similar trends been identified in groups of pygmy blue whales found outside of Antarctic waters, offshore of Madagascar. Fin whales—which emit such deep sounds that their wavelengths can be longer than the bodies of the whales themselves—have also dropped their pitch. (For the human ear to hear a fin whale clearly, their sounds must be computerized and sped up.) In one million recordings of different whales' songs that were digitally analyzed, scale shifts were observed across species, populations, and pods that don't interact or stray within earshot of one another. Whatever triggered the change may have taken place in the ocean or in the whales, but it does not seem to have a localized origin.

My first question, when I heard about the whales' dropping frequencies, was *How are we involved?* But lowered pitches have been identified across populations of whales that live in icebound seas traversed by no major shipping routes, where artificial noise is negligible—as well as in habitats nearer to intermittent, or more significant, sources of noise. Scientists suspect that the underlying, unifying cause is not sonic pollution. The descent in pitch was not so significant that it would alter the distance that these communications travel, refuting any theory that the whales are attempting to speak over longer intervals, or to dodge the frequencies of maritime traffic and military sonar.

One explanation for the deepening whale calls offers insight into how global conservation efforts may function not only to increase protected animal populations, but also to change what individual animals, within these safeguarded species, sound like. A clue lies in the fact that blue whales today also sing more quietly than in previous decades. Anatomically, because of the biomechanics of how blue whales verbalize, the louder the sound they make, the higher the pitch. When the whales soften their voices, they "speak" with more bass. Scientists have speculated that as their numbers have grown, the whales have decreased their call intensity (their volume), and, concomitantly, their pitch has dropped. Today's Antarctic blue whales may be quieter, and lower-toned, than in previous decades merely because more whales are communicating, and over shorter distances.

Less reassuring is the following hypothesis: the whales may not need to waste energy being so loud because sound waves travel farther in oceans made acidic by the absorption of carbon dioxide. In time, if escalating acidity goes unchecked, the sea's new biochemistry will also amplify and extend low-frequency noises generated by mechanical activity: construction, fishing, sonar, resource extraction—compounding our noise, to create ever greater crescendos in the world's coastal shallows, and the mid-depths. So the sea will grow to be a louder place not only because the human bustle within it is multiplying, but also because, as a medium for conveying sound, the sea itself is becoming a better magnifier of noise, permitting certain sound waves to travel further. It is as though some forgotten god on the seafloor were, twitch by twitch, turning up the volume dial to account for their advancing deafness.

Recent monitoring of Antarctic blue whales has shown that, during the austral summer, their pitch rises again. The whales increasingly have to use their most forceful *forte* volume to be heard amid the cracking ice (a natural sound, amplified by unnatural processes as rising temperatures exacerbate ice melt). So the impacts of a warming climate may modulate animal sounds even in remote places where no ships go, where barely any humans subsist, and where the most thunderous notes come from the clatter of breaking ice. As we adapt our storytelling to the conditions of a transforming biosphere, whales likewise are called to modify the qualities of their dispatches, in line with this new reality.

Changes foisted upon the oceanic realm by humankind have shaped how whales understand their world and how whales relate to one another, at the same time as what whale song and silence means, to us, has also shifted to sync with the urgencies of this environmental moment. How they sound *to us* is linked to how we feel about the future of animals, the future of the sea. The blue whales' voices fall between them; and between us and them. We may not yet know what the sounds of blue whales mean, but whether through the intent to preserve their species, or, inadvertently, as a result of refashioning their environment, it seems that our deeds do echo in their voices; that our actions are somewhere there, in their sounds.

The sea streams still, with nebulous sounds, transmitting from as-yet-unknown sources. One recent summer a ping emanating from the Arctic Ocean seabed is reported by hunters in the Qikiqtaaluk region of Nunavut territory, near Igloolik. Earlier: something else in the San Diego Trough is overheard, crooning to its prey. From the Pacific, at a point west of the tip of South America, come bleeps known to be of organic origin, but several times louder than the loudest identified zoological source. The Caribbean Sea whistles in A-flat. Sonic upsweeps, frequent in spring and autumn, have been ebbing off New Zealand for over two decades. The Western Pacific Biotwang—fricative and metallic—is audible all year round to sensitive machines in the deepest tracts of a crescent-shaped trench. Equatorial hydrophone arrays recorded a low trumpeting in the 1990s that NOAA officially named: "Julia." "Antarctic BW29" is the code given to a voice belonging to an animal undiscovered, but made out by equipment at three different locations in the Southern Ocean. Some people are certain that a keening known only as the "52 Blue" is an as-yet-undocumented species of cetacean, or a group of cetaceans, moving south from seas off Alaska to near Mexico and calling at an anomalously high frequency. It, or they, speak within the range not of whales, but at the pitch of the fluttering you might hear near a substation of high-voltage transformers on a summer evening. Whether their call is loveless, lonely, or the last of its kind, their voice has gradually dropped over time, perhaps only as a result of growing larger.

An elected official, later appointed premier of Nunavut, released a statement on the pings coming from the Arctic seafloor. "Sometimes there are mysterious things, and there are people who report those mysterious things," he said. "I want to thank them."

Perhaps the most concrete example of people retuning the voices of whales can be found in the story of Noc, the beluga who sounded like a man. Noc's name was given to him by the Inuit hunters who aided in his capture—it alluded to the tiny stinging flies, colloquially known as "no-see-ums," that still appear on hotter days where Noc was born. Netted as a two-year-old calf off the northern coast of Manitoba in 1977, Noc lived for twenty-three years under the remit of a US Navy surveillance and retrieval program called "Cold Ops," based out of San Diego Bay, in which beluga were trained to respond to hand signals and spoken commands in the open ocean. Along with dolphins and seals, the navy deployed beluga whales for reconnaissance, to identify aquatic mines, and to recover inert torpedoes for reuse. When not in training, or tasked to an operation, the whales and dolphins were housed in enclosures adjacent to one another. It wasn't until thirteen years after his death (from meningitis, in 1999), that a recording of Noc's voice became renowned online.

For a period while in captivity, it seems that Noc learned to mimic the sounds of divers communicating with one another on surface-to-seafloor "wet phones." One diver described being told to "get out of the water" by a voice he mistook for that of a superior. It was Noc. Acoustic scientists, identifying the novelty of this imitative behavior, began to induce Noc, via rewards, to repeat his humanoid vocalizations, which they then recorded and studied, publishing a paper over a decade later. At the time of Noc's impressions (which only lasted until he matured), there were already other unofficial accounts of beluga learning to speak: at the Vancouver Aquarium, keepers held that Lagosi, a fifteen-year-old beluga, could say his own name. Elsewhere, whale handlers documented the animals emitting sounds akin to "Russian, or similar to Chinese." But Noc's voice is, to date, the only verifiable example of human mimicry by a whale. Listening to a recording, the beluga sounds less like a faithful imitation of speech than a helium-ed caricature—burbling and slurpy, more *Looney Tunes* than declarative talk. But Noc's voice is wildly unlike anything within the ordinary range of sounds beluga make, and his diction had an amplitude, and a verbal rhythm, that, according to researchers, matched human communication. You can almost make out words, a *yoy-yoy*ing, a *do-di-do*ing.

If ever I hear myself recorded, I think: a voice is an uncanny possession for any animal to have, especially a recording animal who has tooled itself to ply apart voice and body. The human voice is hewn through an inner karst, the hidden alcoves of the torso, voice box, and teeth, and, in that way, the tenor of your personal voice—its timbre and grain—seems a wholly intimate, unique, and biophysical property. But then: the tongue. The tongue—that pliant and beguiling traitor grown inside your mouth—schools the voice for its reception, dropping vowels, placing stresses and sibilance, anticipating the prosody of class, slang, or the accent of geography, in league with its two cunning elocutionists, the lips.

Your voice, the philosopher Slavoj Žižek once said, is "coming from somewhere in between your body."

Whales are not usually imitative animals: they do not replicate the sounds of other animals with which they share an environment in the wild. Noc's voice fell somewhere short of being human, but perhaps it was just human enough to suggest that this whale thought of its handlers not as the others, but as an "us" he might communicate with.

In late 2018, the *South China Post* reported that scientists from Tianjin had successfully engineered a means of conveying concealed messages within audio-manipulated, digital sperm whale pulses. Sonically, the sea turns out to be a good place to hide clandestine information. The Chinese technicians created the new technology with the intent to evade, or baffle, modern undersea surveillance systems programmed to filter, and ignore, cetacean calls, during sweeps for the acoustic waveforms of covert communications. Contrived to blend in with background biophony, the dispatches could then transmit, unnoticed, between relay stations and submarines. Concealed in whale voices; human data. And as for sperm whales passing through these signals, no one yet knows if the animals recognize the Chinese intelligence smuggled into their calls, or if they can learn to split out, and disregard, navigation coordinates and naval commands from the transmissions of their kind.

————————

Beyond the sounds distinct to a species, an individual whale is singular enough to conceive of it having a personal voice. Some scientists are convinced there have been stand-out soloists throughout the decades. "Today's humpback whale songs are such pale copies of those of the sixties," opines Roger Payne in an open letter from 2005. Payne insists that industrial whaling subtracted major symphonists from the sea. When whalers harvested the largest, mature whales, Payne says, they altered population demographics so that the young never got to learn the music of the old, impoverishing their musical lineage. To Payne's ear, the population bottleneck meant that no matter how sophisticated the songs of humpbacks are now, they remain less striking than those of a former generation.

Once, people thought that whale songs were ancient and immutable—this lay at the heart of their charisma: that the songs were a window into a mysterious world that preceded us, an imaginary place of being parallel to, and outside of, the sphere of humankind. Now we know that the sounds whales make are changeable. The voices of some whales have deepened and quietened; others, perhaps, bear the acoustic legacy of whaling and conservation. Whale voices have been shown to be responsive to extrinsic forces exerted by human societies and industries, but they are also, recent research reveals, an expression of the intrinsic and independently evolving culture of whales.

An infant humpback does not arrive in the world charged up with a self-same song, the way a froglet comes equipped with the croak of its species. A humpback is brought into a historically distinct linguistic community—the conventions of its song must be learned. Which is another reason why, perhaps, some noises made by whales are deemed, by us, to be songs. The songs of whales emerge from a mammalian collective, shaped in an ever-evolving shared score. Whale song is *social*, as much as it is natural. Whales today do not, indeed, sound as they did in the 1960s; just as we, too, by and large, have dropped the argot of that decade from our conversations.

Off the east coast of Australia, the songs of male humpbacks tend to increase in intricacy over time, evolving through incremental ornamentations and novelties. As the humpback population has returned so, too, apparently, has the potential for ingenuity and originality within the population. Humpbacks have regional dialects that change year to year; the animals also express a seasonal preference for certain song structures. Sonic motifs circulate when males from the same population-group reiterate them to other males within earshot, traveling either ahead or behind along their path of migration. So the songs can transport, up and down the length of the route, betwixt the whales.

Humpbacks recycle and vamp off the songs of other, far-off populations, too, over the span of oceans; sometimes, even if a continent has come between them. When the whales drift within audible range of one another, they may loan a theme or fragment from a song overheard in the distance—building a repertoire that they carry away. A single whale's distinctive warble, or a phrase from it, might transfer almost virally, like slang, a joke, or a summer hit, though no one has yet been able to identify what makes a specific sound catchy to humpbacks. Call them samples, these popular phrases. Remixed, they sweep across hemispheres. Sonic samples can pass between populations of humpbacks that never physically meet, as has been documented in songs moving from west to east, out across the Pacific. The jingles—pop songs—travel from Antarctic-Australian waters, to the calls of humpbacks grouped up near New Caledonia, Tonga, Samoa, to around the Cook Islands and French Polynesia. Having encountered one another in an early phase of their migration, humpbacks off Madagascar and Gabon, with the landmass of Africa now between them, sometimes play from the same songbook. The songs do not need to be heard over a great distance, to travel it; there just need to be sufficient whales in the sea to bring them near enough to one another to relay a fragment—before bearing it out, to other whales.

These "cultural transmissions," report biologists, occur "on spatial and temporal scales that have previously only been documented in humans." Humpback whales participate in the largest communication network on Earth—outside of those that we have constructed, kitted out with satellites and fiber-optic cables running under the sea.

Then, roughly once every three years, an event takes place that alters the song structures entirely. Scientific observers have deemed these events "cultural revolutions." When a cultural revolution sweeps across the humpback population, the whales' compositions are stripped of showy details, and simplified—the animals switch to what the scientists call "revolutionary songs." This process could suggest, say a team at the University of Queensland, an upper limit to humpback learning and humpback memory, if sound *is* a means of gauging a mind. When their songs cannot become more ornate and still be remembered, rote-learned, or transmitted adequately between whales, then the degree of complexity crashes. A whale singing a less demanding song stands out against those attempting poorly learned, but more complicated tunes. What follows is a race to uncomplicate.

One upshot of this perpetual revision of humpback songs is that the dreamy recordings of whales from the 1970s, including those aboard the

Voyager probes, no longer accurately represent the sounds of whales in nature. Whatever those compositions conveyed, whatever the meaning of those exact groupings of pitches, most were never heard again in the wild. Perhaps they never will be. As whales have sometimes captured human cultures in their bodies—arrowheads buried in their flesh, plastic products in their stomachs—we have frozen their cultures inside ours.

To sound is to measure within the body, or, anachronistically, it is the action of a whale body plunging, a whale rushing to the deep. How does a whale locate itself in a universe without landmarks, without light, without a way to quantify time?

The answer is found not in the deep oceans, but in deep space.

Out past the ozone, the tropopause, the stratosphere, the sun takes eleven years to cycle through its turn of seasons. The sun is a superheated cloud of dust and gases; it has no surface. But though there is no crust on the sun, no marine or terrestrial environment, no diurnal transition (the sun is utter daytime), physicists report there is still a type of weather there. In its most torrid state, the sun's convective exterior is strafed by channels of charged plasma, traveling at thousands of miles per second. Great whips of plasma lift off the sun's outer layer and lash from its fizzling corona. Sometimes these "ejections," as they're called, "coronal mass ejections," are so immense they spurt beyond the sun's atmosphere and enter interstellar weather—meteorology fanning through space. The ejections cannot sustain their temperature there. But persisting as shock waves, as jets of subatomic, ionizing particles and globs of magnetism freed from the sun's gravitational control, these solar storms move outward. If such a tumult occurs on a quadrant of sun facing our planet, three days later, its effects fall to Earth.

Any astronaut might see a flash or flashes before their eyes when a solar storm descends—even if their eyelids are squeezed shut and their fists are pressed into their sockets. What causes the light is protons, sped up in the wake of the ejection, creating a visual disturbance. On Earth, compasses quaver. Shortwave radios at the poles and on aircraft will hiss, and radars return what is known as "glint noise" in their imagery. There may be blips in the power grid, short-lived brownouts or blackouts. Once, the Canadians had to shut down their stock market when a solar storm disrupted central computers. ("I don't know what the gods were doing to us," said the vice president of the Toronto Stock Exchange.)

At ground level, little indicia. A few fewer bees return to their hives. The senses the insects use to navigate are baffled. The bees drone on, oblivious, over salt pans and in yellow canola, long after stars come out. Homing pigeons take longer to loft, and migratory birds linger overhead, searching for lakes or certain roosts the flocks prefer. The minor deviations of other journeying species—bats, mole rats, lobsters—are harder to observe. These animals, too, align themselves using magnetoception. The magnetic fields of Earth compel and direct their movements.

To humans, the solar weather is most conspicuous as raptures of beautiful night air, appearing in some parts of the world. Auroras spritz long corridors of color across the sky, extending their range from loops at either end of the planet, to appear at latitudes over Tasmania and Victoria, and, descending out of the Arctic, through Canada. Scrolls of emerald green, fuchsia, dark red, and electric blue: the auroras are created in the upper atmosphere by atoms agitated by particles from the sun. After coronal mass ejections, the vibrancy and sweep of the auroras increase, driven by the effects of interstellar weather subtly deforming Earth's ionosphere and its magnetic fields, which exist like netting, or a fine skein of hair, around the entire globe. Sperm whales deep underwater cannot see the auroras, these fluttering, starry drop curtains. Yet the aftermath of solar storms might govern the lives of whales in ways even more dramatic than the churn of the northern and southern lights that turn stunned human faces skyward, far above the water's surface. The consequences of solar storms are believed to come up at whales *from beneath*, off the seafloor in darkness, rather than out of the air.

How is this so? Under the sea there are great mountain ranges made not from minerals but magnetism. They can be sixty miles wide, these natural, unseen features of the planet. The mountains have no specific density or charm. They cast no shade, reflect no light, and obstruct no submarines. An oil rig could pierce one without ill effect. Unlike geological peaks, these are soft mountains, hills of pure energy measured in nanotesla, and their formation, their silhouettes, are a product of interactions between Earth's magnetosphere and the interstellar weather.

A solar storm humans can't feel—which we only have eyes to observe secondhand as shimmers of color in the night, and which our ears might take in, absentmindedly, as the sound of lost birds circling—that far-off ejection of sun can, nonetheless, migrate a geomagnetic mountain. It can tremble that mountain like a jelly, or reoutline it temporarily. A strong sun storm might altogether dissolve such an undersea mountain, and cause an abrupt change in the formation of the seafloor, as it is perceived by a creature arrayed with senses sufficient to apprehend these fields and energies. A creature that, say, navigates long distances in the deep not using its hearing or vision primarily, but according to the presence and relative strength of magnetic topographies. Like bees, birds, bats, mole rats, and lobsters, sperm whales are believed to migrate via the detection and direction of these energetic forces.

Researchers writing in the *International Journal of Astrobiology* propose that magnetic mountain ranges, impalpable to us, act as "thought barriers" to sperm whales. Those are their words: "thought barriers." A cosmic storm, the geophysicists argue, could bewilder a sperm whale by rearranging the invisible aspects of its underwater environment, from the astronomical dimension. In this way, it might turn out to be true, what has before now sounded like fuzzy, new age mysticism: that sperm whales exist in celestial space; that their minds are tuned to the information of interstellar planes, and that whales are animals buffeted by weather coming from between the stars.

Winter 2016: After significant solar storms, twenty-nine erratic male sperm whales entered the North Sea and stranded along the coasts of Germany, Great Britain, the Netherlands, and France, all within a few weeks of one another. The stricken whales died, many cordoned off from engrossed crowds.

Some people speculated that the malignancies to blame for the arrival of these huge cetaceans were emitted from the human world: a biosphere saturated by signals, electrical forces, and other presences of synthetic generation. In Skegness, Lincolnshire, two of the dead sperm whales were spray-painted by activists or vandals. *Mans Fault. Fukishima* (misspelled). *Man killed me. RIP*. The logo of the Campaign for Nuclear Disarmament (CND) was speedily applied to one whale's tail fluke, though the icon was missing a spoke and the CND disavowed any connection to the protest. Whose fault was it, the newspapers demanded, that so many sperm whales were appearing? Could it, indeed, be attributed to heavy water from the failed Japanese reactor, ejected into the Pacific many years prior? Were the spirits that moved, in these animals, not ours?

Where the natural world was once interpreted as latent with meanings installed by some godly power, events in nature now manifest the dread of ecological disasters people have created, and which we fear have grown to roam beyond the detection of our senses and instruments. If animals are a kind of extrasensory perception, we fear that the shadow realm they testify to is one haunted by the elusive legacies of human industry, manufacturing, and warfare.

Most of the stranded animals were necropsied by biologists. A sorting of the stomach contents of these North Sea whales revealed copious beaks, which is as it should be. Sperm whales eat squid, gulping them whole, and squid come equipped with hard beaks, akin to those of parrots, which they in turn use to consume fish, crustaceans, and other squid smaller than themselves. Squid beaks are impervious to the whales' digestive acids, so the sperm whales were taken to be in good health: the animals were well-fed. There seemed, initially, no comprehensive explanation for first, their presence in the North Sea, and second, their deaths. Sperm whales typically avoid this part of the Atlantic—young bulls in groups go west of the British Isles to the Norwegian Sea in winter and spring, where the armhook squid are abundant.

Though it remains only a hypothesis, the notion that these sperm whales may have taken a wrong turn in the North Sea because the magnetic seamarks the animals use to navigate were distorted by solar storms returns, to us, old ways of knowing the incarnate world. *What are the gods doing to us?* asked the Canadian financiers, of the sun, that all-seeing eye, when it derailed their markets. There were two huge sun storms in the months before the sperm whales showed up on the European coastlines. From the droning nocturne of bees, to this deliverance of sperm whales and the reshaping of magnetic mountain ranges, the storms' effects might have ranged from the minute to the planetary. If animals offer a kind of perception beyond our senses, the shadow realm they testify could be universal—almost heavenly.

And to the sperm whales too: Was there a cursed moment of realization when the whales traversed the zone where their geomagnetic mountains had eroded, when the sand rose inexplicably up from the dark below to meet them, and the sun drove the animals over the edge of their world? Did the incredulous whales then ask the same of themselves and each other: *Why have our gods decided now, to desert us?*

06

Sea Pie

DAYS BEFORE FLYING TO TOKYO, I LEARN THAT THE TRANSLATION for the name of a Japanese dish of whale tongue is "twitter." *That can't be right*, I think, shutting down my laptop. It can't be right because whales don't twitter, birds do. A twitter is a winsome sound: a noise (and, at that, a receding noise; trebly biophony, lacking pertinence—cute). I went and found the dictionary where it lay facedown, still paged open to the entry for "sound," and I flicked forward. *Twitter*: "a succession of small, tremulous, intermittent noises," a term "of imitative origin." A form of noisemaking that implies a hesitancy to make noise, then; a suppressed sound, the scandalized tittle-tattle of Second Empire parlor rooms—and, though the dictionary omits this nuance, does the word not also have connotations of self-soothing? The echolalia of sundowning. Birds twitter to themselves in roost.

I returned the book to the shelf. A serving of tongue emits no sound at all, of course. I wondered if the name of the dish referred instead to the diner, chattering with anticipation at the table. Was it a joke, a conscious irony, to order up this specific body part with a word that recalled its postmortem silence? Or did the Japanese people, in fact, interpret whale calls as twittering? If that *was* the case—if the sounds of whales were deemed "twittering" elsewhere, rather than "songs"—then here was a curious question regarding the ways different human languages, histories, and contexts, marshaled the charisma of other species. The contrariness of "twitter," compared to that expansive, grand word "song," implied divergent impressions of a whale's presence and its sociality. I thought that, to declare an animal edible, you likely had to demote it to something that clucked and peeped, or made otherwise meaningless, self-regarding noises. Harder to call "food" that which sings, accumulates memories, and belongs to a multigenerational oral culture.

And why was I headed to Tokyo? If the whale that ate the Almerian greenhouse had set me on the path to reckoning with how the synthetic human world could end up inside wild animals, I was also aware, from the outset, that I would need to consider places where whale meat appeared under plastic-wrap, as an ingredient, a foodstuff; where whales were consumed by people. I was going to Japan to think about that most literal internalization of whales into culinary culture and into human bodies—ingestion. Who devours some of the biggest animals on the planet? Why had the Japanese whaling industry been prolonged into the twenty-first century, on a commercial scale, where other nations fronting opprobrium and the unprofitability of whale products fell away from the pursuit? Have the implications of eating whales changed in the current environmental moment? These were my questions, and though they might seem, on their surface, to be about taste and national identity, perhaps even about cruelty, I understood, even before my departure, that at root they were questions about the fluidity of animal charisma. Claims of loyalty and connectedness to other species can, and do, resolutely divide us—and no more so than where an animal revered or cherished in one culture is categorized as edible in another.

People no longer eat like animals. Our diet has, by and large, been domesticated and corporatized—the planet could never have supported nearing eight billion humans, most of them non-nomadic, on game and forage alone. Today's retail produce is less diverse, but more bountiful, and reliably nonseasonal. Genetic variation in crops has narrowed over time, to give way to the dominance of a limited set of hardy, high-yield cultivars: vegetables and fruits are trademarked products—as human as Teflon or Neoprene. Likewise, meat animals, the preponderance of which belong to a handful of overbred species that are farmed, not hunted or harpooned. The modern pantry is stocked with a greater percentage of processed foodstuffs: machine-extruded, shelf-stabilized, single-serve, and convenient. In a marketplace of artificial abundance and unseasonal variety, we are asked to be judicious—to make choices beyond survival instincts.

Decisions made in the space between the spoon and the mouth are deeply personal, yet the range of possible options that any one of us gives thought to, inside that window, is also a factor of culture, class, geography, science, and story. The origins of what we eat are more than biological. The science of food chemistry, agronomy, and animal husbandry, of health and medicine, shapes dietary preferences. But so, too, do official subsidies, as well as superstitions, fads, fears, advertising, and the difficult-to-describe limitations of "taste." Within the scope of what food is available, what is affordable, what is habitual, and what is anathema, many of us make choices steered by two guide rails: our heritage, and what our peers eat. If what we eat makes us, then it makes sense not to stray too far from the example set by our kin—for so long as that remains practicable. Food is fellowship. Food is also nostalgia: our earliest meals can seed a sweet tooth, spoil an appetite, or instill a craving later in life. The barrier between taste and memory is paper-thin.

National cuisines can speak to the religious or political past of a people, but they also tend to memorialize climatic and ecological conditions: those foods "safe" from rancidity or localized strains of bacteria, for example, as well as the botanical varieties best fostered by regional rainfall, altitude, and soil type, and the livestock that were first able to survive the terrain—before its refashioning into fields, feedlots, fish farms, and ranches. Though many cultures prize it, among food taboos meat is inordinately more likely to be forbidden, restricted, or subject to specific cooking protocols than any vegetable or grain. Ancillary to the category "meat," organs of sense and sentience (eyes, mouthparts, brain tissues) as well as those associated with life-force (milk, gonads, blood, glands) have, across history, attracted the most reproof and cringing distaste. Meat stars in commemoration and celebration dishes: think of "coronation chicken," Christmas gammon, Thanksgiving turkey. But perhaps because it is associated with esteem, caregiving, and the gathering together of clans (familial or fraternal), meat is also a leveler and a mark of class solidarity: fried chicken, the hamburger, and the plowman's pie have all, in turn, signified such lauded values as self-sufficiency, unanimity, and comfort. It is no hyperbole to say the animals we eat—and most especially, the animals we *don't* eat—make a "we."

Eating is intimately individual. Yet what we wish to eat—and what we refuse to eat—can clue us into a circle of intimates, or inculcate us within hallowed traditions that connect back to our culture's past. I believe all this, and at the same time, I know I am two-faced on the subject. My own background has permitted me to shift my diet without estranging myself from my origins. Even though my family is, to a person, meat eaters, no one takes offense at the fact that I have been vegetarian since my mid-twenties: it's scarcely a novelty by now (to all other than my youngest cousin, who, at fourteen, will not be disabused of the possibility I belong to some secret monkey-wrenching outfit, set on liberating beasts of burden: in this regard, I am a letdown). Many of my friends are either vegetarian or they eat animals only rarely, or they eat only some animals. One or two are strictly vegan. Australia's multiethnic and miscellaneous cuisine details few public observances wherein refusing meat is poor etiquette, or gets met with censure. During federal elections voters are rewarded with "democracy sausages," barbecued on a hot plate near the exit of the polling booths and folded into bread: this is the only occasion I can think of where any community I belong to comes together over meat exclusively. Even then, it's more a trend, not a weighty orthodoxy from the distant past.

What it means to be a vegetarian has shifted across the decade or so that I have (mostly) stuck to a plant-based diet. The notion that eating is *an environmental act* has, in recent years, gained traction. The basis of vegetarianism has come to be viewed as an expression of expanding spheres of concern: from the health of the human body and spirit (historically, the earliest vegetarians went meatless for reasons of godliness); to encompass disquiet about the suffering of animal bodies; to consider the condition of regional watersheds and topsoil in areas adjacent to containment agriculture. (Land clearing, effluent pools, and prophylactic antibiotics discharged into waste water feature in accounts of this latter justification for not eating meat.) The most recent set of arguments for plant-based eating represent a further dilation of that concern, to contemplate how meat-eating figures into climate change. Vegetarianism is now viewed, in increasingly vocal quarters, as an ecological commitment that extends to planetary health.

The writer Michael Pollan famously described how the human "food environment" was radically renovated by the ubiquity of refined and vitamin-boosted foodstuffs. So, too, are huge numbers of agricultural animals—species bred into self-same, fast-growing artificiality—raised today on a diet of calorie-rich fodder and pelleted feed wildly unlike the nutriment of their ancestors. It can seem that the food chains connecting people to these "pre-meat" animals today have little to do with nature. Livestock are not "natural" animals, nor do they live under natural conditions, acting out their natural instincts: they are part of the protein production industrial complex, bringing cheap sustenance to market. Yet, however segregated from nature these edible animals are, their existence has ecological—and atmospheric—effects on a world scale. Beef is a flashpoint. The oft-repeated figure is that if cattle were themselves a nation, they would be the world's third-largest emitter of greenhouse gases, principally via the belching of methane. (The opposite of whales, cattle likewise geo-engineer the air, but as an accelerant to global warming.) The downstream impacts of growing, fertilizing, and transporting fodder are also injurious: greenhouse gases from feed production represent 60 to 80 percent of the emission coming from eggs, chicken, and pork.

But is it ever possible to be impartial on the subject of what we eat, and why? The external facets—personal, cultural, and political—are so close at hand, and become so mingled with our ways of seeing the world, that a discussion of what we *mean* by the things we eat seems impossible to stage in all but the most introspective terms. We may not intend to communicate anything at all, to anyone beyond ourselves, in the choices we make about food: though a lack of intention does not prevent those decisions shaping the environment in which people, and other animals, live. How we resolve to draw the line between what we'll do to nourish our bodies, and the interests of others, is neither a solitary, one-off decision, nor is it resolved in a vacuum. To each, the sacrifice can be minor, or it can be costly. In thinking about the implications of eating whales, I cannot help but lump all of this into the frame: both how eating configures a symbolic relationship to nature, and how nature itself is changed by what we collectively choose to dine out on.

With these thoughts in mind, I packed up and walked over to a friend's house for dinner. The others around the table ate heartily, but I scarcely nibbled. The image of a whale's tongue tip, resting on a plate, was still with me: a vision of someone bringing chopsticks to their mouth and biting into the whale tongue, its familiar texture so much like the gristle of their own tongue, *and no pain coming*. "Are you well?" the host asked, his hand on mine when the dishes and pots were stacked in the sink and we had all flopped down on the sofas.

"Yes," I said, "I'm just stuck on thinking about the worst thing you could possibly eat, and if you were forced to, would you eat it very fast, or slow."

The conversation turned to how you should eat the worst things to eat and what they were. Some were very funny, and nearly all were meat. Sambal stingray, serpent soup. A whole lamb's head, appearing from beneath a flourished cloche. Sweetbreads, *animelles*, tripe: pseudonyms for offal (parts that "fall off" a carcass hung in a cold room). A shudder telegraphed around the room when someone said, "'Gator and cooter' in Florida." I mentioned the zebra meat my sister had seen basting in a bain-marie in the Netherlands—oh, no, it was skinless. Incredulity met the claim that fast-food burgers contain mince from up to a hundred individual cows and "laboratory tests have shown"—scare quotes curled midair here—"it's a mix of different breeds." One friend, a filmmaker and actor, then leaned forward to tell the story of how he had thought to play it safe at a lavish breakfast in the Philippines by refusing every dish offered, other than an innocuous-looking broth, which he ate smilingly, with a metal spoon. At the end of the meal the cook expressed great pleasure that he had so enjoyed his *bowl of blood*.

When everyone had wiped the tears of laughter from their eyes and poured the last of the wine, I talked about how the Tudors used whale ingredients in cordials, biscuits, and something called posset: a custard drink. Their version of a doughnut got fried in pilot whale fat. Further back, during the Middle Ages, a whale functioned as an apothecary in Europe—large whales' intestinal secretions were used to ward off epilepsy and headaches; superstitious chemists powdered narwhal horns as a remedy for melancholy, and an antidote to poison. Famous kings had table utensils and apotropaic drinking vessels whittled from the horns, hoping to scotch would-be assassins. During the Black Death pandemic, ambergris from the hindgut of sperm whales was used as a foil to the pestilent air that was supposed to be the origin of infection. Ambergris ("floating gold," as it is still colloquially known) is a rare and expensive gummy substance that forms around a nub of chitin—undigested squid beaks—bonded with a claylike ooze discharged by the whale's entrails, to prevent these sharp stubs shredding its innards. Its use as incense, a perfume fixative, an aphrodisiac, and in cookery continues today. Sperm whales tend to vomit up the residual mouthparts of their prey periodically: only a few (1 percent, biologists estimate) produce compacted ambergris, as a result of the beaks migrating too far along their digestive tracts to be regurgitated (in which case the marbled, pumice-like lumps are either defecated, or rise to the surface after the whale dies and its body decomposes). Expert noses have compared the strong and not unpleasant smell of ambergris to "the wood in old churches" (a penitent place, inside the whale).

"That's all old news," I continued—on a jag now, having spent the past few days reading up on the history of whale eating. Dolphin fillet appears in recipes from the court of Richard II in the fourteenth century, but even deep into the nineteenth century, the painter Henri de Toulouse-Lautrec included stewed porpoise in his personal cookbook. During World War II, the Americans encouraged eating whale as a substitute for beef and mutton, prioritized for military rations. In 1918, a formal dinner held inside the Museum of Natural History showcased the many ways a whale could be served. ("Notables Try the Feast"— this was how the night was reported.) The "deep-sea pie" on the menu, guests were reassured, contained humpback meat; the species of whale most frequently captured off the Pacific coast. "Not such a whale, but a part thereof, was consumed," the newspaper explained (lest any reader had assumed a dead humpback, intact and entire, had been laid out on a high table for the aristocracy to attack with knives and forks). Tinned and smoked whale meat remained available to purchase in America at least until 1973—in the gourmet section of Macy's in New York, among other places. By then whale meat had already become a distasteful gimmick in showboat dinners that gave pride of place to Jell-O salad and canned pineapple.

"Out of all the animals, I would never, ever, eat a whale," said the person who had, minutes before, praised snake soup for its medicinal qualities. An otherwise dauntless omnivore, she punctuated each word with a tap of her ring on the glass she held. "They're just too intelligent," she observed. Everyone nodded. That whales were, in some sense, known to be like human beings—smart, social, communicative, *songful*—made eating them, tacitly, cannibalistic: this was the main reason, people agreed. Plus, killing a whale was extreme. Possessed of central nervous systems and brains partitioned into hemispheres, as whales were, it wasn't difficult to comprehend that when they hurt, it was in many of the same ways humans underwent pain. Their suffering was imaginable: the biggest animals surely died in the biggest agonies.

It had grown late in the cozy living room. No one yet seemed ready to depart. The wind jostled the window frames, and, slouching in from the dark, the tendrils of a climbing vine I know as mile-a-minute weed brushed against the glass, its leaves the shape of ears. Someone soberly pointed out that, before refrigeration in early history, a whale brought down by hunters would have had to have been shared, if it was not to go to waste. Which might more readily inculcate whale meat to ritual; the allocation of its parts reinforcing social hierarchies and forging new alliances in a community. Who got the professedly tasty underbelly, and who got the tasteless fluke? Wasn't that a way to arbitrate *power*? So wouldn't a whale, therefore, get to be charismatic because, via its abundant flesh, social power was revealed and visibly distributed?

The host sniffed at this, and weighed in to point out that some cultures consume animals on precisely the same grounds that we had, just now, turned our noses up at the meat of an intelligent creature, like the whale. Animals with humanoid qualities (cunning, virility) were eaten elsewhere because eaters wanted to take on those qualities for themselves. Eating could be a way to acquire "essence," he said, gesturing inexplicably as though settling a crown onto his head. The silence that followed was broken by a woman who agreed, and said her Chinese mother took a daily spoonful of "Tasmanian squalene"—oil of blue shark—because she believed sharks were voracious and never stopped being hungry. Late in life, her mother had lost her appetite: she thought imbibing a shark elixir would return it back to her.

I hadn't contemplated this before, but it occurred to me then that rather than being relegated to some lowly creaturehood to be classed as "food," an animal's symbolism—its charisma—might arise from, or justify, it being classed as edible. If charisma was a source of reverence, reverence didn't always lead to protection: sometimes, it warranted consumption.

"You won't eat whale in Japan, though, will you?" asked the actor. "Because, of course, you're a vegetarian," someone else rushed in to complete his thought. True then; true now. But, for the briefest moment in the company of several Japanese whalers—stood in dockside shade cast by the world's last factory-whaling ship—that conviction was to become the lie of a diplomatic, and compromising, mouthful.

Today whale meat is eaten around the world, not only in Japan. If you know where to look, you can discover whale on plates in the chilly cities and settlements of Iceland, Norway, Greenland, and the Faroe Islands in Denmark, or pinched at the tips of a *sujeo* set in Ulsan, South Korea. When the late LA food critic Jonathan Gold ate whale—tossed lightly in sesame oil with slivered Korean pear—he called it "leaner than beef, with a rich, mineral taste and a haunting, almost waxy aftertaste." The *New Yorker*'s Dana Goodyear described the raw whale she ate in Iceland as having "an oily flavor that recalled the smell of a burnt wick in a hurricane lamp." In an attempt to reverse a declining trend and grow youth markets, Norwegian companies have sold whale in paper cones at rock concerts and donated excess minke, caught off that country's coastline, to charities for those in need. With low demand in the consumer market, whaling entrepreneurs in Nordic countries have fostered innovation in cosmetics, supplements, nutraceuticals, and medicines. It is possible to buy whale-meat protein powder and whale-oil skin creams, made from the by-products of whaling, in Norway, though scarcely anyone does. Research underway in an Icelandic university aims to develop anti-anemia medications from the iron-rich stuff of whale bodies: a new frontier for whale cures.

Elsewhere whales are hunted for food, in observance of cultural tradition, and using ceremonial and mechanized weapons, in Alaska and Canada, in the Russian Federation, near to the Atlantic islands of Saint Vincent and the Grenadines, and off Indonesian islands including Lembata and Solor. More nations, including the Solomon Islands, Peru, and China, can be added to this list of where people can eat whales, if smaller cetaceans—dolphin species—are counted. In some countries, including South Korea, it is still legal to take, and sell for consumption, whales that have died as bycatch in fishing operations, or which have stranded. (Slender evidence can confirm whether a whale has been incidentally caught or actively pursued, after it has been conveyed to a port. A single animal has been known to net USD$85,000—"ocean lotto," Korean fishermen are said to call the meat.)

Nearly everywhere it is possible to eat whale meat, the quantity consumed is negligible. Typically a seasonal inclusion in the diet, whale may be eaten for a mere handful of months, or by a trivial number of individuals on the occasion of a festival. In Iceland only 3.2 percent of the population report eating whale meat on a regular basis (defined as six times a year or more), and even fewer people (1.7 percent) eat whale at least once a month. That number is slightly higher in Norway; and Norway and Iceland both export whale meat to Japan. Perhaps only on isolated and very icy islands, with barren soil or no soil, and hard weather, is whale meat a staple. These are places small human populations reside, or move across, carrying, among their supplies, strips of dried whale skin, fermented nobbles of fat, or raw *muktuk* (cubed bits of skin and blubber). Their degree of physical exposure, and the elemental privation of their livelihoods, might vindicate this resourceful consumption of whale meat: though, too, it is one context in which people can end up dependent on polluted blubber as a cornerstone of their diet.

Japan is among the last remaining countries where it is possible to consume cetacean meat products all year round, and, though few Japanese people eat whale routinely, many more than those who do, insist upon their right to do so. Polling results offer a clouded picture. A survey conducted by the Japanese newspaper *Asahi Shimbun* over a decade ago found 56 percent of respondents supported eating whale meat, and 26 percent opposed it. Four years later, a Nippon Research Center survey, commissioned by the International Fund for Animal Welfare (IFAW), showed widespread indifference to whaling: 54.7 percent polled had no strong opinion—pro- or anti-whaling—though those who did support whaling (27 percent, and 11 percent strongly) were broadly against subsidizing the industry. A study of more than five hundred fifteen- to twenty-six-year-olds in Japan in 2007 found approval of whale-meat consumption, and demonstrable "anti-anti-whaling" sentiment: the sense that Western interference in Japanese culinary traditions and fisheries practices amounts to an imposition of cultural imperialism. Finally, the Nippon/IFAW poll also projected that less than 11 percent of the population bought whale meat in 2012. The quantity of whale purchased in Japan is very low in comparison to other sources of meat. Per capita, annual whale meat consumption is around forty grams per person—equivalent to one slice of ham.

The amount of whale meat harvested by the Japanese surpasses domestic need, but unlike the excess gathered by the Russian whalers of the 1970s the surplus is not, by and large, diverted to secondary markets in animal feed. Instead, the preponderance of whale meat is stored, uneaten. Numbers are, again, hard to come by, but in 2012 the *Japan Times* reported that around five thousand tonnes of frozen whale was then being held in refrigeration in Japan, in industrial warehouses. In 2019, Junko Sakuma, a researcher at Rikkyo University, estimated the stockpile at 3,700 tonnes. Where once, after World War II, whale was a cheap substitute meat found in school lunches, the contemporary availability of more affordable animal protein in Japan has meant that whale now belongs to "a very narrow niche market, like horse meat, or wild game." (There are echoes here, of decades past, in which whale oil was upgraded from being an ingredient in margarine and soap, to use in luxury goods.) A superior cut of whale meat costs around ¥2,300 Japanese Yen (JPY) (USD$22) for around three and a half ounces, making it far more expensive than other high-end meats in the Japanese market—for example, Matsusaka steak.

Marketing whale meat to a new generation of Japanese consumers, attempts have been made to quantify its environmental impact. Analysis published in the *Sankei Shimbun* newspaper in 2009 posited that the amount of carbon dioxide produced by procuring a kilogram (2.2 pounds) of whale meat, even accounting for the long distances then traveled to obtain it, amounted to less than a tenth of that which was created in the production of the same weight of beef.

At the time that I visited Japan, the nation remained engaged in its Southern Ocean scientific whaling program; an annual undertaking that began in 1987, after the IWC moratorium came into effect, and which continued—seemingly ruggedized against international condemnation—until 2019, when the Antarctic hunt was abandoned in light of Japan's renewed commitment to commercial whaling in territorial Pacific waters. For the three decades prior to Japan's withdrawal from the IWC in 2019, their government capitalized on a provision of the commission's charter that allowed the Japanese Institute of Cetacean Research (ICR) to fulfil catch quotas, granted for the purposes of science. The IWC permitted any whale meat unutilized in testing to enter the domestic market, to be sold for food.

It is inaccurate to say that Japanese whaling science during this period was an outright misnomer. There was nothing covert about Southern Ocean whaling and its purpose—it was, in the broadest sense, "scientific." What *is* true is that the research undertaken was never intended, as a top-line outcome, to improve human understanding of cetacean biology or behavior, or to advance the science of polar ecology. The Japanese analysis was aimed at verifying the population structure and generational dynamics of Antarctic whales—largely, the animals' reproductivity, and their age and sex distribution—so as to provide a defensible basis for future whaling. Its aim was resource management research, as opposed to conservation science.

The species harvested by the Japanese in the Southern Ocean were, for the most part, minke. Trim, narrow whales with baleen, minke have pointed heads and light-gray saddle markings. They are found in two distinct species: the northern (common) minke, seen in the North Atlantic and North Pacific; and those in the waters of the far south, called Antarctic minke. The name, with roots in Norwegian, is believed (perhaps apocryphally) to be a verb: "to minke," to shrink, to contract. Antarctic minke were late targets for international pelagic whaling prior to the IWC moratorium, but as bigger and oilier whale species crashed, and as nations hunting whales for food and extractive products ramped up their technologically advanced operations, these quick, compact whales became expedient quarry. Close to one hundred thousand minke were taken by pelagic whaling expeditions in the Antarctic from 1972 to 1987. Under the remit of scientific research since then, the Japanese have harvested slightly more than eleven thousand.

The highest value components of a minke whale, to Japanese scientists, are its ears, followed by its eyes. The whale's ear channels—sealed off on the exterior of the animal's head, and compacted into the skull—contain waxy plugs that, in the way of a tree trunk's rings, attest to the whale's age. A cetacean earplug looks like a celery stick, long forgotten in the bottom of the fridge crisper: mucky, clotted, and fibrous. Magnified, its layers chronicle periods of growth, as well as intervals of feeding and, in migratory species, recurrent oceanic travel; a darker or lighter lattice is laid down in roughly six-month increments. From an analysis of this substance, scientists can infer how old the whale was at the time of its death. Though any whale may be identified as either mature or a juvenile via other methods, determining its exact age requires the excision of earplugs; or else, its eyes. Eyes, like ears, are clocks. The whales' ocular lenses are also gathered at sea. The proteins in the whale's eyeball, to quote the paleobiologist Nick Pyenson, are "bottled off from the body's circulation," and so the rate of change in these proteins—unreplenished by any process or source outside the eye—can also be analyzed to determine a minke whale's age.

Japanese scientists aren't the only experts interested in what can be learned from whale ears. Researchers at the Smithsonian have called whale earplugs "oceanic core samples," because the organic wax records chemical information and can be assayed to impart a history of contaminants (pesticides and heavy metals) that the whale has been exposed to in its lifetime, as well as periods of high physical stress—registered as internal peaks of cortisol—which are perhaps attributable to an animal experiencing noise pollution or the depletion of its prey. Whales' ears are a plastic document. Biologists say these animal waxes are dossiers not only on whales individually, but on the marine worlds they inhabit.

It may seem, in some ways, poetic, that the physical apparatuses of the whale's senses—how it sees, how it hears—are also the whale's inbuilt chronometers, testifying to its age. No memory resides in the eye, that we know of. Nonetheless, the eye fills with time. The ears chronicle how the world outside, migrates inside, into the whale. The whale's ears are recording devices. As living cells die and are replaced, these two interfaces collect.

Yet the data extracted from whales' sense organs by Japanese scientists (which exclusively told of the minke whales' age, not their exposure to pollutants) was only necessary to obtain *because* some of the whale's kind were destined to be killed for food. The logic of Japanese scientific whaling hinged on meat provision; but it went further than satisfying underlying demand. Commentators have argued it is several orders more reproachable and wasteful to hunt whales in this way, for scientific sampling, than outright for meat. Rather than being responsive to consumer purchasing, the number of whales sought in the Southern Ocean needed to be declared ahead of time (sometimes years ahead), and to accord with the purpose of procuring a representative sample of the overall population—often this meant taking a greater number of whales than matched the appetite for whale meat back in Japan. Because the whales' heads were to be preserved, to test their ears and eyes, these animals had to be harpooned elsewhere, in the body. How long it took for each whale to die was not recorded—the whalers shot them with penthrite grenades on the harpoon tips: whale bombs, designed for the purpose. (David Attenborough: "The hard scientific, dispassionate evidence [is] that there is no humane way to kill a whale at sea.") Down in the Southern Ocean the animals' deaths were more painful—and more whales died—in service of Japanese research than might have been otherwise, had whaling's objectives been purely mercantile.

The Japanese Antarctic whaling program experienced a short-lived interruption in 2014, when a ruling of the International Court of Justice (ICJ), triggered by Australian proceedings, declared the methodology of the ICR insufficiently scientific (nearly thirty-six hundred minke whales were killed by Japanese whalers and examined by Japanese marine scientists between 2005 and 2014, leading to just two peer-reviewed scientific papers). The Japanese government overcame impediments flagged by the international court swiftly, by redrawing the whaling area and the catch criteria, reducing the permits granted, and widening the purposes of tissue analysis: by 2015, the hunt was recommenced.

During the ICJ hearings Japanese representatives pointed out that the expectation whale science ought to be underpinned by environmental values, rather than fisheries management goals, arose from spurious and imperial ideas of animal charisma—a divergence in what a whale was meant to be *for*. "Environmental values" translated, to the Japanese advocates, as emotion, not science. The emotion came out of culture, but it was also manufactured by the anti-whaling movements of the 1980s. Their legacy was to ensure all whales were regarded as sacred, regardless of whether whale numbers rose to a level that made a return to hunting a sustainable prospect. This disparity would become more clear to me when Japan's former chief IWC negotiator in Tokyo, Masayuki Komatsu, referred to a whale not as an animal per se, but as "a future." If whales were, from the standpoint of modern environmentalism, a narrative as big as the world and a motif of shared responsibilities; and if whales, to whale watchers, were emblematic of the oceanic sublime; they were instead, according to Komatsu, representative not of *a place*, but of *a time* in Japan. That time was yet to come. At least in bureaucratic circles, it surprised me to learn that whaling was considered a futuristic—if small-scale—industry, rather than an anachronistic one.

Komatsu is otherwise famous for referring to minke whales, in 2000, as "the cockroaches of the sea." He says he only meant to imply they were abundant, though of course, he is not oblivious to the specific charisma of the roach; a pest adaptable to junk, indestructible.

Several conditions primed Japan's retreat from their Antarctic whaling program in 2019: technological, mechanical, domestic, and international. Where technological innovation in the twentieth century drove industrial whaling to more destructive ends, big tech was a factor in diminishing it in the twenty-first century. By 2014, both Amazon and the Japanese e-commerce site Rakuten had banned the sale of dolphin and whale products via their virtual storefronts, limiting online sales in response to a series of reports from the UK-based Environmental Investigation Agency (EIA) showing both the ready availability, and the widespread mislabeling of, vacuum-sealed and canned cetacean products from porpoises, fin whales, Baird's beaked whales, and minke. This both closed a distribution network for whale meat, and branded it, in a broader sense, a blacklisted product.

Other reasons for the collapse of scientific whaling in the Southern Ocean were less virtual, and more rudimentary. Japan's famous factory-whaling ship, the 8,145-ton *MV Nisshin Maru*, first launched in 1987, and though it was remodeled for the processing and storage of whale carcasses in 1991, the vessel had become potentially hazardous in Antarctic waters. Since 2011, the ship had contravened International Maritime Organization (IMO) rules, which limit the use of heavy fuel oil—ruinous, if ever spilt into the polar ecosystem—below sixty degrees south, as well as the dumping of waste at sea (some estimates put the discharge of offal, bones, tumors, and other inedible and unusable whale parts from the factory ship at around 40 percent of the total catch). The *Nisshin Maru*, from the standpoint of the Japanese government, would do better to be redirected to whaling within Japan's territorial waters and its exclusive economic zone—not only because of reduced risk, but also as it would be cheaper. The government had been subsidizing the whaling industry by around USD$46 million annually.

A return to commercial, localized whaling with a more modest haul would reduce that contribution—while continuing to express, to the international community, a staunch refusal to be dictated to.

The nail in the proverbial coffin of Japanese Antarctic whaling was that the whales themselves were deemed to be imperiled. In 2018, the IUCN revisited its assessment of both common minke (listed as "of least concern," being among the most abundant whales) and Antarctic minke, moving the latter from the categorization "data deficient" to "near threatened." Though the IUCN acknowledged a lack of confidence in the total population estimate for Antarctic minke, given that these whales live in habitats that are hard to monitor and increasingly vulnerable to climate change, the union erred on the side of caution to flag the minke as close to being at risk—meaning that the animals the Japanese hunted in Antarctic seas were now, at least officially, members of a population with a higher profile in circles that arbitrate endangerment.

That same year, the Japanese Southern Ocean whaling program was widely castigated in Western media when 120 of the 333 minke whales the Japanese killed over the summer months were identified by their researchers as pregnant. The sample of minke taken from the Southern Ocean was designed to provide a cross-section of population structure, and cetacean health (within certain guidelines: the Japanese forbade taking young calves, for example), and so it came as a surprise that so many in the haul should be female, carrying fetal young. Activists were aghast to see these expectant whales killed—though, until very recently, biologists have had no benign or reliable way to give a wild whale a pregnancy test (even telling female from male minke whales is difficult).

Many people in Australia and elsewhere held that Japanese scientists should have, somehow, exercised greater selectivity to discern pregnant females and exclude them from the haul. Skeptical observers believed that the high numbers of pregnant minke killed that year exposed a cruel flaw in Japan's survey design: these females are likely slower moving and more averse to ice-pack waters, so they were easier to harpoon. More damning commentary yet speculated that the 2017–2018 catch represented a deliberate attempt to gather data that amplified the perceived fecundity of the Antarctic minke population—giving grist to the argument that the whales were breeding at such a rate that a return to commercial whaling in the region would prove low impact.

Japan's decision to regroup around commercial whaling does not represent a policy change, though it does mean a shift in the whaling area. In 2019, the commercial-catch quotas granted by the Japanese government amounted to fifty-two minke whales, 150 Bryde's whales, and twenty-five sei whales, caught in waters more proximate to Japan. At auctions on the northern coast of Japan the meat achieved "celebration prices," several times higher than the going rate for frozen or preserved pieces of Antarctic minke, sold wholesale. The entry into the market of this new category of whale meat—a fresher product—effectively created a new criteria for valuing it into the future.

Before I dropped in to see Mio Bryce, the head of Japanese studies at Macquarie University in Sydney, to talk about the politics of eating whale, I heard rumors about her office. The academics in the faculty were in the process of clearing out their workspaces—the old multistory arts office block was due to be demolished—and Bryce's office, the facilities' manager acidly confessed, was his greatest challenge. When I entered the room, I could see why. Her many narrow shelves were stacked with hundreds, perhaps thousands, of rainbow-colored manga comics; floor to ceiling, two rows deep, some books packed in horizontal blocks atop of those lined up vertically. Bryce's workspace had become, over many semesters, a shrinking cube of lamplit air. It felt like walking into a three-dimensional smartphone game.

Incidentally then, we began by talking about manga characters and the loanword *kawaii*, a particular Japanese aesthetic mixing cuteness with extreme vulnerability. "The eyes have become too big, Rebecca," Bryce said, showing me illustrations—large-eyed figures with dots for pupils. "Actually, they don't show anything. Before, even the *robot* eyes showed emotion. I want to collect different manga eyes, because these recent ones have become different." She sighed and slotted the volume into a transparent sleeve. "This is probably related to how people relate to each other, because today, it is an internet world," she said. "The eyes"—here she brought her fingers up to rest on her cheekbones, in a "y" shape—"are not in contact." I thought of the Japanese whalers collecting Antarctic whale eyes for their research, and then of my own experience, coming eye to eye with a humpback but being mystified by the animal's emotional inner life. I mentioned the dolphin selfies from Buenos Aires, and the idea of cute aggression.

"I don't think dolphin and whale are the same thing," Bryce exclaimed, pointing out that because people encountered dolphins in the zoo, they felt attached to them in ways they didn't when it came to whales. Dolphins were *iruka*, and very cute. "In picture books, they become family," she said. "If I was told *this is whale meat*, or *this is dolphin meat*, to me, it is a totally different thing."

She meant that eating a dolphin was worse, yet the animal ethic in Japan, Bryce wanted to impress on me, was an ethic informed by Shinto Buddhism: to think of all life as equal, from the whale down to the sardine. Biophilia? Well, better to kill one whale than many, many other sea creatures to satisfy the appetites of the same number of people, seen from this perspective, she said. Bryce recalled temples dedicated to dead whales in Japan, where hunted whales had been given their own names posthumously, and those names were recorded alongside village people who had passed away. In Nagato stood a stone memorial for whale *fetuses* even, an engraved plinth, orientated toward the sea. When a whale was killed, it was important and respectful to use all of it: unlike the Americans, who, for so long, had flensed the whales of their exterior blubber and wastefully abandoned the rest.

Perhaps most importantly, Bryce went on to remind me, whale meat has long been a symbol of historical resilience to the Japanese people. Toward the tail end of World War II and into the 1960s, Japan experienced a food crisis born of the wartime decimation of supply chains and the Japanese agricultural sector. The ecological conditions memorialized in this postwar national cuisine were the aftermath of conflict, when Japan—not irrelevantly, an island nation with a narrow land-base—had to establish a food-security policy less reliant on importation, at the same time as farming lands were in disrepair. US overseer general Douglas MacArthur urged recommencing Antarctic whaling, not only for nutritional reasons but also to retrofit and repurpose Japanese naval vessels, decommissioned as per terms of surrender. The Japanese people were starving and crippled by vitamin deficiencies; whale meat served in elementary and middle schools helped bring young bodies back to health. Though, in time, the Japanese turned away from this substitute meat, its association with the ideals of self-reliance, and restored pride, have never entirely been dropped.

A narrative that establishes whale meat as a mainstay of Japanese cuisine further back in history has sometimes been extended to encompass the entire country, but though there are limited sites dotted around Japan in which small-scale whaling has long supplemented local fisheries, whale meat was never a national dish of tradition prior to World War II. Indeed, a number of coastal communities historically refrained from killing whales because cetaceans were held to signal good fortune. The whale's charisma in Japan has not been historically or geographically consistent—yet as part of a national story, its power dates back only a matter of generations.

But why does Japan *continue* to whale? I put it to Bryce. Are the whale-meat reserves meant as a fail-safe for lean times to come? Or, as one activist had previously suggested to me, was whaling the defensible "outside peg" in a broader Japanese policy setting toward wildlife, extending to include the trade of exotic pets, ivory used in luxury name stamps, and the overfishing of endangered tuna? By this measure, whaling was regarded as a "stalking horse" that diverted attention from other exploitative areas of wildlife commerce: if Japan held out on a willingness to be negotiated down from whaling, as the worst violation of global norms dealing with wild animals, it might provide cover for lesser, but still censured, industries. Bryce doubted both these explanations, and though she said they were, to differing measures, plausible, she favored instead the likelihood that the continuation of whaling was, in fundamentals, an assertion of the Japanese government's right to configure fisheries and food security with no deference to international oversight. Where once whaling had been symbolic of Japan's self-reliance, it had evolved to represent resistance, and the imperviousness of a cultural hegemony that would see it quashed.

At the end of our conversation, I asked Bryce how she had felt about what Australians ate when she first relocated to Sydney. She described the "gruesome" meat aisles and deli counters of her local supermarket. "Whole chicken, roast chicken, you won't see in Japan," she said, splaying her hands with disbelief. "To me, it's really gruesome. Sometimes I see a rabbit, it's a rabbit shape. Lamb meat with bones. Cutlet, that was . . . *horror*. The shape of the animal is still there!" I asked about seafood, wondering if the frame of these animals, these meats, seemed repellent because it suggested the food was, in some way, unprocessed; that the sheep, rabbit, or chicken hadn't been properly disassembled and cleaned. But, as our time was running out, Bryce deferred the question with a different anecdote: "My husband still can't eat shrimp," she said with a shiver, recalling the Australian summertime custom of serving up barbecued prawns. "Lots of eyes, their tiny, tiny eyes. It's just a bowl of eyes."

Having arrived in Tokyo, I am headed by bullet train to Shimonoseki, a port city at the westernmost extremity of Japan's Yamaguchi province, on the southwestern tip of Honshu: a place famous for political power brokering and a ceremonial dish of poisonous pufferfish. For a single day in April the *Yūshin Maru* and the *Yūshin Maru II*—both harpoon vessels—are in port and open to the public, alongside the immense Japanese factory ship, the *Nisshin Maru*. Shimonoseki is the home dock for these three whaling vessels in Japan, where the ships are repaired and repainted with anti-fouling lacquer over the northern hemisphere summer. The upkeep is needed to mend damage rendered by the high seas of the Southern Ocean, the wear and tear of hunting large whales, and—very rarely—because of scuffing interventions staged by the international activist group Sea Shepherd. The Japanese whalers, crew, cetacean scientists, and a retinue of government officials are scheduled to parade off the ships and gather in a flapping canvas tent to acclaim their achievements throughout the whaling season.

The event has not been widely promoted. These homecomings were once flashy celebrations. Before 2011 (the year of the Tōhoku earthquake, the tsunami, and the Fukushima nuclear incident), the fleet's arrival was attended by many local families: children bobbed beneath whale-shaped balloons and were dotted with twinkling minke stickers. On the forecourt snacks of whale meat were offered on toothpicks, and whale pikelets, *okonomiyaki*-style, were sold from carts. But after the natural disaster and its cascading impacts on infrastructure, official events all across the country dialed down their spirit of gala and pageantry. The tenor of the welcome-home ceremony will be dampened, too, because, according to a statement released by the Japanese Fisheries Agency, this whaling season has been Japan's least successful since 1987. The ships have returned, this year, with a haul of only 103 Antarctic minke whales.

Along the way to Shimonoseki sunlight falls in long columns to the green ground, as if from trapdoors flung open in some distant ceiling. It is the tapering of sakura, the cherry and plum blossom season. The last of the pinking trees streak past, sheltered in hillside gullies. Transience is the spiritual theme of spring in Japan. A dozen passengers gather around a handheld radio: a baseball game is being broadcast, they tell me when I ask. Then the stadium flashes by the window, a gulp of green and red between buildings. The Taiyō Whales used to be the local team. Now they're known as the Yokohama Baystars because, in the early '90s, superstitious fans petitioned to change the name after coming to believe that dead whales, sold by the team's parent company—a seafood corporation—had cursed their success.

Disembarking into the train terminal after six and a half hours of sprint travel, I buy a can of coffee from a vending machine and am startled to find it burning hot when it rolls into the slot. I board a public bus to the waterfront. I've come to Shimonoseki without assurances of an audience with the whalers. Such undertakings take a long lead time to negotiate, often using intermediaries, and they can be especially hard to obtain from overseas. The journalist Mark Willacy, the ABC's North Asia bureau chief for over five years, tells me that when Australian reporters have been permitted to ask questions of working whalers, it's been on the nonnegotiable proviso that they're anonymously silhouetted behind screens. The whalers fear reprisals, hate mail, and trolling. All my requests have, thus far, been deferred to processing.

The bus curves through an architecturally eclectic city. We pass a grand sandstone-colored hall near the station with baroque cupolas and pediments, a mansard roof, and lower down, pillars like an Italian palazzo. The Kaikyo Yume Tower—nearly five hundred feet tall, and capped with a spherical viewing chamber—presides above. There are oversize, galvanized puffer fish pouting and blowing their cheeks out on top of every phone booth. I am looking for a landmark; the Kaikyokan Marine Science Museum, near to the dock, which is run by a man who was once a senior bureaucrat in the ICR. The museum houses multiple whale skeletons, as well as an entire floor dedicated to "Balloonfish of the World." The bus turns toward the water near a glass-fronted edifice that scatters a lacework of reflected sun into the street around it. A sign above the automatic doors calls the building "Dream Ship." I assume it is the office of a pleasure-cruise company (in fact, it is a civic services center). Still mildly stupefied from the train journey, I lift my head from jotting *Dream Ship?* in my notes, to glimpse, pulled up at the end of the street, the soaring hull of the *Nisshin Maru*.

The *Nisshin Maru*, more than four hundred feet in length, is the world's last factory whaling ship—for now. Japan's Fisheries Agency is thought to be constructing at least one more to replace or supplement its work in the future. *Maru*, a suffix commonly appended to the names of Japanese merchant ships, means "circle." The word signals both the ship's return to home, and a history of ships depicted as water castles; defensively encircled. The *Nisshin Maru* is black hulled with square, yellow masts rising over it, thick as twin lighthouses. In dock, it looks less like a factory and more like a fortress. Minor scrapes and gouges in the paintwork are the work of icebergs, not thwarted assailants. Sea Shepherd's far smaller ships do not make contact directly with the *Nisshin Maru*, though they're known to get between the factory ship and refueling vessels—and, in the past, seaborne activists have thrown stink bombs of butyric acid at its decks. Globular telecommunications units sprout from the rigging of the *Nisshin Maru*, and water cannons defend the prow. The English word "RESEARCH" in six-foot capital letters runs down its portside, along the line of the wharf. Momentarily I glimpse the whalers on deck wearing bright hard hats and dark jackets. Then the bus passes under a walkway and the ship slips behind a line of waterfront buildings.

I'm the first to step off. It rains in Shimonoseki not as it rains in Tokyo—not sudden and heavy like loquats shaken from a bough. It needles. Drops spike the pavement through the empty air, forming no puddles. The sky above is white and motionless. The Kanmon Strait is black. There is no haze here, no soft shifts of seawater up into sky. The world is as sharp and unitary as the gestural ink of the Japanese *hitsuzendo* artists who line the steps of Ueno Park.

In the ionizing air, a brass band honks through warm-up scales in a marquee on the forecourt. The two smaller ships—the *Yūshin Maru* and *Yūshin Maru II*—are moored ahead of the *Nisshin Maru*. Immaculate in blue and white, the word that swims up to describe them is that childish, naval adjective "jaunty." Up close, a businesslike fierceness overtakes that sentiment. They're little battering rams compared to the *Nisshin Maru*, but still six or seven times as long as a minke whale. There's no sign of the harpoons they're purposed for, but the sharp, yawing bows and array of radars clearly evince their function: fast and guileful ships. The smart colors, I realize, are not some pretense to the sanctioned marine authority of military ships. They're painted so as to disappear. Under the gunmetal sky here in Shimonoseki the two *Yūshin Marus* stand out, but against a backdrop of ice and sleet the high white stalks above their bridges would be invisible and their low bows might seem a rise of ocean. It would be possible to be almost upon them, I think, before picking them out.

On the upper deck of the *Yūshin Maru II*, I can see a family walking around, guided by a man in a highly decorated uniform. Bowing and smiling with the seaman whose responsibility it is to secure the plankbridge leading onto the ship, I ask permission to look around, and I walk aboard. The quarters inside are compact and surprisingly disorderly given the shipshape, pristine exterior. Life vests, logbooks, cable ties, bundled clothes. Every nook is wedged with equipment, as if the crew had grown slovenly on the journey home and in their haste to disembark, abandoned the jumble for on-shore staff to straighten out. Headed toward the black screens in the bridge I meet another sailor in the low hall and introduce myself. Would it be possible, I ask, to see the harpoons? I'm thinking about the whale in Perth; how delicate the methods and negotiations around its euthanasia. He gestures to a woman dressed in civilian clothes behind me, and there I find myself penned into the narrow corridor between them. What follows is an interrogation, but without express intimidation. Who am I with? I'm with myself. Who else? Myself, a writer. This is my publisher's name. No, no green thing, no group. Yes, Australian. It's all convivial actually, though I'm aware that someone has closed the door to the bridge in a hurry.

I'm told I need the captain's permission to be aboard, but the captain is now leaving and the *Yūshin Maru II* will be closed to the public because a ceremony is about to begin. I explain my project, give my name, and offer my business card flat, held out with both hands. There's an odd moment of jammed social etiquette when both people reach for, but then refuse to offer, their own cards. It's a *faux pas* in Japanese culture, not to exchange *meishi*, and I see a fleeting opportunity to take advantage of their hesitant sheepishness, pressing the point that I'm here to interview the whalers, the ship captains, and to see the ceremony. This is rude of me, and though I try not to be insistent or pugnacious, I come off, I'm sure, as presumptuous. The first night I arrived in Japan, sharing a drink with translators and a handful of students, I was assured by all: *You won't get them to talk to you.*

We disembark the ship together, stepping over ropes, and walking toward the marquee, still talking. It's begun to rain, and I forget the compact umbrella in my bag to take refuge in the shared space offered by the woman from the *Yūshin Maru II* beneath hers. She tells me her first name, and says I can't formally interview her because she knows nothing useful to me. But we strike up a companionable discussion all the same, and when I take notes she sometimes spells a word or two for me. Her English is terrific: she was educated in Canada and the United States. She's a marine scientist who worked for five years on the *Nisshin Maru*, on whales. Now she works with fish stocks—sardines and mackerel mostly. Gesturing behind us, she refers to the *Nisshin* as "this ship" ("Now, could you have gone aboard this ship? No."). A small quirk of English—*this* ship *here*, or *that* ship *there*—that reminds me how subtly a language implies a perimeter of intimacy between object and speaker. She won't answer any of my questions about whales, but we find common ground discussing women in the sciences, science education in Japan, long sea journeys, and her visits to Australia. Informally, she's become my escort, and the other sailor drops away. With one eyebrow cocked wryly, the scientist recounts that the last time she was in Tasmania, she bought a kangaroo skin. *It wasn't even expensive*, she says.

When the band set to with a munificently brassy number, the whalers come marching down from the *Nisshin Maru*. Or rather: they come shuffling down, grinning widely, exchanging a few words with the crowd and ruffling kids' hair. Some smile gladly at the marine scientist next to me, happy to recognize a friend on dry land. The men are leathered by glare and gales, a few have missing teeth. They're fishermen, after all— old salts, not naval officers. It seems enough to ask that they don a uniform. They're wearing black gumboots, real coveralls—some with navy or gray jackets against the wind. They all have carrot-orange hard hats on, and pairs of durable, sky-blue gloves point hollow fingers from their back pockets. The whalers together have the semblance of lunch-break mechanics, scuffing the ground as they pass the time of day, wincing in the weather, chewing chemical-strength gum. But the spattering on their boots and the legs of those coveralls is not engine oil, it's fat. Pearly fat, plashed from dead minke. These men have waded thigh-deep through whale.

As the whalers settle in a row before the gathered crowd, a compere introduces the mayor of Shimonoseki and other notaries. At the urging of their parents, children sit up tall and stop swinging their shoes. Agriculture, Forestry, and Fisheries minister Yoshimasa Hayashi tells the assembled crowd that Sea Shepherd's obstructive pursuits of the whaling ships "amount to piracy. The International Whaling Commission has recognized the legitimacy of Japan's research whaling!" He's just warming up. (In fact, an expert panel of the IWC declared it was unable to determine whether lethal sampling was necessary for Japan to achieve its scientific objectives: less a full-throated approval, than a shrug.) The scientist translates some of this on the hop for me. Generous applause matches the loudening patter of rain on the tent. Flowers are doled out, though it seems that some of the whalers are uncertain about their bouquets, holding them awkwardly upside down, or tucked under an arm.

When the formal proceedings draw to a close, a jubilant mood fills the space. Families, recently reunited, are puffed up by all the talk of a crucial national industry, a patriotic charge. No one seems gloomy about the small haul, and no one's going to sea today. On trestle tables inside the marquee, stacks of tinned and vacuum-sealed whale meat are laid out with a cashbox. The whalers themselves are enthusiastic customers, pushing to the back to buy armfuls. I see one man on the forecourt staggering under a wholesale box of whale meat. His arms are straining but his face is suffused with happiness. A younger man rushes out to help him.

Now it's really starting to rain, it's freezing, and we all huddle in under the tent. With so many bodies hedged in, standing around the muddled chairs, microphones, and cameras, it's a confined environment. I'm standing shoulder to shoulder with an older whaler, and no one has told me yet to leave. Two girls in matching corduroy dresses shyly peek from between the whaler's heavy legs. When I catch their gaze and poke out my tongue, they fall apart in a berserk giggle fit, making faces. Trays of hot soup are being shared around, and the scientist whisks two bowls off, passing one to me, saying something quick to the whaler. He laughs and asks me, twice over, "*Vegan?*"

"*No,*" I reply, "*you speak some English?*"

"*No, no,*" he says, scowling.

The plastic soup bowl is warm in my hands, and the whaler motions with his chopsticks, *Eat?* Tannin broth, tiny mushrooms as if from a Lewis Carroll woodland, scallions (parsley?), spots of clear, fragrant oil.

And minke. A few thin pieces of minke, like autumnal leaves, in the bottom of the bowl. Not tongue meat, I don't think; perhaps a kind of bacon.

"Don't eat it if you don't want it," the scientist says.

The two girls are watching now, each with their own small bowl of whale soup, chewing with their mouths open. I remember the blue whale in the Perth Museum, my first whale, running round and around it; a flash of my sister's baby tooth, balanced on the whale jawbone. The strips of minke are a cocoa, brickish red, marinated in a sauce the shiny color of an antique chair. To stall I ask the scientist about the recipe, on which she will only say, it is traditional. To be clear: it was never my intention to eat whale in Japan. For one thing, my research so far had led me to wonder just how apt for human consumption whale meat is. But the accusation *vegan* seemed to carry more freight than a mere inquiry as to my tastes, and I'm still hopeful of asking a question or two of the whaler now hovering nearby. I make a snap decision.

A rich taste, I get more off the marinade than the meat in one bite. Strong, dry, like duck or aged beef, I think—so much as I am able to summon those flavors anymore. The girls are beaming, wide-eyed. I take one more piece, pushing the sauce off into the soup. There's no hemal buzz to it, no whiff of metal or blood. Was it preserved, before being doused in the marinade? While I'm examining the shred with my tongue, I can feel my brain turning over. The whale in my bowl. The whale in my mouth. The whale swallowed into me. This charismatic, personified creature. And yet, at the same time, I am aware that what is going through my mind in this moment is so far away—astrally far away—from the feelings that the whaler, and these children, experience, eating whale soup. For them, this is a meal that signals a homecoming: a reunion.

Some anthropologists have argued that the start point of anthropomorphization—projecting into animal minds—was hunting animals for food. Zoocentric (animal-centric) sympathies otherwise look like an evolutionary misstep. Human hunters in prehistory had little to gain from feeling remorse over an animal they'd killed and eaten, or from identifying with their prey's fear and pain compassionately. On the other hand, inferring how animal minds worked, how animals perceived their environments, "extrospection" advantaged hunters. Steven Mithen, an archaeologist, has argued that humans developed a theory of their own consciousness alongside an understanding that animals had mental processes and sensory lives that paralleled those of humans. This psychological trait or aptitude might have been passed on, through generations, because it benefited those who possessed it; those people were better hunters. Identifying with the experiences of animals, which can now make eating any animal uncomfortable (perhaps especially an animal that has, historically, been elevated to the status of uniquely intelligent and sociable), may have begun, in the imaginable past, as a psychological propensity in humans who sought to make other species their food.

Hurriedly, almost dropping the wooden chopsticks, I thrust the bowl over to the whaler just to be rid of holding it. "*I apologize, sincerely,*" I say. "*It is not my taste, not my appetite.*" He laughs, and offers the rest to his granddaughters. As they guzzle the way cold kids do, he comments to the marine scientist that these girls will grow up strong.

People no longer eat like animals, and so we look for ways to divorce patriotic sentiment from the consumption of creatures that are considered charismatic, or endangered. But do animals even eat "like animals," today? For some wild animals, their diets, too, have become historically modern: from gray whales shifting to eat seaweed in the absence of their customary prey, to killer whales collecting pesticides in their blubber from the seals, otters, and fish they consume—all the way down to the bottom of the food chain and the innards of amphipods, glittery, like fiberglass, with microscopic slivers of plastic. The "food environment" of even undomesticated animals, inhabiting wilderness, is changing.

Much pelagic plastic comes from packaging associated with food-stuffs, and the chemical ballast whales carry is often the outflow of crop agriculture. Oceanic ecologies are changed by escalating global carbon emissions, a portion of which come from industrial meat production, but food waste—how much food goes *un*eaten, including wasted vegetable produce—is another significant contributor to the devolving climate. Worldwide, roughly 8 percent of global emissions come from uneaten food, either junked from homes, rotted in the course of transportation, or rejected from retail sale because of trivial imperfections (by comparison, air travel accounts for around 2.5 percent of emissions). Hydroponics and importation have expatriated even the pulpiest tropical fruits to grocery aisles well outside the original species range, and seasonality, of those plants. What facilitates this is refrigeration—using electricity, often resulting from the burning of fossil fuels—and packaging. Many self-same, leafy food plants are grown, year-round, in single-variety warehouse environments like the Spanish greenhouses—not rooted in the ground, but sprouting from hygienic grow bags of perlite, sand, and loft insulation-like "rock-wool," untouched by direct sunlight.

Though packaging from food products and agronomy may enjoin pelagic rubbish, the trade-offs between food waste and wrapping are complex: the visible plastic that preserves some perishable produce may have less impact than the invisible (gaseous) cost of it going wasted. Either way, many more foods are plastic-wrapped than need to be in the commercial supermarket—where many more comestibles are also displayed than ever will be sold before they start to go bad—and more packaged, refined foods are bought to be eaten than ever have been in previous generations. How this shapes broader ecologies is a growing concern. What we're eating, it turns out, may be eating the world. The Japanese diet is not the biggest challenge facing whales in the twenty-first century, and, perhaps happily, there are concerns that are far closer to home, that we can begin to remedy.

The afternoon before leaving Tokyo I found myself wandering Asakusa, in the kitchenware district, known best for its trade stores stocked with armories of knives, and plastic display food. It was peculiar to stand in shops full of food and smell only plasticizing chemicals—the eye and nose estranged. Peculiar, though in truth, it was also soothing: this aroma of toy-stores, a cosseting scent found the world over. I felt like a doll, entrusted to choose play food for the dollhouse—that is, I felt myself the empty instrument of someone else's pleasure. Bowls of permanent ramen, glossy with egg; vinyl sashimi sprigged with green cellophane; the beauty of a shaft of light crystalizing in a resinous, undrinkable pint of beer. I could find no inedible plastic whale-meat. I thought of whales eating stones in captivity, and of dogs eating bits of a toxic, euthanized humpback, in its long green dream. The paramount, evolutionary question *Is that food?* had gotten impossibly muddled, so it seemed.

Saezuri, twitter. The whale tongue dish; a store clerk in one of the many Asakusa crockery shops explained it to me when I asked. The name comes from the sound it makes in your mouth. The story goes that a Japanese chef gave it that name when he first bit into the fatty, squeaky meat. It sounded, he had said, like birds cheeping inside his head. I thought of all the voices we put into animals, the sounds they make to us—and what happens when we think beyond diet, to consider the intimacies we share with whales, outside of whaling the oceans.

I leave, passing chefs, or chef apprentices, gliding through empurpled backstreets in their whites. They weigh what I make out as wads of anonymous seafood, peeling on blue latex gloves and turning away, momentarily, to sniff for freshness. Fish change hands. Most of the chefs are bowed beneath the kind of see-through, bell-shaped umbrellas that are popular here, in which gather mini thunderheads of vaporous nicotine. Their smokes flare in intervals. Red smidgens, like imagination moving through a school of jellyfish. I watch them for a minute, listening to crosswalks chirruping for no pedestrians in the nearby empty streets.

Kitsch Interior

A Cooler Box in the Arctic — Out-of-Placeness — Goose Barnacles on
a Basketball — Grasshopper Effect — The Ocean Online — Spiraling
— Horse Latitudes — Middleness — The Gyre Is Invisible — Net Flock
— Plastic Is the Weather — Rafts — Micro then Macro — Its Skin Is
Changed — Whale in the Vortex — A Beluga Passes Gravesend — Qi
Qi — White Flag — The Vaquita Is Not a Fake Animal — Neozoons
— Grolars and Wolphins — Outbreeding Depression — Balloons —
The Things Afloat That Will Never Sink — Counterfeit Was Quality
— Newness Forever — Oil Spills Compared — A Conundrum with
Seagulls — Galleoning — Algae Craze — Paleopolymertologists —
Tomb or Time Capsule — Dumped Desire — And Hope

HERE'S THIS MORNING'S NEWS IN *THE GUARDIAN*: A BRITISH-LED research expedition reports "sizable chunks of polystyrene, lying on remote frozen ice floes in the middle of the Arctic ocean." The scientists were surprised to discover the blocks of plastic foam many hundreds of miles from land, in areas that have only recently become accessible to them because of thaw. "It is one of the most northerly sightings of such detritus," they state. Years earlier (perhaps so long as decades ago), cold, languorous rivers are thought to have portaged the polystyrene—a kind of crumbly, inexpensive insulation, used mainly as cushioning for merchandise—into the Arctic basin, where it was seized by the seasonal freeze and churned, as is the habit of ice floes, into deeper, slower, harder ice. Now, out of the newly fluidifying world, this plastic has returned.

Accompanying the article is a photo of a rectangular foam slab, floating in a pool of ice frazil. I look and realize: *That is the lid off a cooler box*. Almost certainly, yes, a cooler lid: pitted with age. Somewhere then, in a more tertiary layer of the ice, hangs the box it belongs to, wide open, storing nothing at a set temperature.

The meltwater, a symptom of climatic warming, keeps on disgorging these sorts of symbols, as though history didn't move in a line, but in a spiral. The poet Tomas Tranströmer once called time a labyrinth, and wrote that in a few places, where the walls had thinned, you could press your ear against the division and listen for the footfall of an earlier iteration of yourself passing by on the other side. Things that were lost or forgotten reappear, strangely altered by context, as we loop back past them on our next involution, moving inward or outward, upwards or downwards—who can tell?—lapping the unknowable eye of our life's epicenter. The cooler box comes up, there where it ought not to be. How much out-of-place-ness should occur before we admit the impossibility of there being any *in*-place-ness left? Though we go on murmuring in our green dream, our former relationship with the wild resists restoration. We are all tumbled together, human and nonhuman, the far and the nearby, deeply in torsion, inhabiting this change of state.

In unoccupied and hard-to-access places—blank parts of the Arctic and Antarctic, the deep sea, and twinkling alpine summits—people are pained, and perhaps blackly awed, to discover human presence bestrewn ahead of their arrival. Packaging bedded into the landscape. Icebergs opaque with wet wipes like browning hibiscus. That tiny seahorse pinched onto a Q-tip. A basketball, sailed all the way across the globe with a resplendent chandelier of goose barnacles growing from its lower hemisphere. In the belly of a whale, the greenhouse; and an unopened tin of Spam found six miles down in the stygian depths of the Mariana Trench (surely the furthest point any swine, dead or alive, has ever traveled to: that pig, an astronaut to its kind).

Meanwhile, what gets called the "grasshopper effect" brings vaporous and persistent chemicals to both ice caps. Pollutants evaporated from warmer regions are blown around the planet as noxious particles. When they reach colder latitudes, the toxicants condense out of the air—falling as rain, or dirty snow—and so the world's frigid realms become landing zones for chemicals used distantly elsewhere. These semi-volatile pollutants (mostly kinds of benzenes and organochlorines: pesticides and PCBs) may hopscotch across mountaintops on their way to the poles, lodging in sedges and lichens, the thin bark of high-elevation trees, or dipping into the drip lines of icicles. P-C-B: hop, hop, hop. The migration of these substances can take years, accreting and dispersing as they spring toward sites of greater accumulation. Like Van Gogh's painting *Starry Night*: the chemicals whorl in pointillist air.

The world as we once knew it has of course disappeared. Now, too, quietly, the world as we don't know it yet—a nature we've barely met— slinks away.

Gray whale (subadult: thirty-seven feet long)
Puget Sound, 2010

Tracksuit pants. Golf balls. Pairs of surgical gloves. Twenty plastic bags. Small towels. Unidentified decayed plastic sheeting.

Into the online ocean, I cast my net for knickknack cetaceans, a global superpod of squishies, throws, and ornaments fashioned from polyester, polyethylene, polyethylene-terephthalate, polyvinyl, plastic. *"Three dolphin heart laser-inscribed crystal LED nightlight gift"* (Hong Kong), *"Rhinestone 'enamel' pendant cut stud humpback"* (Sydney), *"3D bedding quilt duvet cover whale mother—fleece"* (Saint Petersburg), *"Carved whale hair pick"* (Denpasar), *"Whale baby in change color light-up water ball"* (Shanghai). And so on, and on: a myriad drifting on the digital tide, each object's billing a poem unto itself in the global diction of eBay English. I buy none of these decorative whales, only bask in their lurid glow, toggling forward and back. Today, I'm spiraling.

I have in mind a circular momentum, because recently I have been considering the so-called gyres; plastic gyres in the sea. There are several, and they go by different names: "garbage patches," "trash vortexes," and "white pollution" (after the color of single-use tableware, Styrofoam, and thin shopping bags). Belongings now devoid of belonging. The plastic gyres appear where currents converge and create slowly revolving eddy fields. The "horse latitudes," as they were once called, intersect with the gyres: becalmed regions of little rain, where sailors seeking the New World are said to have pitched their thirsting horses overboard, so as to conserve freshwater for the crew. I once thought of the gyres as producing artificial islands; aggregations of fisheries junk, jettisoned cargo, debris drawn into the ocean by tsunamis and high winds, and trash flushed off the land by stormwater. I imagined a pell-mell of spoilage—pontoons of human language atop the languageless sea, rasping softly *Coca-Cola, Sprite, Gatorade, Schweppes, Sunkist, Mountain Dew, Nestlé.* The surface scythed, maybe, by a wingtip from Malaysia Airlines' MH370—crashed into the ocean and still unaccounted for. Or even: a terrible, sucking funnel, at the tapered root of which I shut my eyes and saw dark things, smoothed into jostling, anonymous spheres by the constant, spinning motion of the gyre. Not a void in the seascape, then, but a ghastly commons, a humanmade *middleness.* A buoyant midden.

As it turns out, though, the gyres are, mostly, invisible. They have no plasticized surface, and no definitive boundary. The gyres are better thought of as a force, rather than a giant compound artifact from this, the age of unintended consequences. Largely compiled of microplastics and durable monofilaments, in the gyres' deeper revolutions synthetic nets move in flocks with the tide. Around 60 percent of the plastic produced worldwide is buoyant, being less dense than seawater, and some eight million metric tonnes of plastic enters the ocean each year; a mass greater than that of the Great Pyramid of Giza. (How much is recycled before it is junked? As of 2015, people of the world had generated approximately 6.3 billion metric tonnes of plastic waste, around 9 percent of which had been recycled, with an additional 12 percent being incinerated.)

Seaborne plastic from the land, or fallen off boats, might eventually sink, it might be carried by the wind to be recaptured by a coastline, it might float, or else, it gets eaten. Over time, plastics in the ocean are shattered by wave friction and UV radiation, into a bleached and dainty shrapnel—tinier than krill or a limpet on a whale bone. Because most polymers remain impregnable to water and microbes, it may take hundreds of years, thousands even, for the particulate to disappear. If it ever does. One of plastic's most pernicious qualities is that it doesn't so much decay as divide into littler and littler pieces. Only a microscope would reveal the full extent of the plastic, though it's a ubiquitous and global problem, occurring in every ocean, and in rivers and lakes, as well as, more diffusely, on land. Plastic is a component of dust; it granulates in the farmland soil of Shanghai and falls in rain over the Pyrenees. Plastic is in the weather. Current estimates hold that in one of the largest gyres found in subtropical Pacific waters between California and Hawaii, 94 percent of approximately 1.8 trillion pieces of plastic are microplastics. Within an area of more than six hundred thousand square miles, there are at least seventy-nine thousand tonnes of polymer debris.

In the same way that the fat in whales attracts and stores contaminants, so, too, do these fragments of plastic collect legacy chemicals—pesticides like hexachlorobenzene and DDT, for example—puddled on the sea's surface. Such chemicals are hydrophobic; they repel from water, and tend to pervade lipids and oils where they come into contact with them. The microplastics, pollutants themselves, are tiny rafts, concentrating other chemical pollutants. Some microplastics descend to the seafloor in marine snow, more drift near to, and on top of, the ocean's topside. Ranging from under five millimeters to down to 100 nanometers, oceanic microplastics work their way into the food chain either because they are mistaken by fish for water-insect eggs, or because filter feeders, such as mussels and oysters, passively absorb them—as do zooplankton. Amphipods—small crustaceans—are voracious microplastivores: UK scientists studying life in six of Earth's deepest ocean trenches have unearthed microplastics in the digestive tracts of 80 percent of the amphipods they retrieved. Baleen whales, like bowheads, humpbacks, and minke, are likely to be particularly subject to the sublethal effects of ingesting microlitter because of how they feed, which is typically at the surface, and by capturing, and then straining, huge volumes of seawater. How microplastics impact large marine mammals, like whales, is yet to be concretely determined, but though these pollutants are minute, they leave a lasting imprint. Phthalates, which are trace chemicals left behind by plastics, have been identified in the blubber of dead fin whales off the Italian coast, and may be evidence of microlitter consumed by the whales.

Bigger pieces of plastic (macroplastics between five and fifty centimeters, and megaplastics larger than that) are a more significant concern for marine life: 86 percent of the megaplastics (forty-two thousand tonnes) in the largest Pacific gyre were discovered to be nets. A crueler, tighter confinement than the smallest oceanarium tank—nets and plastic mean captivity, outside. Michael J. Moore, a marine mammal expert, writes of right whales snared in the slow-motion flensers of ghost nets: "The majority of North Atlantic right whales are repeatedly more restrained than any animal in a zoo." Born an obsidian, polished black, almost all of these whales are laced with white scars: the coloration of the species initially scripted by genetics, but scribbled upon by junk, so that the animal's appearance is now far from natural. Some nets in the gyre displayed extensive hand-stitched repairs and sun deterioration, suggesting they had been discarded, rather than being ripped by high weather from active fishing use. If dropped in the open ocean, the tulle of fishing gear may prevent ensnared whales from diving in pursuit of prey—but even if this isn't the case, the flesh and bones of younger whales are cut by nylon netting, as surely as by a pair of secateurs, as they outgrow the cage. Manacled by ropes, these cetaceans are typically deserted by their pods. If they aren't strangled and don't succumb to a sequelae of infections from wounds inflicted by the fishing gear, these whales starve; their bones are sometimes visible through their skin before they die.

Worldwide, there are five gyres that stretch for miles, swirling with the refuse of our modern lives. Do whales know to avoid them? In October 2016, scientists crossing the Pacific gyre aboard a low-flying Hercules C-130 observed and photographed four sperm whales (including a mother and calf pair), three beaked whales, two baleen whales, and at least five other cetaceans traveling through these waters.

Sperm whales (4 stranded together)
Schleswig-Holstein, Germany, 2016

A forty-three-foot-long net from a crab fishery, pieces of
a plastic bucket, over 110,000 indigestible sets of squid beaks
(the whales' natural food), and a car engine cover.

How much out-of-place-ness, indeed. Car parts in a whale, a cooler box lid in the Arctic, and, over Christmas 2018, a beluga, pet-named "Benny" by the public, foraging for fish in a stretch of the Thames passing by Gravesend, Kent. Today individual whales are turning up outside of what scientists have long believed to be their traditional habitats and migration paths, which might signal other changes in those ecosystems, or in the whales. "We don't know where [Benny] went, because we don't know where it came from," said a spokesperson for British Divers Marine Life Rescue to the *Telegraph*, when hydrophones ceased to pick up the voice of the whale.

A bowhead recently appeared off Cornwall, some one thousand miles outside that animal's customary range. Then a narwhal, another Arctic whale species, turned up thirty miles downstream in a sea-fed river in Belgium. Killer whales go plunging after bowheads into quarters of the Arctic where sea ice previously impeded them. Beluga off Canada have accepted a narwhal into their pod. ("[This shows] the compassion and the openness of other species, to welcome another member that may not look or act the same," said a researcher from Harvard, faintly echoing anxieties about immigration in the human world.) Gray whales, too, are altering the choreography of their long transportations. Gray whales are leaving the North Pacific and crossing over the top of the globe through the ice-free Northwest Passage to appear in the Atlantic. Gray whales surface now off Israel, Spain, and Namibia, where they have never been recorded before. Some people say that these whales are *re*appearing—resurfacing underneath stories of sirens that once were whales, back when whales were a storehouse of superstition. Or it may be that during the world's ice ages there *were* gray whale calving areas in the Atlantic. These whales may be returning, rather than migrating, into new habitats, if their range is viewed on a long enough timescale: back to when there were no humans watching them at all.

Cuvier's beaked whale
Bergen, Norway, 2017

Thirty pieces of plastic litter: a filmy sheet more
than six feet long; shopping bags that once carried chicken in
Ukraine and ice cream in Denmark (the logos had not yet
been dissolved by the whale's stomach acid); a wrapper
off a packet of Walkers crisps from Britain,
still printed with a triangular recycling symbol.

The Norwegians called it the *Plasthvalen*. A curator of osteology
at the University of Bergen, Dr. Hanneke Meijer, said to *Sky News*,
"That's when it hit us. We have a plastic whale."

No whale species was ever driven to extinction by whaling, for all its sweeping violence—but cetaceans *have* disappeared from the planet, already, as a result of pollution. The last Chinese baiji dolphin, Qi Qi, the endling, died in captivity in 2002. Though there are persistent, unconfirmed reports of the baiji's rediscovery, it seems certain the animal is "functionally" extinct. Even if a handful of individuals remain in the more turbid corners of the Yangtze River, the species will not rebound.

The baiji is the first cetacean to be killed off by exposure to heavy metals, habitat destruction, electrofishing, and boat strike. Its scientific name—*Lipotes vexillifer*—means "the left behind." *Vexillifer* is "flag bearer." The "flag" alluded to is its pennant-shaped dorsal fin, which lent the dolphin its other common name, the white-flag dolphin. The white flag, left behind: Qi Qi represented an unintended surrender.

The vaquita is the rarest porpoise on Earth (if it has not already, as of printing this, vanished); a sooty-eyed, snub-nosed creature, the size of a wine barrel, of which there are perhaps only twelve left, trailing through the murk of a northern embayment in the Gulf of California. In addition to the gill-net junk that has killed most of them, vaquita are highly sensitive to mechanical noise, and their hearts beat very fast ("We think of them as the hummingbirds of the marine mammal world," said a senior scientist). In 2018, a fisherman in San Felipe told reporter Ben Goldfarb that local people believe the vaquita has always been a hoax, or a mythical animal. "Most people here think they don't exist," he said. One effect of a species fading out may be the discovery of a tipping point, at which the loss of real animals spurs the apprehension that those animals were fakes all along.

Sperm whale
Isle of Harris, Scotland, 2019

In a "litter ball" weighing over 220 pounds: knotted
synthetic netting, ropes (red, blue, green, brown:
some as thick as an adult's wrist), packaging tape,
plastic drinking cups as from a child's birthday.

Whales of a new order appear on the fringe. Discovered but perhaps also, in a sense, created. Neozoons. A 2010 commentary in *Nature* called attention to the increasing hybridization of high Arctic animals as a result of sea-ice loss and ice-sheet retreat. The retraction of the Arctic cryoscape, the authors noted, has not only removed habitat for life-forms that flourish in frozen terrain, but has also, consequently, dissolved physical and climatic partitions between animal populations long isolated from one another. A few of these species are capable of inter-breeding, and some produce genetically viable offspring.

Across the tundra emerging from beneath the ice traipse rare pizzly bears—grizzly polars. The bears' white coats ombré into brown paws, and though they remain long-clawed for seal-hunting, their heads are jowly in the way of Alaskan grizzlies, and their shoulders roll with a fatty hump. Only a few pizzlies (otherwise called grolars: their com-pound name isn't settled) have been confirmed as crossbreeds via DNA testing, but it's very likely there are more, because the bears tend to give birth to cubs in pairs or litters of three. In the sea, too, narlugas have been sighted—narwhal-beluga hybrids—off west Greenland. Beluga and narwhal were once divided from one another by walls of sea ice that have become, in the warming world, seasonal, and may soon dis-appear for good.

Similarly, two kinds of minke—common minke and Antarctic minke—both migrate out of colder oceans and head to warmer waters come winter, but they do so during alternating times of year (the whales are genetically distinct; different in size and slightly, in coloration). The staggered visits of the minke to equatorial seas may have begun to over-lap. At least two hybrid minke, with parents from either end of the globe, have been identified by geneticists. Bowheads from the Pacific and the Atlantic are meeting, as they haven't done for some ten thousand years. Along the coast of British Columbia, Dall's porpoises mate with harbor porpoises. Many species of seals diverge and fuse—spotted, with harbor seals; harp, with hooded seals; ribbon, with spotted seals. There is pres-ently a wolphin (and there have been others in the past) in captivity in Hawaii: half false killer whale, half bottlenose dolphin (whale-dolphin, *wolphin*). A wolphin is a beast humankind created by more direct inter-vention, pairing these two individual animals together in a single pen: In the wild the bottlenose dolphin and the false killer whale would seldom, if ever, encounter one another.

The authors of the article in *Nature* noted that hybridization was not, by itself, an inherently negative phenomenon. As far back as 1886, a bulletin of the US Fisheries Commission made a note of "bastard whales." Hybrids are not entirely novel, though the changing conditions make them more possible. On an evolutionary level, the crossbreeding of animals is a driver of novel mutation, helpful for organisms adapting to pressures and opportunities. But risks to the whales manifest where hybridity reduces genetic diversity in a species over time, lessening the animals' overall adaptability to environmental change (particularly if such changes are quick, or extreme). Or else, interbreeding threatens species if the intermediate genotype of a hybrid alters its ecological interactions to the detriment of that creature. The wolphin, for example, has more teeth than a dolphin, but less than a killer whale: there may be no midsize prey fish in its forebears' natural habitat easily captured by such a mandible. ("Outbreeding depression," biologists call this comparative disadvantage of hybrid offspring.) Other endangered animals, the scientists warned, were at risk of genetic swamping—being overtaken completely by a relative. The North Pacific right whale, for one. Though scarce, North Pacific right whales do encounter the more numerous bowhead whales—and will continue to, as the icy seas they prefer concentrate poleward. Bowheads are the "kissing cousins of the right whales," as one American professor of biology put it. Interbreeding between bowheads and North Pacific right whales could push the latter to extinction in a short time span.

Fantastical in the way of the drolleries that once menaced medieval voyagers from the corners of their incomplete maps, these hybrid animals may be some of the most uncommon creatures on Earth, but the fact that they do not belong to a distinct, identified species means that they lie outside the protection of conservation laws and international enforcement. Their in-between-ness puts them in a chancy position. When a pizzly bear was brought down by a trophy hunter from Idaho with a permit to shoot polar bears, there were no repercussions (had it been a grizzly, the penalty might have been prison time). When Kristján Loftsson's Icelandic whaling company Hvalur hf killed a blue-fin whale cross in 2018, they found they could not sell the meat to Japan—food standards foreclosed on the importation of "mystery meat" from an unrecognized animal (Hvalur hf might as well have been trying to offload yeti steaks). So it turned out the Icelandic whalers had killed what was surely one of the world's biggest and rarest animals, for no gain or profit. They laid it on tarmac; they washed the bluish body down with hoses. Despite the ire of green groups, no legal sanction could be brought to bear upon the company. Dead, the blue-fin fell into a gray area.

Another blue-fin hybrid has been returning to Skjálfandi Bay in Iceland annually since 2012; it is a popular attraction among Icelandic whale watchers, an animal that hovers between worlds.

Rough-toothed dolphin (calf)
Fort Myers Beach, Florida, 2019

Two plastic bags, one shredded balloon.

When I haven't been writing sometimes I have been watching—out of the louvered window of my leased apartment in Glebe—a building being constructed on the skyline, in Pyrmont. For a long while today I tracked a freight elevator beetling up and down on the exterior of the skyscraper, depositing sacks of concrete grit, reels of cable, or whatever else is needed. The cladding isn't finished yet, so you can see the inner structure. The facade only reaches partway up, and the rest is a sort of steel cage. Black pillars jag out into cranes on the open rooftop. Each week I take a felt-tip marker and make a little strike on the inside of the window glass to mark the top of the advancing skyscraper there, on the other side of Blackwattle Bay. Each time I find the building has inched above the previous line.

For a quarter-hour this afternoon I stayed hooked on the slow traction of this construction elevator when, suddenly, an immense cluster of red balloons appeared in the corner of my eye, in the gray sky. Picture-book balloons, sticker balloons. Red! A very literal, Chinese New Year red. The balloons floated along above the fish markets. As I watched they blew right *into* a floor of the windowless skyscraper. I stood up at my desk. What a gift. I pictured the foreman later, throwing open the elevator door with his load of tarps to see the balloons there, a tree on fire in the dark.

Sperm whales (two stranded within a month of one another)
Northern California, 2008

Two hundred and thirteen (dry weight) pounds of debris
total, including one individual net of approximately
182 square feet. Branded cord and netting, bearing the
trademarks of Indonesian fisheries. Types of net identified:
bait nets, gill nets, shrimp and trawl nets, tallying
134 nets in total. Polyurethane and nylon.

Curtis Ebbesmeyer, a famous oceanographer who models dispersions of rubbish in the sea, says, "There are things afloat now that will never sink."

I think: *Never?* I can't think with this: *never*.

What happens to time in the gyre? Vortexes in literature and myth were event horizons: places before time, or after time, or where time stood still (as in a black hole). Dante's hell is a vortex leading down, and heaven, in the *Divine Comedy*, was found by following a spiral path up Mount Purgatory. Describing the Aztec creation myth wherein Quetzalcoatl, God of Wind, blows a shell horn above a heap of bones that leap up to become humankind (swirl before time), Eliot Weinberger writes, "A conch is a vortex you hold in your hand." If the blue whale in the WA Museum carried me into thinking about the past, plastic objects inside whales made me think of the deep future. Plastic extends organic and impermanent things toward a dim, far-off horizon (take: Tupperware, take: cosmetic surgery). Plastic refashions. Plastic is hygienic, inert. Plastic, since its inception, has always counted itself a futuristic material: luminous raw-stock of the space age, protective envelope of new technology. From plastic, people have compiled myriad impossible dreams. Artificial hearts! A nylon flag on the moon! Transatlantic cables, super computers, jumbo jets. Connecting antipodes, plastic makes the world worldly—it permits the *there* to occur here, and inversely.

But plastic equips appetites for the shoddy, too, the throwaway, any quotidian pick-me-up—a flourishing of what has been called the "orgy of the ornament" after World War II. All those temporary things, glitzy tchotchkes. "The future of plastics is in the garbage can," declares an editorial in *Modern Packaging Magazine*, foretelling, from the mid-1950s, the expanding market share of polymers engineered into high demand, cheap, single-use goods designed for rapid turnover by a public for whom the word "plastic" designates disposability.

At first, plastic was considered an industrial improvement on the shortcomings of the "real thing." Horn, tortoise-shell, pearl, and wood, subbed out for Bakelite, celluloid, laminate, and Formica. Counterfeit was quality: natural materials, after all, proved splintery, chip-able, sun-fading, and idiosyncratic. Such items often had to be treated with varnishes, antiseptics, and strong-smelling preservatives, like creosote; maintaining them entailed custom polishes and waxes. So the ersatz product celebrated its uniformity, cleanliness, and fakeness as ultra-modern. Early consumer plastics were made into durable, hard-edged objects. Polyethylene had been synthesized in the 1930s, but Teflon, PVC, and polyester fabrics were wartime innovations; their downstream legacy branched into squeezy bottles, Velcro, LEGO bricks, Lycra, drip-dry no-iron clothing; the plastic bag, the Barbie doll, the silicone breast. These softer plastics made time pass with greater ease; they decreased workload in the domestic sphere, reduced people's exposure to harsh cleaning products, and increased the safety of household goods. Light-weight cars were cheaper to run. Medical plastics saved lives. Mean-while, the plastics industry sold consumers on the fantasy of a lifestyle that was abundant, inexpensive, and swiftly modulated. Fast food in takeout containers. Seasonal fashion; semi-seasonal fashion; forecast fashion: newness, forever. Plastic permitted a person to be one thing, one day, and then the next tomorrow; to slip skins. The luxury was that plastic could be used, and forgotten.

Things look different from the standpoint of a time when there are more ornamental lawn flamingos than living flamingos. Fifty years ago, the ocean was very nearly plastic-free. In the lifetime of a single whale, its environment has absorbed hundreds of thousands of tonnes of polymer litter—much of it not products, but product packaging.

In a 2017 paper coauthored by three American environmental scientists and titled "Production, Use, and Fate of All Plastics Ever Made," the authors note that plastic manufacturing increased from two million metric tonnes in 1950 to 380 million metric tonnes in 2015. Thirty percent of all plastics, ever made, are currently in use.

This is the incongruity of plastic: it can be so changeable and compliant, so *plastic* and expendable, yet we're told that, somehow, it will endure beyond our lifetimes. Omnipresent and cheap, no substance *less* appears to warrant the aura of the eternal. Yet plastic will survive us. Escape us. Plastic *is* geologic in its original state. Polyvinyl comes from coal. The feedstock of polyethylene is crude oil. Other plastics derive from natural gas and its by-products. All are fossil fuels. The ocean may be able to assimilate oil spills across a long time, but never plastic. The scale of the 2010 Deepwater Horizon disaster notwithstanding, over the past four and a half decades, oil spills from tankers have decreased markedly. In the 1970s, there were 24.5 large oil spills ("large" being more than seven hundred tonnes) annually; in the 2010s, the average number of large oil spills decreased to 1.7 oil spills per year (even though the oceanic trade of petroleum products increased). Oil spills rightly arouse people's disgust, provoking strict regulation. Darkening the ocean, or sometimes setting it alight, spills draggle marine creatures and seabirds in asphyxiating grease. A visually alarming event (the sea on fire, great tracts of it blackened), an oil spill's morbidity is all surface. Plastic may *look* prettier—microplastics resemble nothing so much as fine confetti—and, being degusted by animals, smaller plastic may also be less conspicuous: but plastic *is* an oil spill, once-removed. An oil spill modulated, objectified, in fact, into something more persistent, and harder to capture or remedy.

Yet, the durability of plastic remains still, only imaginary, in the sense that it hasn't been experienced. Many types of plastic have been in existence for less time than their projected rate of decomposition, so evidence of their dematerialization lies still to be discovered, in the environments of the future. This is the great, visionary power of plastic: with it, we can think beyond the edge of our own lifetime, into the world of generations to come. Which is what many people, today, are doing: calling for an end to the use of disposable, rapid turnover plastics, having projected a mental picture of what the world of their descendants—and the kingdom of the animals—will look like, plasticized by dross that has been overlooked today.

All of Earth's creatures are called to readjust their bearings and map their behavior to conditions emerging now, but among them, we are, surely, the only species capable of adapting to a vision of the future. As F. Sherwood Rowland, Nobel laureate and chemist of the ozone layer, said to the *New Yorker* in 1989, "What's the use of having developed a science well enough to make predictions if, in the end, all we're willing to do is stand around and wait for them to come true?"

Sei whale, subadult (forty-five feet long)
Elizabeth River, Chesapeake Bay, 2014

One broken DVD case, shards found in the
esophagus and stomach lining.

Can some animals live with plastic? How might that change food webs?

In Argentina, I've read, seagulls are eating southern right whales not after they've perished, but alive. The birds have been in the news more than once. As the whales surface for air, the gulls dive, tearing strips off their skin to obtain the blubber beneath. The seagulls target the right whale calves, which are less agile swimmers than the adults, and being smaller, must more frequently rise to fill their lungs. The infants and their mothers are found nearest to the shoreline and the craggy island outcrops, where the gulls nest in colonies. These are big gulls, locally known as kelp gulls. Heads the size of softballs, wingspans over four feet long. Their bills are the yellow and red of motorway signs. Feeding from the whales' open wounds, the gulls spiral and mob in opportunistic packs. Attacks can last nearly four hours.

Three decades ago this behavior was aberrant. Researchers in the 1970s first observed gulls gleaning thin peelings of the adult whales' skin, sloughed during their year-round molt. Southern rights lose translucent swathes of their epidermis naturally, shedding with it barnacle-larvae, cyamids, algae, and other gummy clings of microscopic, aquatic flora and fauna that stick to their bodies in the water. The kelp gulls, en route from their roosts to the mainland, sometimes plucked the slough off the glassy tabletop of the sea, where it floated. Occasionally the gulls were seen to alight on the whales' backs in nipping flurries, picking at the molt. How their habit changed is unclear, but in the '80s and '90s the gulls began to forcefully snick and dig *under* the whales' skin, into the whales' blubber. They became adroit raiders. They taught their fledglings, or so it's believed.

Along the Patagonian coastline today the harassment of the gulls has become so aggressive and widespread that scientists estimate 99 percent of southern right whales there have gull-raked and cratered backs, right down to the white, adipose tissue. The lacerations become lesions, some as big as dinner plates. Skin loss can lead to dehydration, compromise a whale's thermoregulation, and allow pathogens and parasites to enter in under the dermal surface. What follows is infection. Many whales grow sick. Nursing whales expend great energy evading the gulls, so much so that they lose important fat reserves laid down the season prior. Twenty-four percent of their daylight hours are spent under states of gull-induced disturbance. Stressed, the southern rights stop feeding their calves. Juvenile whales spend less time socializing with mother-calf pairs. Thus maturing females, it's thought, have slender opportunity to familiarize themselves with the practices of infant rearing. Young and old whales both learn to take their inhalations secretively, in gasps. Maternal whales assume a defensive posture researchers call "galleoning"; arching their spines downwards to expose only their heads and fluke-tips above the surface, when they need to breathe. Still, gulls come like lightning strikes, smashing through the air.

At Península Valdés, a whale nursery in the Chubut Province of Argentina, the mortality rate of southern right calves has been declared "globally unprecedented." In 2013, 113 calves died: a third of those born that year. Before 2005, fewer than six calves died annually in the area, on average. Though the overall southern right whale population has increased, the escalating mortality of calves has outpaced the attrition rate of prior years to such a degree it is considered anomalous.

Kelp gulls are only one potential culprit for these premature deaths. Another might be a phytoplankton called *pseudo-nitzschia*, which flourishes in waters enriched with agricultural run-off, a plant booster for crops. The growth of dense tracts of *pseudo-nitzschia* accords with the timeline of the right whale calves' deaths. *Pseudo-nitzschia* emits a neurotoxin that, when unintentionally consumed by people in crabs and clams, can cause seizures, a weakening of the eye muscles, and irreversible memory loss. "Amnesiac shellfish poisoning," it's called. As with Alzheimer's, the effect in humans is that the near past is lost and more distant years become jumbled. One hypothesis is that the whales' development could be affected in utero when the toxin is swept up by pregnant southern rights in mouthfuls of prey fish (in this case, the adults seem too large to be harmed). Too, the algae could be responsible for the crazing of marine birds. Alfred Hitchcock's *The Birds* (1963)—loosely based on a short story by Daphne du Maurier—takes inspiration from one episode of *pseudo-nitzschia* intoxication in California, when a frenzied flock of sooty shearwaters began vomiting anchovies and careening into cars and buildings in North Monterey Bay.

A concert of factors could account for the mortality of young southern rights off Argentina; the gulls and the poisonous plankton acting together in ways yet to be discerned. The infant whale bodies, most under three months old, are swept ashore. The gulls get to those, too, unzipping them from snout to fluke.

When the New York author Diane Ackerman visited Península Valdés in 1990, she described southern right whales gamboling in the sea near the beach, a sight "as wonderful as discovering dinosaurs in your garden." The area is ensconced within a 1,275-square-mile parkland preserve, and a marine reserve extends almost six miles from most parts of the shoreline. Ackerman, profiling that famous cetologist (and whale-song enthusiast) Roger Payne in her essay "The Moon by Whale Light," observes right whales "tail sailing"—thrusting their flukes straight up out of the water to catch the breeze and drift across the bay for no other reason, according to Payne, than to delight in the motion. From her tent in the night Ackerman hears whales snoring and snuffling. All the ordinary intimacies of cetacean inspiration and expiration waft up out of the sea and into the warmed air of Ackerman's writing—at Península Valdés, the whale-world rushes over her coastal one.

Since the publication of "The Moon by Whale Light," human numbers in the region surrounding the Península Valdés nature reserve have grown steadily. A seasonal influx of holiday-makers and eco-tourists (including whale watchers) inflates the residential population further—notably during the months when southern rights migrate through Argentinian waters. Kelp gulls have always found rich pickings in the effluent of slaughterhouses, sewage outfalls, and the silvery gush of gutty discard from the area's fisheries, where it is debouched into the Golfo San Matías: these food sources track the burgeoning human population. The gulls' other significant scavenging ground, sites of open-air landfill, have likewise mushroomed. In the kelp gulls' nests are often found regurgitated pellets of indigestible garbage. Leftover, larid junk food. In short, more gulls appeared *because* there was more rubbish.

A thorny problem. One paper in *Marine Biology* noted that, having fostered gull flocks on municipal waste over many decades, initiatives reducing the birds' access to rubbish (enclosing the tips, increased recycling) could drive *more* gulls to plundering the southern right whales. Dealing with the trash on land could be worse for the whales in the ocean, in the short-term.

There are now around 75,000 breeding pairs of gulls roosting along 2,175 miles of Argentinian coast. A decade ago, government authorities instigated a cull of birds aimed at the elimination of attacker gulls in whale-watching areas. Individual "bad" birds were thought, by some observers, to be leading the gull assailants—but the impact on the wounding of southern rights has not been lessened, the sharp-shooting of kelp gulls notwithstanding. The link between expanding waste heaps and the number of gulls is only a theory, but it's one that makes sense to me, having seen seagulls eat any foul and unfeasibly large thing from a bin. Blubber is a nutritious addition to the kelp gulls' scavenging diet, though it's more difficult for the birds to get whale fat than pickings from the scruff of landfill.

Bryde's Whale
Cairns, Australia, 2000

Sixty-five square feet of plastic, including checkout
bags with pharmacy brand-names still discernible.
Several disposable cigarette lighters.

I sometimes wonder: in the tattered, sun-struck expanse of the Mindarie tip, how much remains of the dead humpback whale, stranded years ago in Perth? What animals have picked it over—what feral cats and filching worms, what fungi, what nascent flies? In the thrum of midday what birds halve its heart? Do they oil their feathers with liquor from its liver?

Do they cosset their nestlings on the indigo bunting of the whale's innards? The smell is surely awful. Or the smell is, by now, indistinguishable amid the whiff and redolence of the waste moraine. Perhaps there's nothing left of the whale at all. Contra to a whalefall: piece by piece it has been lifted off into the sky, by gulls.

I resist this image, though. The whale flung open and carried away. What I enjoy imagining is somewhat deeper, and slower moving—the compression of the humpback's bones and its transition toward being a fossil. I don't know whether this is chemically possible, if the place where the site is located has the necessary tectonics, the right type of low-oxygen sediment, to preserve the bones of dead animals. Can a landfill cache and mineralize organisms the way the tar pits and carboniferous forests of prehistory stoppered the Pleistocene mammals? I like to envisage then, not paleobiologists, but international paleopolymertologists and taphonomists of plastic to come, chemists of the ancient rubbish bag. Women and men peeling apart stiffened skins of baled packaging, tweezering fruit-netting, aglets and buttons, splinters of disposable cutlery and plastic clothing-pegs interlarded around the whale. That humpback mummy, swathed in historically notable clingfilm. Long after a civilization's monuments erode, its garbage heaps remain. How is a tip like a tomb, and how is it like a time capsule?

Cuvier's beaked whale
Sitio Asinan, the Philippines, 2019

88 pounds of (empty) rice sacks.

Many kinds of gull are enthusiastic plastivores—as, too, their seabird relatives: the albatrosses, shearwaters, auks, and petrels. There is no nourishment for the birds in the synthetic dross humans throw out, though very often there is good sustenance in what is smeared onto, or what accrues around, the plastic. Insect larvae ripen in the sticky crevasses of litter, in turn drawing beetles to feed off the grubs. Food remains wedged in unwashed containers, or spilt across and saturated through nondegradable dreck.

The biofouling of plastics adrift in the sea—spun green in a soft coat of algae—produces a sulfurous scent birds mistake for a foraging cue (we are also able to smell it, this brothy odor like bladder wrack and forkweed). Fish eggs stick to floating trash and baitfish huddle in the shade below, pecking at plant-life grown on its underside. For some aquatic species, pelagic plastic has been a boon, providing habitat, a mobile graft for algae, and—for sessile organisms like barnacles and anemones—more numerous rafts on which to cross the open sea and infiltrate new coastal environments. The birds, meanwhile, eat rubbish not always because they confuse it for the appearance of prey, but because it smells like food and has food gelled around it.

On islands off Australia the bodyweight of some birds is 8 percent plastic—the equivalent of a 137-pound person with a bolus of diaphanous trash almost eleven pounds heavy in their stomach. The albatrosses of the Midway Atoll in the North Pacific die so utterly stuffed with plastic that their carcasses famously look like a magician's stage props, splitting open to reveal ribbons and straws within. These birds' bones will eventually decay and disappear, but they leave behind little brightly colored cairns shaped by being tamped up in their innards. Fulmars in the North Sea, meanwhile, so reliably eat floating rubbish off the surface they are officially used as biological instruments to measure marine debris, their stomach contents examined in order to estimate the abundance and dispersal of plastic pollution in rough oceans, difficult to monitor.

To their nests, gulls bring things people once needed, or thought they needed, and then discarded. The effect of the gulls, then, is to remind us not just what those objects were and what aspirations propelled their purchasing, but how unconscious this act of throwing away was; how willfully we believed what was put out in wheelie bins got rolled into the sickly twilight of landfill, evaporated into smoke, or was left to be eaten by organisms too small to taste anything, or too feral to care about.

In a particular mood, I like to whisper aloud the Dutch term for the pellets the gulls spit up: *voedselmonsters*. An evocative, gothic word, itself a compound of collected bits congealing into the English tail, *monster*, (though translated it means only "food sample"). A 2008 study found, in *voedselmonsters* examined on the island of Texel in North Holland:

bottle tops, bits of elastic, the legs of a doll, an army of toy soldiers, a medal ("WINNER") on a ribbon, and a small Sony Ericsson mobile phone. Elsewhere since, bits of aluminum have appeared in ducks. A toothbrush in a living albatross. A glow stick was found entire, inside a shearwater. The *voedselmonsters* are more than a conglomerate of disused items. The pellets amass dumped desires.

Cuvier's beaked whale
Biscarosse, France, 1999

Three hundred and seventy-eight individual
inorganic items (a record).

I've lost count of how many times someone has asked me, on the subject of this book: Is there cause for optimism; is there cause for hope? How do you sit with this terrible, sad news from the ocean, day after day? So here, I want to clear some space to speak quite directly and plainly to those questions, as if you were sitting beside me.

There is hope.

A whale is a wonder not because it is the world's biggest animal, but because it augments our moral capacity. A whale shows us it is possible to care for that which lies outside our immediate sphere of action, but within our sphere of influence—we care deeply, you and I, about the whale *because* it is distant. Because it speaks to us of places we will not go. Because it magnifies the reach of our humanity, and reminds us of our collective ability to control ourselves, and of our part in a planetary ecology. Because a whale is a reserve of awe and humility. You might take hope from the movement around plastic pollution—the shopping bag bans, the campaigns against drinking straws—but this, to me, looks like low-hanging fruit. What I mean to say is, there are many beings not proximate to ourselves that we will have cause to extend our compassion to in the decades to come: the future generations, the vulnerable people overseas, the creaturely life, and you might ask yourself, *How should I care for that which I do not know, that which I have never met?*

Do you care for the whale?

Could you act on behalf of the whale?

Being hopeful follows from being *useful*; this has been my experience, and to be useful, it matters that you identify a part of the problem that you might see change in, using the talents and the resources that you possess. Hope is fellowship. Hope is in the doing. We may be the only species capable of imagining a future robbed of the wonder of encountering other species. This knowledge, in the end, gives us cause to start.

I went to get a cup of tea. I drank the tea, washed the cup, and when I came back something made me look again at that unfinished skyscraper. Impossibly, at that very instant the balloons blew out the other side of the building. They carried on through the Sydney air, red. Red as the red in the mouth of a bear in winter. Did anyone else see this? I sat down, stunned, happy, and found my head was completely empty.

When I shut my eyes I could still see those red balloons, flipped electric blue. My hands rested upturned in my lap. It was not an unpleasant sensation, to feel so uninhabited. The night grew dusky, then dark. I kept the lamps off. What of the skyscraper? When I opened my eyes, the skyscraper's upper-stories had dissolved into night. Only the lower levels of the building were yet fitted with power and lighting.

At last I tasted rubber, as though the balloons had passed through me on their way across the world; as though I was a part of the world balloons could pass through.

Scantling

The People Discovering Dead Monsters — Egg and Eye — Coral: Is It a Thin Animal? — "Animalia Paradoxa" — Kraken, Owl — Of the Fauna Americana, *Basilosaurus* — *Hydrarchos*: An Enormous Mass of Petrified Snake — Their Heads, Being Too Small to Nourish Their Tails — Mythomania — Virginia Woolf Reports He Is Constantly Seen — Not Ignoring Bermuda — Or Things in the Thames — Cryptozoologists — Whales: Are They Islands? — How Much of It Was Plastic? — Are They Sensitive to Noise? — Killing the Animal in Ourselves — Cetaceans Never Seen Alive — A Naked Name — Seep Genera — Defaunation Is Terraforming — Minor Orders of Hermitic Monsters — Lice, Suckers, Worms — Gothic-Domestic Aesthetic — An Essence that Flickered and Ran — Million-Faced Beast — To Be Parasitized Is to Be Pluralized — The Swarm, the Squirm, the Sucker, the Spiral — Lice Riddle — Self as Circus

FROM THE STICKIER CORNERS OF THE INTERNET YOU READ REPORTS of the people who discover dead monsters. They're up early, or out in the hours on either side of midnight, with or without a dog. This is a person not given to conspiracy, but often—because they are found walking at a time when others don't dare to—they're immigrants or itinerants, shift workers, rough types, failed artists, insomniacs. The world has already become so strange to these people, it can easily get stranger. What we're talking about is: the broken-hearted. A group to which conspiracy attaches against their design.

I have watched such souls from a distance, but I have also been one of their kind: a figure the size of your thumbnail, wide awake and restless. Wherever you are, let's say that you saw me, or someone like me, at the edge of the ocean, tracing out brutishness. An otherwise empty beach. Scooped with shadow, and shifty. To be so impressionable: a quality to circle, to covet. Yet, at toe-point, nothing reveals itself as a personal talisman. The seaweed doesn't scrawl like a sentence. A rockpool's only a rockpool—at night it collects no reflections. Any octopus has long since turned into stone; its skin color changed, as biologists say, "at will." No gluey jellyfish halos, no limpets huddled beneath holdfasts either, no life capable of regrowing a limb or several. The shoreline pocks with bubbles on a wave's retreat, but there's nothing down there you can touch. It's only pressure.

Time. Time, and such weight.

Wait. Who's that? What's there? Out of the blue, now: a lumpiness in the shallows. I can picture it. Fearing a body, but being drawn closer, step by step. The obscured bulk grows larger. Sea fleas shiver its outline, reeking with punctuality. Its mass is rocked by fractional waves.

What it is, is then: bewildering, surely. A whole hut of flesh. Something like hundreds of jellicate teeth in linen. An immense fleece coat on struts, oddly snapped—like a garment off the back of Genghis Khan. A rill of organs inside weird, bog-man leather. And macro bones, too huge. The bones of a behemoth, a titan, a troll, all those giant things brought into existence to assume the excess of our selves. The mega-hominoids. Or, is it a crate of poached tusks, maybe, flung overboard or returned from a shipwreck, bundled in sea chunk. "Junk": the term people once used to describe the tissues inside the head of a sperm whale—*the junk*. Eggs there, or maybe eyes? Shouldn't the ability to discern an egg from an eye be drilled by primeval instinct? How hard is it to separate the unborn from a look?

The monster has no face. But if there is a dog, the dog's face is barking.

During the era in which Carl Linnaeus set about compiling his sorting system for life on Earth ("Linnaean taxonomy," the basis of modern biological classification), the demand for a global approach to recognizing and organizing natural organisms arose from two impulses. The first, to Linnaeus's mind, was a need to discipline the language of botany and zoology. In the eighteenth century many common plants and animals were known by several names, sometimes cobbled out of multiple dialects—or else they had ludicrously verbose titles that either described an organism's qualities at length, or reveled in flattering patrons and peers esteemed by the namer. Linnaeus's solve was to assert the primacy of Latin or Greek nomenclature—indicating likeness between organisms in the linguistic roots of their binomial naming. The second reason was colonial. As European powers overran foreign territories to exploit for trade and resources, settler scientists were encountering flora and fauna that challenged their understanding of life's natural order by defying existing categories. Were the corals of the tropics, for instance, living stones, a vegetable "bark" grown over an animal core of bone, or a very thin kind of animal, stretched across a branching plant? What defined the essential traits of a mushroom, now that the halls of the equatorial rainforests had been found decked with so many bizarre new fungi—and why *wasn't* a very large fish, after all, a whale? Attempts to emplace freshly described organisms within a schema of relatedness necessitated a hardening of the rules.

Linnaeus may have been a stickler for standardization, but he left space open for the possibility of unverified creatures, many of which had a longstanding heritage in lore—even if they'd never been resolutely identified. The *Systema Naturae* (1735), Linnaeus's formative taxonomy, concludes with an itemized miscellany titled the "Animalia Paradoxa." The unicorn, the satyr, the phoenix, and the siren; the Animalia Paradoxa contained supernatural creatures, as well as a few that couldn't yet, on the evidence, be declared true discoveries. By the early nineteenth century this category of chimera had largely fallen away. The only real dragons yet found were *Iguanodons*, *Plesiosaurs*, and *Ichthyosaurs*, clamped motionless in rock. After Charles Darwin's *On the Origin of Species* (1859), the physical features and behaviors of living creatures came to be viewed as plot points in the narrative of their evolution, rather than being indicative of any animal's moral or allegorical significance to human destiny (as in the bestiaries and fables that had informed Linnaeus's Animalia Paradoxa). With few exceptions, even the world's most charismatic and storied species were divested of myth through the fingertip toil of zoological science. Yet Linnaeus's category of arcane, unbelievable beings had lingering holdouts, well into the 1800s. Alongside the professionalization of natural history came a mid-century surge of belief in monsters, and—as when the ancient Greeks had believed whales to be the terrors skirting the void—so cetaceans again equipped the imaginary menagerie.

Spurred by the notoriety of fabulous beasts, swindlers in dank backstreets and taverns learned to articulate whale leftovers as parts of freakier animals yet. A colossal rib, a narwhal's spiral tusk, a gray whale's eyeballs, bristles of baleen stripped from a humpback's jaw, or armfuls of its spooling tongue—how disquieting these discards from the whaling industry must have appeared to those who had never seen a whale whole, in the flesh. Scraps retrieved from the decks of harpoon ships, or sold by savvy beachcombers, could function as credible props to mobilize a mythical beast. The rest relied on a story. Here, *see*, were oddments of mermaid, ocean centipede, sea swine, and saltwater salamander; remnants of turtles as large as houses, and of aquatic owls once believed to ambush boats in the Northern hemisphere. Before a spellbound audience, a sperm whale's penis (as pale and hefty as a daikon tuber) transformed into a segment of the kraken's mortifying tentacle.

What a whale was, then, was a kind of enclosure in which invisible—but still widely recognized—animals were impounded.

The stories were infectious. Members of the public in America and Europe found themselves caught up in a sea-serpent craze. The monsters branched into species. Gloucester sea serpents and Atlantic humped snakes were so frequently sighted (and circulated as prints and in pamphlets) that the Linnaean Society of New England formed a committee to standardize their reporting. William Hooker, a director of London's Kew Gardens, wrote that even "the most ultra-skeptical" individual should no longer doubt the sea serpent's existence. From 1817 to 1819, over three hundred people claimed to have witnessed a serpent with an equine head lurching about in surf off Cape Ann in northeastern Massachusetts. Its baby was found on the beach. A naturalist declared the baby was an unfortunate black snake with several tumors. This finding did nothing to discourage the public from scouring the surrounding dunes for sea-serpent eggs.

In 1832, twenty-eight immense, fossilized vertebrae washed out of marl on the banks of the Ouachita River in Louisiana, prompting locals to ask: *What animal is this? How did it get here? Are any more of them left alive?* A judge parceled up a single fossil and sent it to the American Philosophical Society in Philadelphia in the hope of intriguing lettered men of science. Having traveled south to retrieve more of the vertebrae (some had been repurposed as fireplace andirons), Richard Harlan, a skilled anatomist and author of the *Fauna Americana* (1825), imagined a gigantic marine reptile enveloping the spine: "a sea monster" of a bygone epoch. *Basilosaurus*, Harlan named it—*basileus*, "king"; *saurus*, "lizard." The animal was, in fact, neither a reptile nor, strictly speaking, a dinosaur, though the misleading *-saurus* suffix has remained attached to its name (it lived in the Late Eocene, a time when primitive horses and huge terrestrial birds walked the land). *Basilosaurus* would turn out to be one of the many ancestral forebears to modern-day cetaceans. Possessed of the most powerful bite pressure of any animal ever known to exist, *Basilosaurus* extended, anguilliform, to sixty-six feet long behind a maw of saw-edged teeth. Harlan was correct to believe that it had died out, but in the early days of its discovery *Basilosaurus* was a creature speculative and malleable enough to permit it to be resurrected by another man, elsewhere, as a different variety of monster.

During the 1840s, the ancestral whale proved a handy scaffold for misinformation and infamy when its unearthed bones were muddled to produce the implausible *Hydrarchos sillimani*—a sea-serpent ruse. Albert C. Koch, a German-born impresario and fossil collector, compiled the 114-foot-long "Hydrarchos" from parts of at least six different *Basilosauruses* exhumed in Clarke, Choctaw, and Washington counties. He purchased one segment of the serpent from the steward of an estate. Other bones he traded or dug up himself. After mounting his hoax in the fashion of a sinuous monster, Koch topped off the work of proto bio-art with several coiled ammonites, or maybe they were nautiluses. He wrote a promotional screed declaring that the animal must once have "reigned as a most tyrannical, cruel and unconquerable monarch." The skeleton toured saloons throughout the United States (an "enormous mass of petrified snake," blanched the *Southern Patriot*), before being shown in Dresden and Berlin. Everywhere the Hydrarchos went, it was met in equal measure by hyperventilating acclaim and scientific ire. Koch confected a second of its kind. The original Hydrarchos went on to be acquired by the Prussian Royal Zoological Cabinet in 1847 before coming to rest, pieced out into boxes, in the basement of the German *Museum für Naturkunde*. There, the disassembled sea serpent was substantially obliterated during bombing raids in World War II—its younger sibling having already been exterminated by the Great Chicago Fire of 1871. The matter of this monster, in Koch's configuration, is now found almost entirely in books of frauds and dupes.

After Koch's Hydrarchos was retired from public exhibit, scientists didn't cease to puzzle over the appearance, and disappearance, of the *Basilosaurus*. The animal long continued to be a wonder: noun and verb. Specifically, this stands out: "Four feet of head, ten feet of body, and forty feet of tail." The words of the American zoologist Frederic Augustus Lucas, writing of the *Basilosaurus* at the turn of the twentieth century. Lucas goes on to lend credence to a novel explanation for the *Basilosaurus*'s extinction. The creatures perished, he suggests, because their heads proved too small to gather and consume enough food to nourish their gigantic tails. Prevalent opinion held then that prehistoric animals amounted to nature's faulty experiments; poorly adapted, ungainly, and of low intelligence, they were thought, in time, to have been outcompeted by superior species. Today *Basilosaurus*'s extinction is attributed to a cooling of the Earth's climate some thirty-four million years ago—a planetary shift that triggered changes in the circulation of the ocean currents these ancient whales inhabited, and in the availability of their prey. The onset of an Ice Age minted new megafauna, like the mammoth and the cave lion: animals we are more familiar with today.

As stagy charlatans positioned whale parts inside imaginary beasts and old tars vouched for tall tales; while zoological taxonomies confirming the classification of whales as mammalian were established, reiterated, detailed, and debated, more spurious reports arrived of—

1808, the Orkney isle of Stronsay, Scotland: a fifty-five-foot-long serpentine clot of tissue with a small head, a long mane, and three pairs of legs. On each foot are five (or six") toes. It's rumored that what are feet could be wings, paws. The carcass glows in the dark. Its stomach contents are bright red. Someone who touches it says, "smooth as velvet." The vertebrae resemble cotton reels. The footed, winged, pawed animal is named by a regional natural history association for a Scandinavian bishop, *Halsydrus pontoppidani*—Pontoppidan's sea snake. It does not exist, this sea snake. Who can say whether once it did? Only two drawings and a handful of fragments remain. A few bones circulate in private collections; others are stored at the Royal Museum of Scotland.

1890, Great Sandy Island, Queensland, Australia: "The flukes were semitransparent," wrote a schoolteacher of the "turtle monster fish" she spotted in the shallows. "I could see the sun shining through them, showing all the bones very forked." She waded alongside it for half an hour. Asked to comment on the claim, one Belgian-French scientist declared that "the whole thing smells far too fishy." "A tissue of contradictions and absurdities"—he called the teacher's accounts—"evidence not of zoological fact but of psychotic behavior, characteristic of mythomania." (This condition of habitual lying, mythomania, often availed to pathologize women's testimony.) "Native blacks call [the creature] Moha-moha," the teacher insisted, "and they like to eat it, and [say] that it has legs and fingers." The fins, not fingers, she saw in the rippling water were chocolate-brown. The thing had scales that shaded silver through to white on its body.

1896, Saint Augustine, Florida, United States: a fibrous pink mass, twenty feet long with several "arms." "The creature could not have been an octopus," proclaims an expert from Yale (though the curator of mollusks at the National Museum of Natural History disagreed). Four horses and six men work to unfurl it in a pit—"no beak, or head, or eyes remain," they note. Some years later, two biologists retrieve stored samples from the Smithsonian: "White as soap . . . the connective tissue dulled four blades," one reports. It is "so tough that when it is exposed to the air, an axe makes very little impression." The flesh had petrified into something impervious.

1922, Margate, Natal, South Africa: Two whales are observed close to the coast "fighting" another "huge sea monster" covered in white fur and sporting an elephantine trunk. "I took [it] to be a polar bear, but of truly mammoth proportions," a witness recounts to the press. The tide sweeps ashore forty-seven feet of gristle and snowy hair, the creature dead and macerated. The pale, anonymous carcass is never scientifically sampled.

1938, Loch Ness, Scotland (naturally, Loch Ness): The novelist Virginia Woolf corresponds with her sister Vanessa, relating what she's heard secondhand from an Irish couple who are "in touch" with the monster: "He is like several broken telegraph posts and swims at immense speed. He has no head. He is constantly seen." Woolf reports that another woman, a renowned golfer and aristocrat, died on the loch when the speedboat she was traveling in burst into flames. She tells Vanessa an insurance company sent divers after this woman, since she was wearing "three thousand [British] pounds of pearls on her head." "They dived and came to the mouth of a vast cavern, from which hot water poured; and the current was strong, and the horror they felt, so great."

1942, Gourock: In the seaward River Clyde in the west of Scotland there appears, on a day of unremarkable weather, a twenty-eight-foot glutinous, bundled accretion from the ocean. Military personnel prevent the public taking photographs (there is a war on—who knows what this discovery betrays?). One council official, Charles Rankin, describes the remains as being "like a large lizard, although its body seems to contain no bones other than its spine. It has smooth skin with thick, knitting-needle-like hairs." Rankin pulls a few bristles out of a flipper and stores them in his desk, where they dry out and coil up like bedsprings. Whatever it was, it was buried in pieces under the town's football pitches.

1959, off the Scottish island of Soay: Two men shark fishing report a head, rising out of the sea to look at them, with "no visible nasal organs, but a large red gash of a mouth which seemed to cut the head in half, and which appeared to have distinct lips." It had "two, huge round eyes like apples." They heard it breathing. (*They heard it breathing*, a horrifying phrase!) When it opened its mouth there were tendril-like growths hanging from its palate. It disappeared beneath the boat.

1960, in the Interview River, western Tasmania: a hairy mass, maybe eighteen feet long, dark with a hump. "A strong acidic reek came off the flesh, very similar to battery acid," a cattle roustabout tells *The Mercury*. Over eighteen months the mass does not appear to break down or decay, this organism of "general outline as like a huge turtle, without appendages" (but perhaps with "quills"). The federal government task an expedition to investigate superintended by John Gorton—later to become Australia's prime minister. On return, the expedition's statement is ambivalent, though it refutes the word "monster."

1988, Bermuda (who could ignore Bermuda?): "Almost like wet wool. It was very dense, and inside of it was white all the way through, and [it] had cells almost like a honeycomb, but not regular"—a salvage diver and historian.

1990, Benbecula (again, the Western Isles of Scotland): "It had what appeared to be a head at one end . . . There was a curved back and it seemed to be covered with eaten-away flesh or even a furry skin, and was about twelve feet long . . . It had these shapes like fins along its back, like a dinosaur," said a child-minder.

1997, Ototoka Beach, New Zealand: "It looks like a bedspread that has fallen off a clothesline."

"Paddling with flappers."

"It purred like a cat."

"The Beaked Beast of Bungalow Beach."

"A snake threaded through the body of a turtle."

"About it was a hard shell like a lobster's."

"Local people were in the process of cutting off the head to sell."

"There were thousands of flies."

"It had also been attacked by something in the water."

"Two things like teeth were seen in the lower jaw."

"Teeth. Teeth everywhere."

"They shot it. It sighed."

"Part of it stayed on the beach for months because no one dared to come near it."

"It," "*That*," "it," "what—"

"Off Newfoundland,"

"Off New Zealand,"

"Off Japan,"

"Off Chile,"

And, at last: "Hoaxes do exist."

Resorting to the scientific literature in the State Library, I thumb through *Biological Bulletin* in search of a paper from 2004: "Microscopic, Biochemical, and Molecular Characteristics of the Chilean Blob, and a Comparison with the Remains of Other Sea Monsters—Nothing but Whales." This forthright title ("Nothing but Whales"!), and the dry, factual methodology the bulletin describes—how refreshing it suddenly seems. I envy the ease of the collective first-person, these findings delivered by a united and definite "we," "*the lab*," and all trace of individual uncertainty, dispute, and hierarchy dissolved.

That the many eyewitness reports of sea monsters are, to date, unconfirmed does not mean that they don't divulge a great deal about the widespread conviction that the ocean remains populated by furtive species people are yet to come to grips with. The belief they testify to is actual, though the creatures, it appears, are not. The oceans have never been more surveilled, but sea monsters aren't endangered animals (even if most of them turn out to be decayed whales, or pieces thereof). You won't find these bizarre beings on display in reputable museums. They persist in the back drawers of public archives, in self-published e-books, and all over the web—particularly where communities resistant to credentialed research and officialdom congregate.

I have dipped into these groups online. They have names like "Things in the Thames" and "UMOs: Unidentified Marine Objects."

Some members identify themselves as cryptozoologists—self-taught scientists of the secret animals. The mode of forum discussion veers wildly between anti-authoritarian cynicism and credulous belief. Conversation tends technocratic, though a cold edge of self-sufficient pragmatism frequently shows through; this is the mentality of frontiersmen, eschewing bookish knowledge for sense. What the cryptozoologists feel driven to investigate, broadly, are suspicious peripheries. Creatures on the edges of being known. Blurs that elude the camera. In the digital age, these conspiracies are catalyzed not by the absence of information but by a surplus of it, siphoned by algorithms and boosted toward the erasure, disaggregation, and distrust of expertise. Images are posted lightened, color-shifted, zoomed into a haze of pixelation or with arrows arrayed around lurking ambiguity. The firsthand reports of hunters, trackers, and experienced game-fishers get kudos. More than once someone submits that what others deem "monsters" would be well-known animals today, if ordinary citizens were not so alienated from nature—if city-folk could even *tell* what *the hell* they were looking at.

Where and when sea monsters have turned up is illuminating. Bermuda, the Scottish Isles, Tasmania, New Zealand. Their distribution inclines to islands—to isolated sites where firsthand accounts can circulate with gravitas, growing ever more cemented ahead of the lagging, absent (or distrusted) verifications of professional scientists, journalists, and camera crews. Writes James Hamilton-Paterson in *Seven Tenths: The Sea and its Thresholds* (1992): "The effect of islands is almost wholly regressive." Islands in the Western imagination remain spaces sopped in libidinal energy; loci latent with "utopias and ideal commonwealths" (though the fantasy of self-reliance and seclusion, as Hamilton-Paterson points out, can equally become one of lairs and prisons). Without being able to pin down exactly how, the appearance of the fantastical and the recognizably extraterrestrial seems induced by the extra-territorial, the extralegal qualities of islands. And what could be more extraterrestrial than a whale, being both not of the land and yet, like us, full of air and warmth, red-blooded and bony?

Yet whales have been mistaken for islands, many times. The *Physiologus*—a Christian bestiary widely distributed in the Middle Ages and translated into numerous vernacular languages—had it that "hoary" (old and graying) whales would deliberately deceive sailors by appearing to be a landscape on which to take refuge. Saint Brendan of Clonfert alighted on such a terraforming whale named Jasconius, which was said to have pulled a coast of beach sand up over its back and become bestrewn with greenery. The saint and his companions lit a fire, which caused Jasconius to dive, nearly drowning the holy men. Whales-as-islands rise and fall to phenomenal destruction in the first book of *One Thousand and One Nights* (1704); in a famous letter from a general to Alexander the Great; in accounts by other saints including Isidore of Seville; through the Babylonian Talmud; in J. R. R. Tolkien's "Fastitocalon"; and lore in Croatia and Chile (the latter a nation adjacent to many whale migrations). Sometimes the island is a turtle or an octopus or a creature called the Midgard-worm, but most often it's a whale.

Today researchers are revisiting these stories of sea monsters and turf-topped whales. Not for evidence of real mythical beings or evolutionary relics in the holdouts of inland lochs and desolate seas—instead they seek the prehistory of a future that preoccupies many: the plasticization of Earth's oceans. What people may have seen, when they saw and reported on these mysterious beasts in the nineteenth and early twentieth centuries, so the thinking goes, were "string-of-buoys" monsters: animals, living, dead, and decayed, trussed in premodern fishing gear. This contention counters the widespread presumption that the lethal entanglement of marine life is a distinctly modern woe.

Prior to the 1950s, fisheries equipment was commonly constructed out of natural fibers and degradable kit—materials that, abandoned or tossed, withered and came apart in salt water. Twisted hemp cording was a mainstay. Linen or cotton fishing nets were reinforced with daubs of resin (ships could carry miles of these nets, with excess for mending). Floats consisted of corkwood, lightweight glass spheres, or inflated pigskins, rather than today's aerated foam or hard plastic, and wooden poles marked the position of fish traps set out on the seabed. Sailcloth, tossed overboard when rent and irreparable, or shredded by high weather, was made from waxed canvas, not—as sails are in the twenty-first century—polyester and heavy laminated Dacron.

But though these simpler raw materials were quicker to perish than synthetic tackle is now, the humanmade impedimenta still fell apart over a longer time period than it took for any snagged creature to decline and die. Hemp fibers swelled in water and withstood rot, particularly where stiffened with pine tar or a mordant dye—or if braided through with wire, as was the case with anchoring cables. Compared to the fluorescent ropes, fine-mesh nets, and industrial oyster lanterns, lobster crates, and fish traps tossed out or lost by today's fisheries, this older gear was inconspicuously colored, wheaten stuff. As it aged, it frayed into shaggy idiosyncrasy. Such dreck would have proved difficult to discern as artificial and of human origin. Bound and dragged by the flesh of an animal, the junk could be confused for an organic extension of that dying creature's body—a longer tail or tendrils other limbs. Shinning glass orbs on the surface could look like buoyant, empty eyes.

Up until the mid-twentieth century, maritime laws and protocols preventing torn or defunct netting and tackle being abandoned near reefs were meager and unenforceable. Tossed onto coral, the nets would be less likely to obstruct out-voyaging vessels, but, being so overlooked, they could roll across whole atolls, picking up a multispecies farrago of turtles, large reef fish, squid, and starfish. A pair of dolphins enveloped in such ghastly maritime bric-a-brac may well have appeared like a double-headed, teratological nightmare afloat on the sea or swept, thrashing, up onto the beach. A dead whale could look like its serpentine ancestors. Which raises the obvious possibility: were the monsters and sea serpents that flourished in the nineteenth century in fact trash amalgams of human making, birthed by expanding maritime ventures? Is there evidence, in the literature of firsthand encounters with sea monsters, to support this claim? If yes, then the history of how marine debris has shaped ocean ecosystems reaches back into the prepolymer age—and the threat pelagic pollution presents to marine life has a past more extensive than supposed.

The more interesting matter, to me, is how identifying animals fettered by waste *as monsters* in the nineteenth century echoes our contemporary fear of the unfolding scale of humanity's reach, our terrifying impacts on the aquatic world. Throughout the mid-1800s, the trackless Atlantic was bridled by telegraph cabling. Even as fishing operations expanded, fish hauls had already, portentously, begun to drop. American catches declined 10 percent between 1880 and 1908—a statistic that surely would have been higher had the steam-powered trawler not advantaged the industry, sieving up many more bottom-dwelling species than in previous periods (and indiscriminately destroying unsalable populations of fry and fish spawn). Increased sightings of rare and mysterious creatures met the rising apprehension that the sea could be—was being—emptied. The biographies of monsters surely contain the emergent anxieties of any age, though ours seems more convergent than most.

Deeper down in the web today can be found the writings of a few fringy natural historians and cryptozoologists who decry human activity as the very reason for the rarity of real ocean monsters. They say the stranger animals have been extirpated by pollution, the depletion of their prey via overfishing, and the insidious strew of pelagic plastic. What has been incompletely described has vanished before it ever came to light. Or else, likely being biosonic creatures, the monsters have retreated to loiter in underwater trenches away from the clamor of trawlers, naval sonar, and diesel-powered ships.

At the nadir of these arguments are appeals for action plans to protect unseen, unidentified, and undocumented creatures, and to extend our compassion to a vaguery of organisms yet to be discovered. Restricting ourselves, in service to other species—creating zones sanctioned for the mysterious in a world where mystery grows daily rarer. Here, surprisingly, appears a provisional model for assuming responsibility for that which is beyond knowing and outside of our control. The call is to *stay mystified*. An inverted conspiracy, this thinking assumes no hidden order in things, no classified or suppressed knowledge; it preserves, instead, nature's innate chaos. Undergirding these proposals, as I've come to understand them, is an ethic not rooted in any green principal, but originating from a capacious sense of privacy. Who else protects the right of alien wildlife to be left alone, unknown? Who will be the savior of their freedom *from us*? Squint, and these questions begin to look like siblings within a broader, conservative-libertarian prerogative to live free from the monitoring of science, the overreach of federal governments, or both.

To believe in the continued existence of large, incognito animals restores nature's richness and its capacity for surprise; a glimmer of magic in a time of humdrum environmental despair. Beyond the brink of our surveillance, Earth seems feasibly wilder, more deeply, strangely textured. People are made smaller and less all-knowing. The appeal of this sentiment makes utter sense to me. What jadedness, to assume all the planet's wondrous beasts have been disclosed! I, for one, long to dismiss the shame and pessimism that makes it easier to envision the existence of monstrosities of human creation—plastic-animal storm clouds, scudding on the seafloor—than verified mythical beings.

The more time I spend thinking about it, the harder it gets to tell whether admitting the possibility of sea serpents is an act of humility or hubris. But I do know it matters, the preservation of the wild as an imaginative space, and not only because what we call "the wild" is an old and precious idea. Losing a connection to the notion of an untamed natural world estranges us from our own species' evolutionary origins, and, too, the instincts of our language-less infancy. The animal *inside*, the creature that quakes at what twitches beyond the campfire light; that animal, too, needs protection. When there are no places unfathomed by technology, when wilderness is excoriated in its entirety, something inherent to our humanness is also lost. We kill off the animal in ourselves, the part that belongs to the wildlife. We extinguish a species of wonder.

In the end, I can't decide if I ultimately side with the cryptozoologists, who would protect what we cannot see, or the conservationists, who call us to pay attention to the partially disappeared. But even if I can't put my faith in monsters, I still believe there is useful thinking here, or at least the beginnings of it—a fragile bridge between conspiracy theorists, scientists, and environmentalists. How to care for unmet things would seem to be a key question of this political moment.

In the indigo of the ocean's depths, in sea-canyon habitats, below the cold, plinking ridgelines of continental slopes—and in more temperate seas, misty with microscopic life—there exist cetaceans never seen alive. The earliest identified spade-toothed beaked whale (not even a complete skeleton: a clutch of distinctive bones) remained the only spade-toothed beaked whale known to science for one hundred and forty years. Then, in 2010, a pair washed up on a beach in New Zealand's Bay of Plenty. After years of DNA testing, their cadavers were certified as members of the species, though no one has reported seeing another spade-toothed beaked whale since. The Shepherd's beaked whale, meanwhile, was only seen underwater for the first time in 2017. A relative to both these whales, the male gingko-toothed beaked whale, has two teeth in its otherwise toothless lower jaw. These teeth are, per the animal's name, yellow and shaped like gingko leaves. The teeth are so oddly formed that they are probably not used for eating—they might be for fighting, but no one has ever seen these whales fighting or eating. A handful of gingko-toothed beaked whales have been caught in longline fisheries and gill nets, and less than twenty worldwide have washed up in places where people can examine them. Preferring cold, very deep waters, these whales stay far from land. The animal's skin is known to darken after death, like underexposed film, so the color of a living gingko-toothed beaked whale remains unknown.

A jet-black beaked whale known to Japanese sailors as *karasu*, or "raven"—a whale American scientists long held to be a "big fish story," or a misrecognized juvenile from another species—was formally identified in 2016, after scientists matched the genetic analysis of skin from a number of decayed cetaceans, with tests done on a whale skeleton hung in the gymnasium of an Alaskan high school. ("We knew it was not any whale we knew," an ecologist involved with the study said—and yet, the creature was a local team's mascot.) This black whale, over twenty-four feet long, doesn't have a name yet, not even a *nomen nudum*—the "naked name" given to an organism incompletely described in the official annals of biology. "We don't know how many there are, where they're typically found, anything," said Phillip A. Morin, a lead researcher from NOAA, to the *Washington Post*. "We're going to start looking," he added.

There may be as many as twenty-two different species of beaked whales, the most enigmatic, elusive cetaceans (elusive, that is, to us and our tracking technologies). Perhaps they amount to nearly a quarter of all known species of cetacean. Likely squid-eating, maybe moving in pods of thirty, the most verifiable data-point about these animals comes from a single satellite-linked dive-tag, showing a Cuvier's beaked whale plunging nearly two miles in pursuit of prey—a depth it remained at for over two hours. These whales have torpedo-shaped bodies. A defining feature of their genus is a muzzle that narrows into a snout like a dolphin, most with an underbite. One, True's beaked whale, has slight hollows under its pectoral fins where it tucks its flippers when it darts or dives. Another, the male strap-toothed whale—with black and white markings not unlike a panda's—possesses a pair of flat tusks that lengthen, over time, to wrap together and gradually bind the upper jaw, narrowing the whale's gape. As they age, these whales must therefore hunt smaller and smaller squid, which they suck through the tip of their beaks the way people eat noodles. To mature, for the male strap-toothed whale, is to grow hungrier and, of necessity, to acquire the agility needed to pursue littler, faster cephalopods. (*The creatures perished*, the zoologist suggests, *because their heads proved too small to gather enough food to nourish their gigantic tails.*)

Populations of beaked whales could decline or boom or disappear entirely, and we would never observe or quantify the change; they are "data deficient" species in the parlance of the International Union for Conservation of Nature (IUCN), responsible for holding data on the relative abundance of the world's species. These animals, many as long as buses, exist offside of natural history in a space between speculation and specimens. Much more is known about their ancestors, from the fossils stored in museum archives, than is known about the living creatures. They have been called "the least understood group of large animals on Earth."

The word "species" lifts off out of "*specere*": to see. A species is determined by examining it and separating it, categorically, from alike others, to fit within the characteristic parameters of a certain name. In the ordinary course of fieldwork, and in laboratories, biologists identify one animal from another by relying on their own senses, and by using the articulations of technology. Cameras, submersibles, microscopes, telemetric tags, and DNA sequencers have all proved instrumental in describing individual animals and defining species. Yet the limits of our perception still confine not just what becomes known, but also what is noticed when it recedes, shrinking toward extinction. What we cannot see, or possess no equipment to apprehend, slips past unobserved and un-quantified, sounding no alarms. This, the as-yet-unwritten world.

Scientists modeling the effects of industrial whaling prior to 1985 have observed that the removal of so many cetacean bodies out of the sea by vessels at the surface almost certainly caused unknown and unclassified bivalves and saprophytic worms (the tiniest sea serpents) to die out when the whalefalls that functioned as their seafloor oases dwindled, and the distances between whalefalls increased, proving unendurable for their minuscule, sometimes luminescent, larvae. Not only were there fewer whales in the sea as a result of industrialized whaling, but commercial fleets selectively pursued the biggest animals, as well as the older, slower individuals, those closest to a natural death, deleting these whales from the undersea cycle. Invisible in the then-unvisited deeps, the consequent disappearance of highly specialized "whale bone fauna" and florid "seep genera" may have been some of the earliest marine eradications ever triggered by human activity. No one saw this happen, though ecologists are certain it did. Things were lost, that is, before they were ever found.

What we fail to see, and fail to name, may nonetheless still fail to survive us.

Though many more charismatic creatures have since evaporated off the Earth, the lessons of these eye-blink extinctions remain pertinent. Defaunation—depleting and displacing animal biomass; in this case, the immense bodies of whales—proves tantamount to terraforming and habitat loss for less conspicuous, subtending creatures that humans may yet encounter. Life existing outside of language is predicted to be abundant, immeasurably so, and especially in the insect kingdom, the microbiome, and slinking over the seafloor. In some parts of the world new species are uncovered and named today at a far greater rate than existing species go extinct, which can create the misleading impression of local biodiversity inching up when, in fact, it is simply the lists of names and the volume of databases that grow. Indeed, the same drivers of animal discovery may be generating animal endangerment, as when development expands into wilderness and the ocean deeps, unveiling previously secluded creatures now at greater risk of homelessness, exploitation, or exposure to new pathogens and pollutants.

A grave turnabout: in the eighteenth century Linnaeus was willing to preemptively include uncorroborated beasts in his taxonomy—their inclusion was a testament to the faith in undiscovered abundance that defined his age and his people. Today, the IUCN's data-heavy lists of species include an increasing percentage of animals and plants that *certifiably* do not exist anymore: an anxious quantification of the loss of creaturely plenitude that characterizes our environmental moment.

This is most troublesome because, not only after death as a

tumbledown feast but also throughout its existence, each animal is an ecosystem and a home. Inside whales, minor orders of more hermitic monsters eke out their existence. Take, if you will, the twenty-six-foot internal whale worm *Placentonema gigantissima*: a being known only by the noble horror of its Latinate name, for it lives apart from any common familiarity. *Placentonema gigantissima*—white, blind, putty soft, the width of a pencil—lives spooled within the dark recesses of its female sperm whale, latched in her uterus. Heartless, bloodless, and lungless, the worm is nonetheless alive (and twitchy). Stretched out, *Placentonema gigantissima* can grow to a length as long as four queen-size mattresses laid end to end, or over three-fifths as long as the sperm whale's own body. No one has yet been able to confirm how the worm reproduces or where it comes from. Other parasitic whale worms have been found to absorb heavy metals from the tissues of their hosts, creating intensified hot spots of contaminants within the animals' bodies. It is thought they have the potential to function as biological "tags"; early warnings of pollutants escalating inside offshore animal populations.

Revolted, I have tried to resist picturing the gigantic *Placentonema* worm—this proves impossible. The worm in the whale's unlit interior inverts charisma so completely, it engrosses me. Mucilaginous and mixing in its own elixir. An abhorrent jack-in-the-box. The worm makes me shudder. For whole days I can't get it off my mind. What is the expression? I can't stomach it.

Whale suckers, the scarf-length remora fish that hitchhike on the outside of whales and dolphins, are not bloodsuckers like leeches (as I had, at first, imagined). Instead, the suckers vacuum up tiny copepods that gel to their host's skin, and they are also known to forage in puffs of cetacean scat and vomit. A whale sucker looks like an eel that has been stepped on by someone in sneakers—their heads are waffled like a shoe-sole. No one is really sure where whale suckers come from to junket on whales; it may be that, when the skinny fish spawn, their fry and fingerlings live in some whales' mouths, sheltered by sheaths of baleen. The suckers prefer warm waters and are more plentiful there, so some scientists study them as indicators of rising ocean temperatures. In the eastern Pacific, whale suckers are thriving.

Smaller yet, but still visible to the human eye, are whale lice. The lice are crustaceans more closely related to shrimp than the lice found on furred, terrestrial mammals (which are insects). Whale lice tend to gather around their host's eyes, and find footholds in nostrils, genitals, and wounds, in the grooves beneath the jaw and throat, and the corners of the whale's immense mouth: here they reduce their chances of being washed off. In clusters, the lice appear orange or pink, which can sometimes create the uncanny impression of lipstick imperfectly wiped off a whale. Aboard right whales, whale lice congregate around barnacles and what are called callosities. Callosities, horny knobbles that form unique patterns on the animals, are present even on right whales in the womb, though the lice roughen them over time, making them more striking. Whale skin is the lice's diet: louses lever up flakes to nibble. Do they irritate the whale, the lice? Who can tell? A single whale louse to a whale is a pennyweight presence. A whale to a whale louse, meanwhile, is a meal of preposterous magnitude: a lifelong, planetary banquet. The lice that eat the whale-skin flakes are eaten by the whale suckers, then the whale suckers are eaten by sharks and tuna (which may be consumed by humans), so that the edge of a whale is the starting line of life for many predators.

Everywhere whales go, whale lice go with them. A whale is yet to be found not bearing whale lice: they are prolific. Seven thousand whale lice can inhabit one whale, which researchers have described as a "living island" for louses. Whale-lice species are distinct to baleen whale species—humpbacks have one kind; gray whales another. When whale species decline, so do their unique lice. The lice may even be sex-specific (one type of louse found living on bull sperm whales, for example, does not colonize females, though both sexes additionally share a third, species-specific species of sperm whale louse between them). Ergo, there can be multiple types of whale louse aboard a single whale, inhabiting its many different crenulations the way different seams of plant life sometimes flourish in the troughs between desert sand dunes.

Whale lice have no free-swimming, larval phase, so they have to be passed from animal to animal during birth, lactation, sex, and other contact. Sometimes a whale louse from one kind of whale will briefly turn up on another kind of whale, which raises questions about how closely these cetacean species interact. When a type of louse loyal to humpback whales was identified on a southern right calf, for example, researchers suggested it was possible—though this has never been witnessed—that mother humpbacks could be nursing infant southern rights off the southeastern coast of Brazil.

Biologists, seeking to refine their estimates of the size of vulnerable whale populations prior to pelagic whaling in the twentieth century, have found a valuable data set in the genetics of whale lice. The lice are retrieved off the skins of surfacing whales using long-handled pincers, or else they are collected from stranded and decaying whales, which the whale lice tend to abandon, scampering away over the sand (sometimes they go up the forearms of unwary collectors to nestle beneath shirt cuffs, though they cannot live long without locating another whale host). Populations of whale lice undergo many generations on the body of a single whale during its far longer lifetime—whale lice live a mere two to three months. The genetic differentiation of the lice can show how frequently their host's ancestors encountered other whales (and thus, in light of the social habits of the whales, the size of the populations they existed within).

In right whale populations, studies of whale lice mitochondria have helped not only to calculate past whale abundance, but also to pinpoint the time period during which southern, North Atlantic, and North Pacific right whales branched from one another along an evolutionary scale, separating into three species (as, in time, each right whale species acquired a unique louse species of its own). In this way, the whales' parasites function as an index card to the longer history of their animate homeland. Researchers from Utah write that whale-lice gene sequencing has shown that "at least one right whale crossed the equator in the Pacific Ocean within the last one to two million years," transmitting parasites to a new hemisphere. Today, right whales avoid Earth's tropical oceans, where the temperatures prove too hot for them.

That whale lice, these scuttled smatterings, can document the migration of a single right whale over a million years ago is, you must admit, astounding.

Before I started looking into all this—what I have come to think of as the gothic-domestic aesthetic of marine parasites and decomposers—

I couldn't understand how, or why, anyone should care for whale lice, horrifying whale worms, or anonymous seep creepers, barely embodied on the seafloor. These inharmonious, inaccessible, thoughtless, frankly ugly things—they seemed, to me, to occupy a categorical wall cavity between two, more beloved groups of animals: wild species and tamed ones. Belonging to neither category, the countless cold-blooded, hair-raising beings that lived off whales could only repel human affection, just as surely as their hosts attracted the same. There seemed to be an inverse relationship between a host species' charisma, and the weirdness, the repugnancy, of its parasites and dependents.

Though I saw that parasites and oceanic detritivores existed in places deemed wilderness, to me they elided many of the qualities that defined a wild animal. These were not self-governing, uncultivated, ferae naturae. Could a whale worm, internalized, be considered a creature of the ocean when likely it never touched, nor knew of, the water? And wasn't a whale louse's loyalty to one type of whale only a step short of domestication? If, in saying the word "nature," people mean to summon, as I think they do, "the backdrop" or "the environment," well then, parasites did not arise out of nature at all. Parasites came to life in the crucible of other animals, bituminized into their flesh. They exploited larger animals' quiescence by being too small, too evasive, too deeply lodged or dispersed for their host to remove. Yet, however persistent their species, parasites were, of themselves, appallingly passive. Their driving instinct was not for freedom. Parasites truckled to dependency, to conjunction and symbiosis in bodily niches where no other organism deigned to inhere, and only bacteria and microscopic life otherwise bloomed. To survive at all, whale worms needed to be unwillingly corralled by larger animals to which they remained captive for their entire lives. If they were removed from whales, they soon limply died. There was something awful about the pressure that produced parasites. As though evolution had achieved a force of creativity so powerful, weird, and severe, it had turned aberrant and tuberously ingrown. At the heart of big life, this twist of death-y-ness.

I thought back to the whalefall, too: to the tiny creatures weaving a lurid carpet of decomposition over dead whales on the seafloor. Their lives were lived in a state of such impotent contingency, that, though I could imagine them as wild compared to parasites, I couldn't think of those marrow-nourished annelids, isopods, clams, and sea snails as independent animals. They were more like fairy rings of toadstools that rose up after rain; possessed of a directionless persistence belonging to vegetables, or crystals encrusting a cave. They were an essence that flickered and ran.

After dwelling, for too many hours, in back issues of the *International Journal for Parasitology*, a line from Annie Dillard surfaced unbidden within me. I heard it like a voice. "Something is everywhere and always amiss." I located the fragment in a tattered copy of one of Dillard's early books and, forgoing yet another edition of the parasite journal, read on:

"Something is everywhere and always amiss. . . . It is as though each clay form had baked into it, fired into it, a blue streak of nonbeing." The blue streak of mortality: this was what Dillard was referring to. Yet whale worms, whale lice, whale suckers are nonbeings, too, set against the whale, the massive being that is their birthplace, habitat, and the genesis of their labor. For parasites are very like death: the faceless thing amiss, everywhere, finding purchase in even the most infolded, sunless surfaces. Long considered indicative of sickness, derangement, and decline, the most unnerving quality of any parasite is its profuse, persistent *vitality*, the wiggling in the womb, the itch on the skin, clinging, consuming, breeding. Worse than a ghost somehow. A parasite will typically reproduce faster than its host, and may exist in greater plenitude; the million-faced beast upon the beast. Being more profoundly alive, according to these measures, than their hosts, parasites can, in contrast, make the larger animal itself seem ghostly and partial—for less of it consists of itself.

Oh, but the plenitude! Didn't the whale lice, specifically, have a machine-like quality to them? How they proliferated, reduplicated, simple bits of code-making copies, extruding one another. Their teeming fecundity also repelled me: how closely their natural accretion resembled a computational process. Could you view the lice as *a whole second animal*, netted thinly across the skin of a whale? Of course not. No louse is a fragment, each is a whole. Though every louse "preys on less than one animal" (E. O. Wilson's famous definition of a parasite).

Connected to this notion in my mind was the idea of ecological superfluousness. Functionally I could not see the whale worms as *useful* to the whales, or to any other organism. Perhaps the worms were of value to science, of course; to whomever was tasked with picking through worm-innards in search of trace amounts of polluting copper (like a haruspex seeking divinations of the future). The whale lice, too, expanded our understanding of whale evolution. Whale lice even enabled scientists to better quantify the harm humanity had done to whale species, and whale suckers provided evidence on a broader remit of damage (being, to some extent, symptomatic of rising ocean temperatures). But to the whales, the parasites were surely only a burden. Primitiveness, plenitude, partiality, purposelessness. I wrote a horrible list.

I had long believed, as I later came to see, that nature is supposed to work in harmony. The reality of parasites conflicted with that conception. Celebrating the return of whales after whaling; fostering compassion for theoretical sea monsters: both these sentiments, the way I understood them, sustained positive visions of the natural order of things. Whales rebound—the force of nature is resilient; nature cures, restores. Sea monsters exist—nature is a source of awe and humility. Yet when I looked at my list of parasitic qualities ("primitiveness," "plenitude," "partiality," "purposelessness"), I could not deny that this, too, was nature. An integral underside to the wild; a "depth ecology" that the writer Gary Snyder once spelled out in his own list that went something like—"moldy, cruel," "crooked," "irreverent, inharmonious," and "virtually inaccessible." But *necessary*.

Some biologists argue that parasites and their hosts should be viewed as a whole, as united entities, because the evolution and health of all animals is produced in interaction with both their outer and inner environments. Relationships of mutual benefit between parasites and hosts are called "symbiosis," and though not all parasites function to improve their host's welfare, recent close study of these dynamics has shown that symbiotic interchanges are far more common than once believed. Gut parasites can aid immunity in some animals, for instance. The presence of other parasites can ensure that potentially harmful microorganisms and bacteroides are kept in check. Though whale suckers create water drag and, in doing so, impose an energetic cost on whales, they almost certainly also help their hosts by preventing the overgrowth of smaller parasites flourishing on a whale's skin (and infectiously, around wounds).

Even where a parasitic species afflicts and impairs individual animals, the overall effect on moderating population, and creating functional food webs, can be consequential—the range of these secondary impacts are only just beginning to be understood. A few terrestrial parasites have been called "ecological puppeteers" for the ways they change the behavior of their hosts and alter the ecosystem—for example, by prompting crickets to jump into streams where insect-eating fish eventually thrive; or by enfeebling moose, therefore supporting more wolf packs. In the longer timescale, evolutionary science has demonstrated that, beyond competition, parasitism can also act as a stealthy driver of novel gene mutation. Sexually reproducing, parasitized animals sometimes have more diverse offspring, presumably to out-evolve the parasites that exploit them. Genetic recombination fuels diversity within a species, which in turn can make that species more resilient to environmental shocks (a reduction of ice cover, say, or a sudden decline in prey), by providing a greater breadth of different possible adaptations to a new set of conditions. In a warming, acidifying, lonesome ocean, these shocks are likely to arrive more frequently. What a parasite inadvertently prepares any animal for, may await it, mistily, in the future.

Before any threatened animal species tips into extinction, its populations often become divided into undamaged regions, inaccessible places, or environments separated by the exigencies of distance. Subsisting in reduced or partitioned territory, or with their numbers depleted because of a pathogen or vanishing prey, fewer mature individuals tend to encounter one another, less frequently. This is particularly true of creatures that are solitary in adulthood, or which spend an interval of their development living segregated from any pack, herd, or pod—as do many of the great whales. Just as deforestation, pollution, and the construction of roadways can accelerate the demise of larger creatures by erecting barriers between their populations, so, too, does defaunation amount to the fragmentation of livable territory for parasites. When the numbers of a parasite's host species change, and particularly if the host tips toward becoming an endangered species, then opportunities for parasites to migrate to new hosts, and to interbreed, lessen.

For some parasites, the failure to be transmitted by contact between animals will imperil their existence: species of whale worms, for example, the life cycles of which depend on being conveyed into mature individuals in a variety of host species. Many parasites die with their hosts at the end of their life, being unable to live freely from them. If a species of whale worm—not "primitive," as it turns out, but highly honed to exist within a specific nook inside an animal—fades away alongside diminishing numbers of its host, and their less frequent interactions, so an opportunity emerges for that niche to be colonized by something else, something noxious, or nothing at all. This is another kind of dewilding—taking the wildness out of the whale—for even though extinction may never be visited on the host species, the depletion of its parasites has the potential to reduce the pool of biodiversity contained within it. We have only a very dim awareness of how this might transform whole ecosystems. In recognition of the importance of parasites, there are conservationists and parasitologists who argue that endangered animals taken into breeding programs should not be deloused, for fear of eradicating one threatened species in the course of attempting to save another.

A parasite is a monster we would not wish to meet; all the more so for undoing the bodily integrity (the purity) of its host. The existence of parasites makes evident the porousness, the uncontained qualities, of life; these things that move under the traction of their own will, into and out of, and across all animals. In contemplating this kingdom of dependent organisms, I have found myself stepping away from my own senses to some zone of thought eerily outside of the human scale on which I experience the world. By which I mean, many of these things exist in the gray space of what is perceptible, and a great number of them simply cannot be seen: they occupy the internal organs and interstitial cavities of whales. These creatures are not knowable, are scarcely imaginable through the faculties of our species (twenty-six feet of worm scrolled *inside* a cetacean?). Yet, in a similar way to discovering that cetaceans could accrue pollution and industrial residue, the parasites also inched me toward thinking of animals themselves as environments, and nature as a process—rather than a series of animate entities in natural places. The irksome and rasping things; the swarm, the squirm, the sucker, the spiral. They were the shape-shifters of our Earth, blurring the edges, kicking the insides out.

In the end I realized that a whale body, seen up close, *was* an incubator, a zoo. It housed many different ways of being alive, which might have been somewhat magical, if it weren't also so spooky. If we could learn from the parasites that everything is *not* quite itself, and that it never was—that there is deathliness and irascible vigor, and plurality and plunder, pushing and pulsing within each creature—then we might undo the charms of charisma and expand the boundaries of our care. We might embrace responsibility for even those things that lie now under our perception, and beyond our control.

Heraclitus—an ancient Greek thinker so intractably melancholic he was referred to as "the weeping philosopher"—once described a group of boys who talked of delousing one another before a long journey. Heraclitus repeated the boys' statement about their body lice as a riddle. In his retelling, the boys declared:

> What we have caught and what we have killed, we have left
> behind. But what has escaped us, we bring with us.

For a long time, what whales there were for whale lice to thrive on had to evade whaling ships and remain unnoticed so that their descendants might continue into this world, the seven seas of the twenty-first century. The convictions of some cryptozoologists echo this history: zones of sparing are conduits through which unknown animals can pass, to appear in the future (*we are attempting to survive our time so we may live into yours*). When the indirect impacts of human activity escape our imagination, and when we are insufficiently far-sighted, then we risk the lives of unmet and ill-considered animals. If restraint *can* still leave environments unaffected today, then we preserve the possibility of many more creatures beyond our ken.

The *beyond our ken* part; that's the part, of course, that keeps me writing.

One recent summer I headed off with a group of artists and scholars to camp at Ingar Dam in the Blue Mountains outside of Sydney. We intended to write about the weather in Australia—which, after a series of vast, almost supernatural, heatwaves, had left us jittery. It made sense to take ourselves away from the insulations of city living to a place where we could really feel the conditions, where the moon was huge and cool, and nearly touchable.

For four days we talked and wrote together. We observed the scribbling of ants in their frenzied forecasts of cloud cover; we discussed the sky that stutters and chatters when ionized before a lightning strike. One among us kept a record of the diurnal shift-changes of cockatoos and yabbies. I wrote with charcoal. The artists called this "a forced restraint." What I wrote about, in slow, childish cursive, was the wind. The point of the charcoal had little to do with its nuance as an artistic tool—the charcoal, in this case, was more an instrument of metaphor. I had wanted the alchemic potential of using an object that had been vaporized—literally, an implement that had had all the air and water burnt right out of it—to describe the wind, or the heat. In the evenings, we swam in the dam, its water like iced tea. We drifted over a toyscape of waterweed and small crustaceans that my mother, only half-joking when later I described them to her, suggested we ought to have skewered and baked in the campfire.

A few days after we all returned home one of the participants— a woman from Ontario, whose academic field was theories of walking— phoned with alarming news. A microbial guest had climbed aboard her during the retreat, an amoebic parasite called *dientamoeba fragilis*, most likely gleaned from the dam. Over the next week, we all came down with symptoms of this common, single-cellular pest; we suffered abdominal pains, had medical tests, were prescribed strong courses of antibiotics— which we then met with various probiotics and tinctures, sachets that were whipped into a salty froth and swallowed hastily before bed. A member of our party, being pregnant, was told she had to wait to be treated until after she returned, notionally, to living as a single organism with only one heart beating inside her.

The idea of interiority that I knew best was one from fiction—the sense of an inner consciousness that the voice of a character mobilized in a novel. But I was beginning to think that this literary construction of interiority—a wafty mental realm—ran roughshod over a more real and important, if insensate, inner life that we ought to be paying keener attention to. That is, the inner life that consists of parasites, microscopic pollutants, micro-plastics, and microbes. The real drift of the world as we create it, and it creates us, as it passes through and settles deep into our bodies. The human, too, is the stable of other species. ("Welcome to the circus of the self," someone wrote in an email exchange that followed my time in the mountains.) These enigmatic microflora and fauna that influence our moods, our appetites, our outlooks, and may even shape our senses, though we usually can't discern them until they are in some way disrupted. Egocentrism turns out to be zoocentrism, in the fine print.

"Anyone who lived in Sydney in the late 1990s remembers the *Cryptosporidium*, the *Giardia*," my neighbor, the poet and essayist Aden Rolfe, said to me when we met for a glass of wine one night around this time. "We boiled the water, or we just lived with all those parasites. They were in the catchments, they were in our friends, they were inside of everyone." A kind of communality, too: a city's populace is something else's reservoir. Giddy on the wine's implication with the antibiotics I was taking, I wondered about all the amoeba I had already killed inside me, and what stories, of what environments, had been knotted into the genes inside them.

Deep End

THE SAND ON NEWPORT BEACH IS COARSE AND REDDISH, LIKE crumbs of stale loaf cake. Poor matter for sandcastles, but people have been trying to build them nonetheless. Waves run rumpled over their efforts, then fan and slide glassily back, unpeeling a surface that hisses. A cabdriver, wearing a blue uniform with epaulets, stoops in front of me. She unlaces her thick-soled shoes and steps out like a sleepwalker leaving the shoes behind, the tongues pushed up. Sand freights the cuffs of her trousers. A few strides further on she strips her socks off and pushes them, balled, into a pocket. More school brogues, slip-ons, heels, and boots with loosened laces are stashed in the swale. With over a hundred others, the cabdriver and I follow a churned channel of beachfront beneath homes with private trails threaded down through the dune grass. It is late in the day. Sea peas are scurried around by the wind, leaving clock-faces on the ground. Both the sun and, there, a faint, fingerprint moon, are in the sky.

Earlier my friend Bird Chisholm called to tell me that a humpback, days dead, had been swept overnight into one of Sydney's ocean pools. Sunrise swimmers were phoning in to the radio stations, reporting what they'd seen. The announcers dedicated a gospel song, "for this great life travels into the light now," or so they'd said. Bird caught me at work, coloring in a patch of my thigh with a black sharpie to mask a rip in my tights. We talked about the whale while I finished this small act of guile.

Chiseled into the city's coastal shorefront, Sydney's saltwater lap pools and public baths are nightly sluiced by the tide. Big waves are known to sometimes dump pieces of coastal life into the pools: turtles, octopi, bluebottle jellyfish, the occasional, small wobbegong shark (sounded out as *woe-be-gone*, befitting the tattered faces of these harmless reef dwellers). Mostly, though, the breakers plow dark masses of seaweed over the barriers—clumps that loom frighteningly into view, like decapitated heads, as you slog along, eyes down, mid-lap. I'd never heard of anything so large as a whale body being lifted over a sea pool's wall by the tide.

Bird wondered aloud if there might have been a storm surge in the evening. Dry lightning quivered over our quarter of the city; the sky taking and retaking its own photo, fretful and adolescent. Thunder shouldered across the lawn. Somewhere it is always raining on the surface of the sea, but in the night there had been no release above the suburbs; the weather only built, the pressure turning the city's inhabitants edgy, then drowsy.

I decided that I wanted to see the whale for myself, though it was hours away on the other side of Sydney. Thanking my friend, I plotted to skive off, feigning a migraine.

More than a few of those headed back to the ridge road at Newport seem to be (yes, they are) softly crying. One or two people stand with their arms crossed in the dimpled access paths that lead to the beach, reluctant to walk further, or perhaps not ready to leave. A recitative I hear echoing along the queue when I arrive is: *Are you coming or are you going?* The tall rocks at the headland in the distance tower over the pool; a 164-foot concrete rectangle, roughly hewn, with a chain-link railing around its perimeter. An onshore wind overruns the ridge, shaking the leeward salt grasses to their root tips and jittering the low foredune, so that the surface sloping down to the water is all vibration and buoyancy; its topside less terra firma than billowy. Above, the sky is stacked with sculptural cumulous clouds, their appearance of fixity belying the wind-brushed earth below. I think of ancient Chinese paintings: midair mountains from the Tang dynasty, afloat in the transparent nothingness of mist or a lake.

After the juvenile humpback in Perth expired and its eyes turned smoky, the crowd (smaller than the turnout at Newport) looked away. It was as though a whistle had sounded. *Time!* People were fast to gather their things from the beach. They stamped over the chaff of shore plants, stopping momentarily beneath the acacias at the entry to the car park where they shuffled into rubber flip-flops and shook out their towels, before jogging to their vehicles. *Vamoose*, the collective resolve, *scram*.

No eye contact or chat in this disaggregation. The mood, I thought, approached guilt, but fell to the side of that word. A kind of humiliation. No, narrower: embarrassment. Embarrassment to have gotten so caught up in the spectacle.

We were relieved, then, to be relieved of the remarkable. Enough pageantry. Amazement proved a finite resource, and at its terminus was weariness. We turned our backs and put both dread and wonder behind us.

But though the humpback whale in the Newport pool is already dead (days dead), people have been arriving since the morning, ecstatic, enthralled and stricken. I'm interested to learn what I can from observing these other whale watchers; whale *carcass* watchers. There was no hope for this sole humpback, so could there be anything hopeful in the spectatorship that attended it? Can a dead humpback whale still conjure the biophilia of ecotourism, or does it become, in decay, an object of biophobia, of disgust?

A train, a bus, another bus, this one smaller—it took three hours and three modes of transport to get here, to Newport Beach, with long stints of waiting between. I arrived at the beachfront wearing my tights, now socked with damp and frigid sand. Others are also still in their office clothes. To enter the beach people have streamed through the thoroughfares of Mona Vale, Clareville, and Newport. These are Sydney suburbs where the lawns run right to the curb and kids knock cricket balls down the driveways in the early evening. Steep suburbs, quartered by tall fences, where the houses are as white as cruise ships rode up onto land.

Today, the season winding up is an oversteamed spring. In the afternoon the cloying scent of trumpet flower swung in warm cupboards of air that I walked into, and out of, along the street, dragging the perfume behind me like some sheeny, liturgical garment. "Butter bomb," I read, turning over a plant variety swatch twist-tied to a leafy fence. The whale in the Newport Baths—I wondered if it would explode. Or would it be towed out to sea before the situation got too messy? When a whale dies on a beachfront in the United States, I had read, "frequently, the body cavity is filled with heavy material: scrap metal, heavy chains, [and] train wheels," to ensure the carcass sinks without attracting too many sharks. And what would researchers in the far future think of these whalefall sites, replete with cetacean bones and pieces of metal haulage? That the whales of our age had fed in human wrecking yards? Or that Americans were a people who so deified their machines, they ritually buried them inside the sarcophaguses of the biggest living animals on the planet?

Vans branded for the television stations were double-parked around the access route to Newport Beach, some sprouting huge satellite dishes and dozens of antennae. No journalists or camerapersons attended the vans, many with their doors thrown open. Power cables had been taped to the asphalt. A few residents turned on their sprinklers to stop people drawn by the whale walking across their front lawns. Heedless, the out-of-towners had continued, in growing numbers.

In his book *Wild Ones: A Sometimes Dismaying, Weirdly Reassuring Story About Looking at People Looking at Animals in America* (2015), the author Jon Mooallem recounts how crowds drawn to the aid of a humpback whale diverted into the Sacramento River in 1985 unknowingly trampled the fragile habitat of an endangered butterfly. The Lange's metalmark butterfly population, already in a downward spiral, diminished to a count of just under 250 individual insects in the seasons that followed—long after the whale ("Humphrey" to its fans) had left the river, guided by an armada of vessels broadcasting whale-song back into the Pacific. "[A] tragedy of charisma," writes Mooallem, "because of their affection for a single celebrity whale, hordes of people had jeopardized an entire species of anonymous butterfly." What beings were we unseeingly stepping on as we overran the Newport gardens and the dunes? Did it matter if they were endangered or not? How much did it matter? Whales are the largest animal we know of on our planet, but to smaller, subtler life-forms, a human footprint might be that of a creature so big as to be beyond imagining; a world crusher.

From the choreography of the crowd at Newport and the postures of individuals within it, you might assess the atmosphere to be one of pilgrimage. Cold-footed strangers murmur muted greetings. There's something of the supplicant in the column's shoeless inching, one or two people on bent knee, or gazing out into the sea, scooping up handfuls of sand to trickle from a fist.

The queue that has formed on the beachfront is not single file; it's more opportunistic. Clustered three hundred feet or more along the coast are groups five and six abreast, against a wrack line of dotty, broken shells. Further, some people have entered the shallows, a few up to their knees, trying to see around the crowd. Theories of a rip current draw all eyes to the deepest chancers. I can just make out the whale in the distance—a pale lozenge rolling in the pool's quadrangle of seawater, up where the shore begins to curve. There's no circumventing the line, and no question of quitting it. I am at the back. Soon others join behind me.

Are the people at the front compelled to observe any specific time limitation when at last they shuffle up to obtain a better view of the whale? Forty minutes later, I am developing a healthy antipathy for those strangers, dragging my toes through the scurfy sand in frustration. They could stand by the ocean pool ogling for a minute or an hour, depending on how they construe their obligation to the onlookers jostling behind them, and the importance of staying with the vision of the whale. Will the tide come in, and take the whale away from us again— if so, how soon? I don't want to have to wait so long that the whale is gone by the time I've arrived at the other end of the beach.

The cabdriver turns and says to anyone who will listen that she'll need to work through the evening to make up the hours. She's around forty. A wet stippling of eczema extends from the back of her neck up to her chin, and kicks a tenderness in me. I confess my fictional headache to her, and that I too will have to account for the time here in other ways, working over the weekend. The arrival of the whale hasn't entirely loosened people from their commitments, though it has broken up routines and brought us together into a spectatorship agitated by intrigue and commiseration.

Waiting huddled in the long procession on Newport Beach takes me to mentally recounting famous names in a lineage of glass coffins. Stalin, the popes, Kim Jong-il. Mandela, the ayatollah Ruhollah Khomeini. Castro. Other enigmatic figureheads and statesmen. The scene here is a reminder of those toiling crowds; a riot slowed down. Children underfoot. Looking upon the illustrious body is intended to corroborate death ostensibly to quash conspiracy, yet it also works to elevate a fallen leader to myth. The whale, on the other hand, needs no memorial cortège to be met as enduringly charismatic.

How the animal came to its end is the subject of conjecture. For a while I listen to the puff and natter of whale talk hovering along the line. *Forty tonnes, oh, at least thirty, took out part of the railing, the swell will lift it. What do they eat anyway—he says sea bugs? Killer whales are the natural enemy; it's a fight to the death. Dumped fuel, you know, humpbacks accidentally swallow it. Or a virus. Gone by morning, will it? Look, that kid's brought his snorkel.* There is something notable, I think, about how facts move through a queue: not radial as in a crowd kettled at a protest, or gathered before a stage, but by way of a kind of channeling. Information turns florid in the hand-to-hand. The more orderly the formation of people, the more bent and deviant is the knowledge passing through, or so it seems. It's not the priestly or political body, lain in state, that cultivates conspiracy—it's the queue. The queue is the agar of conspiracy.

To the nonscientist, the study of ecology requires us to look at the world and apprehend a shadow kingdom of subtending connections. Webs. When a creature like a whale turns up and people suspect it's corrupted, we're drawn to seeing systems of influence where before we registered none. Sometimes these systems surprise us in their beautiful strangeness, as when shimmering auroras in the night sky predicate whales stranding. Other times, it is the sheer expansiveness of the web that startles; the bellyful of a single whale that encloses wrappers from food shopping once purchased in Ukraine, Denmark, *and* Britain. But perhaps most unsettling is the system that connects our lives intimately with the fate of the whale—as was once the case with the whalebone corsets women wore; and as is exemplified, today, on discovering whales girdled by toxic blubber (the chemical legacy of our industries). This much closeness can be painful—and yet the infolding of our lives with those of remote wildlife has much to teach us about how interlinked the future will be. Though we may believe in the reality of being materially connected to many, many far-off things, it is only when we hear of these connections breaking, we can confirm that it's true. Which might be the ultimate value of all these stories: to underline how large our lives are, when they can sometimes feel small and short, slotted into ever narrower silos and categories. The sea is not eternal and unchanging as once we imagined. But neither are we condemned to be changeless. After all, to say that our impacts are global coaxes us toward seeing that our powers to affect positive change are too.

The cabdriver notes two men carrying pints of beer. They are walking up to meet the straggling end of the line on the beach, likely having come direct from the roadside pub. Their mood is jubilant. Children slip out of the knotted neckties and boater-hats that mark them as students of private schools in the Eastern suburbs. The dead whale calls to us all, though it has apparently drifted voiceless for two to four days. Representatives from the Organisation for the Rescue and Research of Cetaceans in Australia (ORRCA) attempt to funnel spectators into a tighter column as the beach narrows between the incoming waves and the rocks beside the ocean pool. A temporary fence of fluorescent netting has been erected between the crowd and the water. The fence keeps on being knocked over and re-staked, each time to afford a slightly better view for the re-staker. The queue is moving much more rapidly where it begins to taper.

Though it has felt ad hoc from ground level, the way we have gathered now begins to demonstrate a satisfying geometry. Something overhead surely clicks into recognizing the momentum of it, the streams of people flowing toward and away from the whale. We move together with the propulsion of grains spilt from a sack cut with a penknife. We move together like baby salmon called smolts, drawn upstream by their brittle skulls, by instinct, by imprinting, by magnetic fields, by collective memory, the presence of polarized light, salt or no salt, and the chemical odors of old, stone riverbed rivers. What is known as pilgrimage might just be some lingering brain-stem compulsion for migration. Why do humans tend to believe in deities at all? To justify our animal instinct for moving in herds. To queue.

The crowd is studded with gleaming smartphones and a few film cameras on tripods that we part and re-join around. Photojournalists, wearing lanyards with press passes, clamber onto the roof of the Newport Baths changing rooms. Ahead I can see two pubescent boys pulling an iPad between them, less interested in the hulking concretion of whale in the ocean pool than in the whale they can catch in their electric frame. Each tap scrunches another image. The animal is compressed to a thumbprint's acreage, pinched or swiped aside. Someone waves a lumination monitor overhead: a device set with a white bauble. I want to explain that the "lumen," as a measure of the strength of light, was first standardized in the nineteenth century using candles made from the spermaceti wax found in the heads of sperm whales. How much light there is, the technical unit for quantifying lumination, tracks back through the history of how humans have manipulated whale bodies, not unlike this one.

The sun drags its fingers along ferrous runnels in the sand. Soon it will be too dim to take good pictures. The race is on. I push closer to our immense animal interloper, and though the crowd still obscures it I can begin to see it in parts between people. The leaving light turns copper in the spray. Like a glowing periscope, the boys' tablet allows those behind them to see a little further ahead: they hold it up. Floodlights are being assembled to afford a greater opportunity to observe the whale and take its mortuary portraits.

The ORRCA personnel hand out a fact sheet on humpbacks to passing children. *Whales are mammals, as we are.* This whale is a male humpback. Not yet fully grown, though it's almost sixty-six feet long and, the ORRCA volunteers confirm, the animal is estimated to weigh over thirty tonnes. The line drawing of the whale on the flyer has a queer sort of smile, benevolent as a sphinx. I hope someone has taken it upon themselves to break it to the small boy with his snorkel that he won't be able to swim with the dead whale.

Ahead of the sight arrives the smell. Volatized molecules loosed from the animal, touching the inside of our noses. The whale is in us already, and it reeks; vinegar, and a haemal stink. "An entire fish shop gone off," one onlooker tells a reporter. Wave sets—cylinders rolling across the ocean pool—are cresting a sudsy, sulphurous yellow in the middle-distance, emulsified with fat and blood. Wafting down the queue comes cigarette smoke to mask the stench of decay. No one complains.

Gary Snyder, laureate of the parasites, would get into this scene, I think. "Life is not just diurnal and a property of large interesting vertebrates," Snyder petitioned, "it is also nocturnal, anaerobic, cannibalistic, microscopic, digestive, fermentative: cooking away in the warm dark." Who was only, ever, themselves? The whale, not alive, is becoming aliver with micro-biota.

What, out there, is newly riven with appetite? I wonder. What is coming to life inside the whale?

Maybe you, like me, know that what is most lifelike about a thought is the way that it winds and unwinds through several events and retellings. Rewinds to graininess, the thought that runs over itself backwards. And maybe you, like me, know yourself to be inwardly shifty as a beach at night, turned cavernous or steep or glossy without pattern. New parts of yourself appear unbidden, then it's on you to understand them. But what ideas alive inside of you now, will outlive you? What can each of us kindle, that will go on, after we are done?

When at last a line of sight clears I'm about sixty feet away. The whale is not a shocking sight; it has filled my head for hours already, a future actuality. The animal is on its back, in the way fishes go belly-up, hard against the thin metal railing that partially divides the ocean pool from the sea. It has been washed to the wall nearest to the beach. Behind the backrush, the surf is heaping. The tide is almost at its highest point. Gulls can't haggle with the wind. They're bunched like plastic bags, stuffed between rocks. On a second glance some of what I take for gulls *are* plastic bags.

The whale looks geologic, marbled and cracking. Decomposition's inward subsidence doesn't scotch the solidity of its dimensions. It is like something dragged up from far underground, chalk and ash streaked— or a meteorite fallen, against all the odds, into our backyard. The pleats of the whale's throat tessellate with the grill-like texture of Sydney's coastal shale formations, the rocks that surround the beach. A sheath of innards extrudes from its mouth. Bits of railing have torn off where this boulder of a body rolled into the pool. More of the railing has been removed by council officers and staff from the National Parks and Wildlife Service, in the hope that the dead whale will be tipped back out into the sea by the surging ocean.

A handkerchief of cloud races across the rising moon.

The incoming swell rotates the whale. An immense fin swings upwards, straight as a joist, animatronic. In the pungent air a girl dressed as a fairy starts grumbling to her father. She's waited so long to discover that the ORRCA people won't allow anyone to go right up to the humpback in the pool, and now she insists on knowing why a dead whale should command so much attention, being still as dead as it was when she arrived in the back of the queue hours ago. The whale's fin swings up again. Slaps back. The fairy wraps herself in the orange mesh-plastic of the un-staked fence, and gets to stamping down molar sandcastles.

———————————

The scene: a date at a Japanese restaurant in Glebe, a few nights before the arrival of the whale in Newport. I order a sampler of saké, clear Japanese rice wine, and it comes to us in several third-filled shot glasses—some warmed, some ice cold. We share the saké in mouthfuls. At the end of the meal we're ruddy with alcohol and the heat of our new attraction. Our knees touch. When the other tables have emptied, and the lights have dimmed, the server approaches us with a bottle of plum wine, *umeshu*, the final measures of which he pours out into three glasses standing in *masu* boxes. A little overflow fills the *masu*, and, putting away the emptied bottle, he shows us how to drink with our fingertips from the glass while it brims.

The wine is good but it makes the server sad, he says. This is the last plum wine the restaurant has stored from Fukushima Prefecture—*umeshu* distilled before the nuclear accident. Since then Japanese farmers have stripped their orchards with high-pressure hoses, shaved the bark off their trees, and replaced truckloads of soil with earth they've imported. But the restaurant's owner and the server are both in agreement: they will never buy or drink wine from Fukushima ever again, no matter the cleansing efforts of the producers there. It's not the risk of radiation. It's that the place called "Fukushima," and the cultivar that was the Fukushima plum, doesn't exist anymore, not really. The *umeshu* we've finished had the terroir of an environment that's vanished. Any fruit grown in Fukushima in the years to come will have none of these qualities, having risen from a scrubbed world.

When the table is cleared, I think: just now, a second ago, we devoured something into extirpation. We disappeared an endling. At that same instant I feel, intensely, that we have to *thank* the server for his generosity, for choosing us, of all people, to be the special two to taste the last Fukushima plum in this part of Sydney. We tip generously.

Can you forgive me, that I wasn't disgusted, that I wasn't ashamed? I thought that my date and I were the deserving recipients of a luminous and decadent gift, and later: it was arousing to me, to kiss him on the street with the sweet taste of the *umeshu* still faintly in his mouth.

I leave Newport as night sinks its feet into the sand. The ocean is a prickling amethyst, black and purple. How the whale in the ocean pool died will be the subject of a necropsy conducted over the coming days, and though the swell will eventually lift and convey the whale out of the pool, the carcass will be returned again, by the sea, half a kilometer down the beach. A group with the National Parks Service—men on-call for clearing fallen trees from highways—will be summoned up with chainsaws and trucks, dressed in white biohazard suits, alongside biologists who take tissue samples and DNA. The carcass flies to pieces. The flies remain a problem for days afterwards.

In wedges, the Newport whale will be transported to the tip at Lucas Heights where, their website describes, there is an organic waste yard and a compound for biohazardous materials. I form a picture of the whale settled into the organic composting facilities at Lucas Heights, surrounded by mounds of grape waste from wine producers, sawdust dumped by timber mills, and the old mash of feedlot vegetables. I see it all cooking under the sun into the pale, dry grist of the blood-and-bone fertilizer people sometimes shake onto their garden beds. But of course, the Newport humpback was not "organic waste." It was "biohazard waste," and therefore incinerated.

There are whale cemeteries in secret locations around New South Wales, though the Newport humpback didn't end up there. The main one, established in the mid-1990s, is somewhere in thickest bushland in Garigal National Park. The sites aren't made public because of the risk that they will be disrupted by trophy hunters and vandals. Here are interred whale bodies of particular interest to scientists; rare species, those that are underrepresented in museum collections, and those that died for reasons that are interesting or inexplicable. Such whales are not cut up and put into landfill. They are wrapped in chicken-wire and muslin, and driven into the forest. One such place is a resting ground for nearly thirty individual animals. Each buried whale, ticking with putrefaction, will be marked by a small sign and an official reference number registered with the Australian National Museum. When research funding becomes available the whales are sometimes exhumed and studied. Or else they're left to leech into the water table there, in glades where the bush glows greener, brightened by the nitrogen.

No definitive cause of death was attributed in the Newport humpback's necropsy, though the whale was described as thin—it had about a third as much blubber as would be expected for an animal its age and size. The Newport Baths were sieved of bigger clumps of flesh and skin over the coming days, and shut for a week to allow the sea to flush through.

On a clouded morning in September I returned to swim laps. The water was chilly: I flinched with each deeper step, barnacles and sea-lettuce scrunching underfoot. A woman wearing a yellow nose-clip laughed at me, openly *hawed*, and then I gasped and went under, kicking off in the familiar stroke.

Midway in my third lap of the pool, as I turned my head to take a breath, I thought I glimpsed a puffy organ, the steel-colored liver maybe, of the whale, bumping along the bottom. But when I looked again it was gone and the water was hazy with sand.

Whales are animals that live in vastness. With an evolutionary history running back fifty million years, not only are whales today the biggest single animals on the face of the planet, found in all major oceans, but some cetacean species roam annually between the tropics and the poles, while others yet call across thousands of miles, or respond to a magnetosphere convulsed by solar storms, rippling through the cosmos. Sensitive to distant phenomena, whales also effect change on an earthly scale. Large whales have altered the chemical composition of the air. The voices of humpbacks sweep underwater soundscapes, yawing into the dark to be passed onwards by other whales that they will never know, far away. Having died mid-ocean, whales become launching pads for a deep-sea bestiary so bizarre and bewitching that even at his most gormless Carl Linnaeus surely would not have credited its existence.

I had gone in search of how it was that whales encountered us, wanting to understand how the triumph of saving the whale could have been undercut, within my lifetime, by what had happened to the whale's environment. I expected to discover uncomfortable concordances of the machine-turned world and whales' lives, which I did—though, too, from the outset I longed to cross the sight line of sovereign, wild animals, existing in marine mystery, out past the end zone of human influence.

What I had not anticipated was that whales could embiggen far more than my sense of wonder. In their breadth of connectedness, do whales not show us how to be conscious of environments we ourselves cannot see, environments beyond our habitation where crisis is being staged?—in the atmosphere, say, at the ice caps, in the far oceans, and in the bodies of animals. Across the course of writing this book I realized how *near* the preservation of nature could be: that our obligations to other species were found not in the wild, but in the increments of ordinary life; in eating, shopping, commuting. To protect any wild animal now, the task is not to look for it, but to consider what it might depend on: the abundance of food, of shelter and paths of migration, the preservation of biophony, of oceanic chemistry and temperature within ranges tolerable to species other than our own; freedom from being crowded out by pollution. We must think about the sensory realities we wish to sustain for animals, and those we wish to protect them from. Both the suffocating love evoked by charisma, and the project of taxonomy—finding, naming, labeling—must give way before a duty to ecology. These are responsibilities to one another, as much as to wildlife, for what we lose when we lose animals is a way to imagine the world as larger than we experience it.

Where humans have harnessed the enormity of whales it has been to both destructive and compassionate ends. Whaling—the first globalized extractive industry—provisioned the dawning of consumer culture, and demonstrated how intractable a proto-energy market could be, even as the resource-base undergirding it shrunk. When I went back to look at the history of whaling, what I saw was that whales had shaped our inhabitancies, our industries and art, all the while we tore them asunder. We live in their wake, as they live in ours. The anti-whaling movement that followed established the moral infrastructure for a story of care that encompassed the world, as well as establishing an environmental citizenry that sought to safeguard nature, beyond national borders. I started out thinking hope had only extrinsic origins; that hope hinged on finding things to be hopeful *about*, rather than things to be hopeful *for*. Now I think it is within our power to find a cure in nature, not a grave site; to connect with our better selves in salvaging a future in which many wild things also feature.

Yet, we are between worlds. In Perth the beached humpback died burning up on the inside—back then, we could not see the heat that it suffered in. Now the fires are terrifyingly visible as we confront a new, globalized zoological disaster. During the weeks that I was putting the final touches on this book, the town of Eden was evacuated as the east coast of Australia experienced its worst-ever bushfire season on record, exacerbated not only by the unseasonably hot conditions, but a forcibly shorter "hazard reduction" period (when flammable undergrowth is safely ignited in a series of controlled burns). Hundreds of residents, some with their pets, were driven out of their homes, and down to the wharf where the *Cat Balou* whale-watching tour docks. They took shelter in tugboats; they armed themselves with on-board fire extinguishers, and held dry woollen blankets to their faces to stop from inhaling the smoke. The sky above turned a scorching crimson red, then black, as daylight lost its purchase on the daytime. By the time the fire front had passed, and the weather in the region changed, an estimated one billion Australian animals had been burnt alive in the eastern bushlands. Koalas, kangaroos, platypuses, and wombats, along with creatures that are less well recognized internationally; quolls, pademelons, potoroos, antechinuses, dunnarts, bandicoots, bilbies, many types of bat, native rats, mice, and several very beautiful and abundant species of moth. Where it took near to a hundred years to remove three million cetaceans from the sea in the twentieth century, this defaunation event lasted scarcely a matter of weeks.

When the mother humpback whale I saw off the deck of the *Cat Balou* swam away from Eden, I questioned where she went to and how those Antarctic seas would fare in the decades to come. I did not consider that the Australian coastline, closer yet than any polar environment, would be so utterly stripped of wildlife before I had a chance to return to that corner of the country to look upon the sea again. Though the fires reared up in the Australian summer, in the southern hemisphere, they were pre-programmed by shifts in the climate worldwide—a force less discriminating and less conspicuous than a whaling fleet, but with a time signature that beats faster. A force we also perpetuate—or resist—with our energy politics, our consumption, our choices.

After the fires, the high winds carried ash-clouds and fragmentary charcoal to carpet the topside of the coastal South Pacific. Then the light rain that finally arrived flushed charred topsoil into estuaries, and out to settle in the kelp, and on top of reefs. In time, waves returned the floating debris as thin, black tide lines, like the fluttering read-out of electrodes monitoring a distant nightmare.

There are two types of environmental haunting, or so it seems to me now—and the decisions we make will determine which of these we will come to be inhabited by, in decades to come. There is a hauntedness that stems from regret; words unsaid, the inward doubt that stilled a connection, or the big gesture that passed us by, leaving us strangers to our neighbors. To be tormented by regret can make us feel more dead than alive as we go about our everyday tasks—for being haunted by *what we did not do* places us in the limbo of never knowing whether we, in fact, lacked the power, or only believed ourselves to be so limited. Such a hauntedness turns a blind eye to the nature that might yet flourish—and fails to imagine how much can be revived in us, when we acquit ourselves, not in service to returning the lost beloved, but in extending our care out to its surrounding world. In this is the start-point of the second, more fruitful variety of haunting—one evoked by an inability to let go of the animals and their worlds, despite the dimensions of the crisis. It is not always unhealthy to *stay attached*, for this haunting compels us in the opposite direction: to envision what stands to be lost from the future, and what might never return, were we to remain changeless. This haunting once saved whales. It is a haunting that creates and guards reserves for hope; that pivots from regretting the past, to anticipating feelings of sorrow, if the nature people love most of all becomes confined to memory, to pictures and stories. It is a haunting that moves us from indecision, to action.

Ash columns eddy through the blackened trees, and it is true to say that things will never be as they once were in and around Eden. But it is also true that where we stand today is grayer than it will be tomorrow. Each of us now sharpens the focus-dial on the future of the ocean, of the weather, of the whales and their kin.

One more thing to relate from the Newport Beach. Toward the end I was separated from the cabdriver. She had been following me, but when we finally arrived by the side of the ocean pool, as near as I would ever get to the dead Newport humpback, I looked for her and found she was nowhere to be seen. I think the cabdriver saw the whale for barely a minute, then turned on her heel and left without saying a word. Was the whale too much for her, or not enough? I wanted to ask, but I'd missed my chance. She was out on the road now, negotiating the trap of traffic generated by the whale, which would take many hours to unsnarl.

I wandered back along the seashore. ORRCA and the council had lit the wrack-line with solar-powered beacons to guard against the visitors twisting their ankles or becoming lost as they sought the route to the road. Laced through the seaweed I saw all the usual taintedness and treasure; blue bits of twine, bait packets. Biohazards, I guess. I am always looking for the shiny pockets of rays' eggs, which look like they could be made of vinyl, or undeveloped film. I knelt down above one patch of dross from the sea and found it dotted with dozens of tiny, transparent fish. These were sushi fish, the kind meant for soy sauce. I held one up to the artificial light, its nose still capped with a little red, screw-on tip. There was nothing inside it now, but air.

What floats after falling is
flotsam, and what floats when thrown is jetsam.
Whatever sinks is lagan.

Whatever is cast up
is yours.

Shevaun Cooley
"Soundings" in *Homing* (2017)

Acknowledgments

CREDITS

FOUNDATIONS

Drenka Andjelic, James Bradley, Julia Carlomagno, Christina Chau, Sarah Chisholm, Sam Cooney, Ben Cubby, Claire Hargreave, Elmo Keep, Cassie Lynch, Robert Moor. Aden Rolfe, Aviva Tuffield, Sam Twyford-Moore, Geordie Williamson, Fiona Wright. Jennifer Mae Hamilton, Astrida Neimanis, and the Weathering Collective. Pony Express, for leading my thinking on affection and nature. To my extended family, such love. David Finnigan—who makes me a bolder writer, and a better person—gets this heart at its fiercest.

Thanks to Brett Zimmer at the Western Australian Museum, Robert Harcourt in the Department of Biological Sciences, Macquarie University, Tim Watters, Mio Bryce, Masayuki Komatsu, Junichi Sato, Geoff Ross, Ed Couzens, and everyone else who took the time to speak to me about whales, whaling, and the ocean. Very early advice from David Marr only later made invaluable sense, with gratitude. Sophie Cunningham—whose writing and creative life has long been a lodestar in my universe—provided mentorship at the end. The preface to this book was published in an abbreviated form in *Granta* magazine—I am thankful to the editors there, and to Eleanor Chandler most particularly, for finding this part of the work a readership.

RENOVATIONS

My friend and first reader, Nick Tapper: a wellspring of confidence, and a backstop at points of doubt. Marika Webb-Pullman at Scribe; and to Henry Rosenbloom, and everyone at Scribe UK, too, for their continued support. The wonderful Bonnie Nadell, from Hill Nadell Literary Agency, who saw the appeal of this project and found the book a home in the United States. Simon & Schuster, and the two brilliant editors without whose insights I would have been lost: Jonathan Cox and Emily Simonson. And to the rest of the team at Simon & Schuster, including Alison Forner, June Park, Morgan Hart, Lewelin Polanco, Alicia Brancato, Erica Ferguson, Anne Tate Pearce, Stephen Bedford, Jonathan Karp, and Kimberly Goldstein: my boundless appreciation.

PROVISIONS

AsiaLink and the Association for the Study of Environmental Literature (Japan). The JUMP initiative of the Australia Council for the Arts. Macquarie University. Varuna: the National Writers' House.

Fathoms was written on the un-ceded sovereign lands of the Darug Nation, and the Gadigal Peoples of the Eora Nation, in New South Wales, and on Noongar Country in Western Australia. It was partially edited in Melbourne, within the lands of the Wurundjeri and Boonwurrung peoples of the Kulin Nation. I recognize the continuity of knowledge that nurtures country and community, and offer respect to the custodians of the land and ocean I work on.

Australian Indigenous words can have multiple spellings. Where possible, this book relies on the Marribank Orthography (Noongar), and the *Macquarie Aboriginal Words* dictionary (1994).

Chapter headings were inspired by Mathew Fontaine Maury's *The Physical Geography of the Sea and Its Meteorology* (1855).

Abbreviations

ABC	Australian Broadcasting Corporation
BBC	British Broadcasting Corporation
BWU	blue whale unit
CITES	Convention on International Trade in Endangered Species
CND	Campaign for Nuclear Disarmament
CSIRO	Commonwealth Scientific and Industrial Research Organisation (Australia)
EIA	Environmental Investigation Agency (United Kingdom)
ICJ	International Court of Justice
ICR	Institute of Cetacean Research (Japan)
IFAW	International Fund for Animal Welfare
IMF	International Monetary Fund
IMO	International Maritime Organization
IUCN	International Union for Conservation of Nature
IWC	International Whaling Commission
NASA	National Aeronautics and Space Administration (USA)
NOAA	National Oceanic and Atmospheric Administration (USA)
ORRCA	Organisation for the Rescue and Research of Cetaceans. (Australia)
SOFAR	Sound Fixing and Ranging
UN	United Nations
WWF	World Wide Fund for Nature/World Wildlife Fund (US and Canada only)

Further Reading

Texts to which the work of *Fathoms* is particularly indebted, as well as those that have been of unique interest to me across the course of writing this book, are asterisked. Some texts may be relevant to multiple chapters.

WHALEFALL

Altman, Rebecca. "How the Benzene Tree Polluted the World." *The Atlantic*, October 4, 2017.

Beeley, Fergus, prod. *Planet Earth: The Future*. United Kingdom: BBC, 2006.

Bischoff, Karyn, Robin Jaeger, and Joseph G. Ebel. "An Unusual Case of Relay Pentobarbital Toxicosis in a Dog." *Journal of Medical Toxicology* 7, no. 3 (September 2011): 236–39.

* Carson, Rachel. *The Sea Around Us*. London and New York: Staples Press, 1952.

———. *The Edge of the Sea*. London and Boston: Staples Press, 1955.

Chen, Tânia Li, Sandra S. Wise, Amie Homes, Fariba Shaffiey, John Pierce Wise Jr., W. Douglas Thompson, Scott Kraus, and John Pierce Wise Sr. "Cytotoxicity and Genotoxicity of Hexavalent Chromium in Human and North Atlantic Right Whale (*Eubalaena glacialis*) Lung Cells." *Comparative Biochemistry and Physiology Part C: Toxicology & Pharmacology* 150, no. 4 (November 2009): 487–94.

* Clough, Brent, presenter. "The Night Air: Whales." ABC Radio National, Sydney, May 10, 2009 [audio].

Das, Krishna, Govindan Malarvannan, Alin Dirtu, Violaine Dulau, Magali Dumont, Gilles Lepoint, Philippe Mongin, and Adrian Covaci. "Linking Pollutant Exposure of Humpback Whales Breeding in the Indian Ocean to Their Feeding Habits and Feeding Areas Off Antarctica." *Environmental Pollution* 220, part B (January 2017): 1090–99.

Demchenko, Natalia L., John W. Chapman, Valentina B. Durkina, and Valeriy I. Fadeev. "Life History and Production of the Western Gray Whale's Prey, *Ampelisca eschrichtii* Krøyer, 1842 (Amphipoda, Ampeliscidae)." *PLOS ONE* 11, no. 1 (January 22, 2016): e0147304.

292 - *Further Reading*

Desforges, Jean-Pierre, Ailsa Hall, Bernie McConnell, Aqqalu Rosing-Asvid, Jonathan L. Barber, Andrew Brownlow, Sylvain De Guise, et al. "Predicting Global Killer Whale Population Collapse from PCB Pollution." *Science* 361, no. 6409 (September 28, 2018): 1373–76.

* de Stephanis, Renaud, Joan Giménez, Eva Carpinelli, Carlos Gutierrez-Exposito, and Ana Cañadas. "As a Main Meal for Sperm Whales: Plastics Debris." *Marine Pollution Bulletin* 69, no. 1–2 (April 15, 2013): 206–14.

Dietz, Rune. "Contaminants in Marine Mammals in Greenland." PhD diss., Department of Arctic Environment, National Environmental Research Institute, University of Aarhus, Denmark, 2008.

Earle, Sylvia. *Sea Change: A Message of the Oceans*. New York: Ballantine Books, 1995.

———. *The World Is Blue: How Our Fate and the Ocean's Are One*, reprint. Washington, DC: National Geographic, 2010.

Ford, John K. B., Graeme M. Ellis, and Kenneth C. Balcomb. *Killer Whales: The Natural History and Genealogy of Orcinus Orca in British Columbia and Washington*. Vancouver: University of British Columbia Press, 1994.

Geraci, Joseph, and Valerie Lounsbury. *Marine Mammals Ashore: A Field Guide for Strandings*, 2nd ed. Baltimore: National Aquarium in Baltimore, 2005.

Goffredi, Shana K., Charles K. Paull, Kim Fulton-Bennett, Luis A. Hurtado, and Robert C. Vrijenhoek. "Unusual Benthic Fauna Associated with a Whale Fall in Monterey Canyon, California." *Deep-Sea Research Part I Oceanographic Research Papers* 51, no. 10 (October 2004): 1295–306.

Hammond, Philip, Sonja Heinrich, and Peter Tyack. *Whales: Their Past, Present and Future*. London: Natural History Museum, 2017.

Hewson, Ian, Brooke Sullivan, Elliot W. Jackson, Qiang Xu, Hao Long, Chenggang Lin, Eva Marie Quijano Cardé, et al. "Perspective: Something Old, Something New? Review of Wasting and Other Mortality in Asteroidea (Echinodermata)." *Frontiers in Marine Science* 6 (July 2019): 406.

Highsmith, Raymond, and Kenneth O. Coyle. "Productivity of Arctic Amphipods Relative to Gray Whale Energy Requirements." *Marine Ecology Progress Series* 83 (1992): 141–50.

Holyoake, Carly, Nicola Stephens, and Douglas Coughran. "Collection of Baseline Data on Humpback Whale (*Megaptera novaeangliae*) Health and Causes of Mortality for Long-Term Monitoring in Western Australia: Final Report." Department of Environment (DEC) Animal Ethics Committee (AEC), January 2012.

Iverson, Sara. "Blubber." In *Encyclopedia of Marine Mammals*, 2nd ed., edited by William Perrin, Bernd Würsig, and J. G. M. Thewissen, 115–20. San Diego: Academic Press, 2002.

Kraus, Scott, and Rosalind Rolland, eds. *Urban Whale: The North Atlantic Right Whale at the Crossroads*. Cambridge, MA: Harvard University Press, 2007.

Lacy, Robert C., Rob Williams, Erin Ashe, Kenneth C. Balcomb III, Lauren J. N. Brent, Christopher W. Clark, Darren P. Croft, Deborah A. Giles, Misty MacDuffee, Paul C. Paquet. "Evaluating Anthropogenic Threats to Endangered Killer Whales to Inform Effective Recovery Plans." *Scientific Reports* 7, no. 1 (October 26, 2017): 14119.

Le Roux, Mariëtte. "'Stinky Whale' Whiff Wafts Over Whaling Talks." Phys.Org, October 26, 2016.

Lertzman, Renee. *Environmental Melancholia: Psychoanalytic Dimensions of Engagement.* New York: Routledge, 2015.

Little, Crispin T. S. "Life at the Bottom: The Prolific Afterlife of Whales." *Scientific American,* February 2010.

McFarling, Usha Lee, and Kenneth R. Weiss. "A Whale of a Food Shortage." *Los Angeles Times,* June 24, 2002.

Meyer, Wynn K., Jerrica Jamison, Rebecca Richter, Stacy Woods, Raghavendran Partha, Amanda Kowalczyk, Charles Kronk, et al. "Ancient Convergent Losses of *Paraoxonase 1* Yield Potential Risks for Modern Marine Mammals." *Science* 361, no. 6402 (August 2018): 591–94.

Milton, Kay. *Loving Nature: Towards an Ecology of Emotion.* London and New York: Routledge, 2002.

* Moore, Michael J. "How We All Kill Whales." *ICES Journal of Marine Science* 71, no. 4 (May/June 2014): 760–63.

Motluk, Alison. "Deadlier than the Harpoon?" *New Scientist,* no. 1984 (July 1, 1995): 12–13.

Rauber, Paul. "What's Killing the Whales?" *Sierra,* June 5, 2019.

* Roberts, Callum. *Oceans of Life: How Our Seas Are Changing.* London: Penguin, 2013.

———. *The Unnatural History of the Sea.* Washington, DC: Island Press/Shearwater Books, 2007.

Ross, Peter, and S. C. H. Grant. "Southern Resident Killer Whales at Risk: Toxic Chemicals in the British Columbia and Washington Environment." *Canadian Technical Report of Fisheries and Aquatic Sciences* 2412 (2002).

Rouse, Greg W., Shana Goffredi, and Robert C. Vrijenhoek. "*Osedax*: Bone-Eating Marine Worms with Dwarf Males." *Science* 305, no. 5684 (July 30, 2004): 668–71.

Sinclair, Elizabeth, and Erika Techera. "Dead Whales Are Expensive: Whose Job Is It to Clear Them Up?" *The Conversation,* February 26, 2015.

Smith, Craig R., and Amy R. Baco. "Ecology of Whale Falls at the Deep-Sea Floor." *Oceanography and Marine Biology: An Annual Review* 41 (2003): 311–54.

Stewart, Brent S., Phillip J. Clapham, and James A. Powell. *National Audubon Society Guide to Marine Mammals of the World.* New York: Knopf, 2002.

Visser, Ingrid. "Killer Whale (*Orcinus orca*) Interactions with Longline Fisheries in New Zealand Waters." *Aquatic Mammals* 26, no. 3 (January 2000): 241–52.

"WA Fisheries Department Refuses to Pay for Whale Carcass Removal As They Are 'Mammals, Not Fish,' Mayor Says." ABC Radio Perth, November 20, 2014 [audio].

Wise, Catherine F., Sandra S. Wise, W. Douglas Thompson, Christopher Perkins, and John Pierce Wise Sr. "Chromium Is Elevated in Fin Whale (*Balaenoptera physalus*) Skin Tissue and Is Genotoxic to Fin Whale Skin Cells." *Biological Trace Element Research* 166, no. 1 (July 2015): 108–17.

Wise, John Pierce, Sandra S. Wise, Scott Kraus, Fariba Shaffiey, Marijke Grau, Tânia Li Chen, Christopher Perkins, et al. "Hexavalent Chromium Is Cytotoxic and Genotoxic to the North Atlantic Right Whale (*Eubalaena glacialis*) Lung and Testes Fibroblasts." *Mutation Research/Fundamental and Molecular Mechanisms of Mutagenesis* 650, no. 1 (January 31, 2008): 30–38.

Wise, John Pierce, Jr., James T. F. Wise, Catherine F. Wise, Sandra S. Wise, Cairong Zhu, Cynthia L. Browning, Tongzhang Zheng, et al. "Metal Levels in Whales from the Gulf of Maine: A One Environmental Health Approach." *Chemosphere* 216 (February 2019): 653–60.

Yong, Ed. "Once Again, a Massive Group of Whales Strands Itself." *The Atlantic*, March 23, 2018.

———. "An Ancient Lost Gene Leaves Whales Vulnerable to Pesticides." *The Atlantic*, August, 2018.

PETROGLYPH

Aguilar, Alex. "A Review of Old Basque Whaling and its Effect on the Right Whales of the North Atlantic." *A Report to the International Whaling Commission* 10, no. 10 (1986): 191–99.

"A New Cure for Rheumatism." *New York Times* from the *Pall Mall Gazette*, March 7, 1896, 3.

Baker, C. Scott, and Phillip J. Clapham. "Marine Mammal Exploitation: Whales and Whaling." In *Encyclopedia of Global Environmental Change*, vol. 3, edited by Ian Douglas. Chichester: Wiley, 2002.

Bale, Martin. "Bangudae: Petroglyph Panels in Ulsan, Korea, in the Context of World Rock Art," edited by Ho-tae Jeon and Jiyeon Kim. *Journal of Korean Studies* 20, no. 1 (Spring 2015): 229–32.

Barlass, Tim. "Bizarre Whale Treatment for Rheumatism Revealed." *Sydney Morning Herald*, March 30, 2014.

Basberg, Bjørn, Jan Erik Ringstad, and Einar Wexelsen, eds. *Whaling and History: Perspectives on the Evolution of the Industry*. Sandefjord, Norway: Sandefjordmuseene, 1993.

Bayet, Fabienne. "Overturning the Doctrine: Indigenous People and Wilderness—Being Aboriginal in the Environmental Movement." *Social Alternatives* 13, no. 2 (1994): 27–32.

"Beached Dead Whales Can Alter the Ocean's Carbon Footprint." *Marine Wildlife Magazine*, May 12, 2015.

Berzin, Alfred. "The Truth About Soviet Whaling." *Marine Fisheries Review* 70, no. 2 (January 2008): 4–59.

Bird Rose, Deborah. "Multispecies Knots of Ethical Time." *Environmental Philosophy* 9, no. 1 (2012): 127–40.

Bird Rose, Deborah, Thom van Dooren, and Matthew Chrulew. *Extinction Studies: Stories of Time, Death, and Generations*. New York: Columbia University Press, 2017.

Birnie, Patricia. "The Role of Developing Countries in Nudging the International Whaling Commission from Regulating Whales to Encouraging Nonconsumptive Uses of Whales." *Ecology Law Quarterly* 12, no. 4 (1985): 937–75.

Boli, John, and George M. Thomas, eds. *Constructing World Culture: International Nongovernmental Organizations Since 1875*. Stanford: Stanford University Press, 1999.

* Burnett, Graham D. *The Sounding of the Whale: Science and Cetaceans in the Twentieth Century*. Chicago and London: University of Chicago Press, 2012.

Burns, John J., J. Jerome Montague, and Cleveland J. Cowles, eds. *The Bowhead Whale*, special pub. no. 2. Lawrence, KS: Society for Marine Mammalogy, 1993.

Chami, Ralph, Thomas Cosimano, Connel Fullenkamp, and Sena Oztosun. "Nature's Solution to Climate Change: A Strategy to Protect Whales Can Limit Greenhouse

Gases and Global Warming." *Finance and Development* (International Money Fund) (September 2019): 34–38.

Clark, Doug Bock. *The Last Whalers: The Life of an Endangered Tribe in a Land Left Behind.* New York: Hachette, 2019.

Clode, Danielle. *Killers in Eden: The Story of a Rare Partnership Between Men and Killer Whales.* Crows Nest, NSW: Allen & Unwin, 2002.

Couzens, Ed. *Whales and Elephants in International Conservation Law and Politics: A Comparative Study.* Abingdon, Oxford: Routledge, 2014.

Creighton, Margaret. *Rites and Passages: The Experience of American Whaling.* Cambridge, UK, and New York: Cambridge University Press, 1995.

———. "'Women' and Men in American Whaling, 1830–1870." *International Journal of Maritime History* 4, no. 1 (June 1992): 195–218.

Cressey, Daniel. "World's Whaling Slaughter Tallied at 3 Million." *Scientific American*, March 12, 2015.

D'Amato, Anthony, and Sudhir K. Chopra. "Whales: Their Emerging Right to Life." *American Journal of International Law* 85, no. 1 (January 1991): 21–62.

Darby, Andrew. *Harpoon: Into the Heart of Whaling.* Crows Nest, NSW: Allen & Unwin, 2007.

Dirzo, Rodolfo, Hillary S. Young, Mauro Galetti, Gerardo Ceballos, Nick J. B. Isaac, and Ben Collen. "Defaunation in the Anthropocene." *Science* 345, no. 6195 (July 25, 2014): 401–06.

Douglas, Marianne S. V., John P. Smol, James M. Savelle, and Jules M. Blais. "Prehistoric Inuit Whalers Affected Arctic Freshwater Ecosystems." *Proceedings of the National Academy of Sciences* 101, no. 6 (January 26, 2004): 1613–17.

Dow, George Francis. *Whale Ships and Whaling: A Pictorial History of Whaling During Three Centuries with an Account of the Whale Fishery in Colonial New England.* Salem, MA: Marine Research Society, 1925.

Drew, Joshua, Elora H. López, Lucy Gill, Mallory McKeon, Nathan Miller, Madeline Steinberg, Christa Shen, and Loren McClenachan. "Collateral Damage to Marine and Terrestrial Ecosystems from Yankee Whaling in the 19th Century." *Ecology and Evolution* 6, no. 22 (October 19, 2016): 8181–92.

Eber, Dorothy Harley. *When the Whalers Were Up North: Inuit Memories from the Eastern Arctic.* Kingston, Ontario: McGill-Queen's University Press, 1989.

Ellis, Richard. *The Book of Whales.* New York: Knopf, 1980.

———. *Men and Whales.* London: Robert Hale, 1992.

* Epstein, Charlotte. *The Power of Words in International Relations: Birth of an Anti-Whaling Discourse.* Cambridge, MA, and London: MIT Press, 2008.

Estes, James A., D. F. Doak, Alan M. Springer, and Terrie M. Williams. "Causes and Consequences of Marine Mammal Population Declines in Southwest Alaska: A Food-Web Perspective." *Philosophical Transactions of the Royal Society B Biological Sciences* 364, no. 1524 (June 27, 2009): 1647–58.

Estes, James A., J. Terborgh, J. S. Brashares, M. E. Power, J. Berger, W. J. Bond, S. R. Carpenter, et al. "The Trophic Downgrading of Planet Earth." *Science* 333, no. 6040 (July 15, 2011): 301–06.

Estes, James A. *Serendipity: An Ecologist's Quest to Understand Nature.* Oakland: University of California Press, 2016.

Fielding, Russell. *The Wake of the Whale: Hunter Societies in the Caribbean and North Atlantic*. Cambridge, MA: Harvard University Press, 2018.

Freeman, Milton M. R., and Stephen R. Kellert. *Public Attitudes to Whales: Results of a Six-Country Survey*. Edmonton: Canadian Circumpolar Institute; New Haven, CT: School of Forestry and Environmental Studies, 1992.

Freeman, Milton M. R., Lyudmila Bogoslovskaya, Richard A. Caulfield, Ingmar Egede, Igor I. Krupnik, and Marc G. Stevenson. *Inuit, Whaling, and Sustainability*. Oxford, UK: Altamira Press, 1998.

Friedheim, Robert L., ed. *Toward a Sustainable Whaling Regime*. Seattle: University of Washington Press; Edmonton: Canadian Circumpolar Institute Press, 2001.

Gaskin, David. *The Ecology of Whales and Dolphins*. New York: Heinemann, 1982.

Gillespie, Alexander. *Whaling Diplomacy: Defining Issues in International Environmental Law*. Cheltenham, UK; Northampton, MA: Edward Elgar, 2005.

Hannesson, Rögnvaldur. *The Privatization of the Oceans*. Cambridge, MA: MIT Press, 2004.

Hayley, Nelson Cole. *Whale Hunt: The Narrative of a Voyage*. New York: Washburn, 1948.

Hearings Before the Subcommittee on Fisheries and Wildlife Conservation and the Environment, of the Committee on Merchant Marine and Fisheries—House of Representatives, 94th Congress, First Session, on: Protecting Whales by Amending Fishermen's Protective Act of 1967 by Strengthening Import Restrictions; to Release Sperm Oil from National Stockpile; and Scrimshaw. Serial No. 94-7. US Government Printing Office (May/June 1975).

Hess, Bill. *The Gift of the Whale: The Iñupiat Bowhead Hunt: A Sacred Tradition*. Seattle: Sasquatch Books, 1999.

* Hoare, Philip. *Leviathan or, the Whale*. London: Fourth Estate, 2009

Homans, Charles. "The Most Senseless Environmental Crime of the Twentieth Century." *Pacific Standard Magazine*, November 12, 2013.

Hoskins, Ian. *Sydney Harbour: A History*. Sydney: University of New South Wales Press, 2009.

Ingold, Tim. *Lines: A Brief History*. London and New York: Routledge, 2007.

Irion, Robert. "Whale of an Appetite." *New Scientist*, no. 2157 (October 24, 1998): 25.

"Is Whale Ban Rotting Cars?" *New Scientist* 66, no. 947 (May 1, 1975): 262.

Ivashchenko, Yulia, and Phillip Clapham. "Too Much Is Never Enough: The Cautionary Tale of Soviet Illegal Whaling." *Marine Fisheries Review* 76, no. 1–2 (Winter–Spring 2014): 1–21.

Jackson, Gordon. *The British Whaling Trade*, new edition. London: A. and C. Black, 1978.

Jarvis, Brooke. "The Insect Apocalypse Is Here: What Does It Mean for the Rest of Life on Earth?" *New York Times Magazine*, November 27, 2018.

Johnston, Sarah. "*Te Karanga a te Huia* | The Call of the Huia." *Gauge: The Blog of New Zealand's Audiovisual Archive*, May 12, 2016.

Jopson, Debra. "Hands Across History," *Sydney Morning Herald*, March 9, 1996.

Kalland, Arne. "A Concept in Search of Imperialism? Aboriginal Subsistence Whaling." In *11 Essays on Whales and Man*, edited by Georg Blichfeldt. Lofoten Reine, Norway: High North Alliance, 1994.

———. "Whose Whale Is That? Diverting the Commodity Path." In *Elephants and*

Whales: Resources for Whom?, edited by Milton M. R. Freeman and Urs P. Kreuter. Basel, Switzerland: Gordon and Breach Science, 1994.

Katona, Steven, and Hal Whitehead. "Are Cetacea Ecologically Important?" *Oceanography and Marine Biology Annual Review* 26 (1988): 553–68.

Keck, Margaret, and Kathryn Sikkink. *Activists Beyond Borders: Advocacy Networks in International Politics.* Ithaca, NY: Cornell University Press, 1998.

Kingdon, Amorina. "Playing Viking Chess with Whale Bones." *Hakai Magazine*, September 25, 2018.

Kuehls, Thom. *Beyond Sovereign Territory: The Space of Ecopolitics.* Minneapolis: University of Minneapolis Press, 1996.

Kurlansky, Mark. *The Basque History of the World: The Story of a Nation.* New York: Penguin, 2001.

Kutner, Luis. "The Genocide of Whales: A Crime against Humanity." *Lawyer of the Americas* 10, no. 3 (Winter 1978): 784–98.

Langlois, Krista. "When Whales And Humans Talk." *Hakai Magazine*, April 3, 2018.

* Lantis, Margaret. "The Alaskan Whale Cult and Its Affinities." *American Anthropologist* (New Series) 40, no. 3 (July–September 1938): 438–64.

Lavery, Trish J., Ben Roudnew, Peter Gill, Justin Seymour, Laurent Seuront, Genevieve Johnson, James G. Mitchell, and Victor Smetacek. "Iron Defecation by Sperm Whales Stimulates Carbon Export in the Southern Ocean." *Proceedings of the Royal Society B: Biological Sciences* 277, no. 1699 (November 22, 2010): 3527–31.

Lutz, Steven, and Angela Martin. "Fish Carbon: Exploring Marine Vertebrate Carbon Services." Washington, DC: Blue Climate Solutions, the Ocean Foundation, 2014.

* Macfarlane, Robert. *Underland: A Deep Time Journey.* New York: W. W. Norton, 2019.

Mackintosh, N. A. *The Stocks of Whales* (The Buckland Foundation). London: Coward and Gerrish, 1965.

Mazzanti, Massimiliano. "The Role of Economics in Global Management of Whales: Re-forming or Re-founding the IWC?" *Ecological Economics* 36 (2001): 205–21.

McCauley, Douglas J., Malin L. Pinsky, Stephen R. Palumbi, James A. Estes, Francis H. Joyce, and Robert R. Warner. "Marine Defaunation: Animal Loss in the Global Ocean." *Science* 347, 6219 (January 2015): 247–54.

McDonald, Jo. *Dreamtime Superhighway—Sydney Basin Rock Art and Prehistoric Information Exchange.* Canberra: ANU Press, 2008.

McKenna, Mark. *Looking for Blackfella's Point: An Australian History of Place.* Sydney: University of New South Wales Press, 2002.

Melville, Herman. *Moby-Dick.* London: Strato Publications, 1851.

M'Gonigle, Michael. "The 'Economizing' of Ecology: Why Big, Rare Whales Still Die." *Ecology Law Quarterly* 9, no. 1 (1980): 119–237.

Miller, Gretchen, presenter. "Hindsight: A Living Harbor: Extended Interview with Dennis Foley." ABC Radio National, June 20, 2010 [audio].

Miller, Robert J. "Exercising Cultural Self-Determination: The Makah Indian Tribe Goes Whaling." *American Indian Law Review* 25, no. 2 (November 2, 2002): 165–273.

Minteer, Ben A., and Leah Gerber. "Buying Whales to Save Them." *Issues in Science and Technology* 29, no. 3 (March 2013): 58–68.

Mitchell, Edward D. *A Bibliography of Whale Killing Techniques.* Cambridge: International Whaling Commission, 1986.

Monbiot, George. "Why Whale Poo Matters." *Guardian*, December 12, 2014.

Newton, John. *A Savage History: Whaling in the Pacific and Southern Oceans*. Sydney: New-South Publishing, 2013.

Nicol, Steve. "Vital Giants: Why Living Seas Need Whales." *New Scientist*, July 6, 2011.

Nihon Kujirarui Kenkyujo. *Whaling and Anti-Whaling Movement*, edited by the Institute of Cetacean Research. Tokyo, Japan: Institute of Cetacean Research, 1999.

Oremus, Marc, John Leqata, and C. Scott Baker. "Resumption of Traditional Drive Hunting of Dolphins in the Solomon Islands in 2013." *Royal Society Open Science* 2, no. 5 (May 6, 2015): 140524.

* Pascoe, Bruce. *Dark Emu: Black Seeds—Agriculture or Accident?* Broome, Western Australia: Magabala Books, 2014.

Pash, Chris. *The Last Whale*. North Fremantle, WA: Fremantle Press, 2008.

* Paterson, Alistair, Ross Anderson, Ken Mulvaney, Sarah de Koning, Joe Dortch, and Jo McDonald. "*So Ends This Day*: American Whalers in Yaburara Country, Dampier Archipelago." *Antiquity* 93, no. 367 (February 2019): 218–35.

Pershing, Andrew J., Line B. Christensen, Nicholas R. Record, Graham D. Sherwood, and Peter B. Stetson. "The Impact of Whaling on the Ocean Carbon Cycle: Why Bigger Was Better." *PLOS ONE* 5, no. 8 (August 26, 2010): e12444.

Pringle, Heather. "Signs of the First Whale Hunters." *Science* 320, no. 5873 (April 11, 2008): 175.

Proulx, Jean Pierre. *Basque Whaling in Labrador in the 16th Century*. Ottawa: National Historic Sites, Parks Service, Environment Canada, 1993.

Reeves, Randall, and Tom Smith. "A Taxonomy of World Whaling: Operations, Eras, and Data Sources." Northeast Fisheries Science Center Reference Document 03-12: National Oceanic and Atmospheric Administration, 2003.

Rocha, Robert C., Jr., Phillip J. Clapham, and Yulia Ivashchenko. "Emptying the Oceans: A Summary of Industrial Whaling Catches in the 20th Century." *Marine Fisheries Review* 76, no. 4 (March 2015): 37–48.

* Roman, Joe. *Whale*. London: Reaktion, 2006.

Roman, Joe, and James McCarthy. "The Whale Pump: Marine Mammals Enhance Primary Productivity in a Coastal Basin." *PLOS ONE* 5, no. 10 (October 11, 2010): e13255.

* Roman, Joe, James A. Estes, Lyne Morissette, Craig Smith, Daniel Costa, James McCarthy, J. B. Nation, et al. "Whales as Marine Ecosystem Engineers." *Frontiers in Ecology and the Environment* 12, no. 7 (September 2014): 377–85.

Scarff, James. "The International Management of Whales, Dolphins, and Porpoises: An Interdisciplinary Assessment." *Ecology Law Quarterly* 6, no. 3 (1977): 571–638.

Schneider, Viktoria, and David Pearce. "What Saved the Whales? An Economic Analysis of 20th Century Whaling." *Biodiversity and Conservation* 13, no. 3 (March 2004): 543–62.

Shattuck, Ben. "There Once Was a Dildo in Nantucket: On the Wives of Whalers and Their Dildos, AKA 'He's-at-Homes.'" *The Common*, no. 10 (October 16, 2015).

Sheehan, Glenn. *In the Belly of the Whale: Trade and War in Eskimo Society*. Anchorage: Alaska Anthropological Association, 1997.

Sherwood, Yvonne. *A Biblical Text and Its Afterlives: The Survival of Jonah in Western Culture*. Cambridge, UK, and New York: Cambridge University Press, 2000.

Shoemaker, Nancy. *Living with Whales: Documents and Oral Histories of Native New England Whaling History*. Amherst: University of Massachusetts Press, 2014.

Slijper, E. J. *Whales*. London: Hutchinson, 1982.

Smith, Laura V., Andrew McMinn, Andrew Martin, Steve Nicol, Andrew R. Bowie, Delphine Lannuzel, and Pier van der Merwe. "Preliminary Investigation into the Stimulation of Phytoplankton Photophysiology and Growth by Whale Faeces." *Journal of Experimental Marine Biology and Ecology* 446 (August 2013): 1–9.

Springer, Alan M., J. A. Estes, Gus B. van Vilet, Terrie M. Williams, D. F. Doak, Eric Danner, K. A. Forney, and B. Pfister. "Sequential Megafaunal Collapse in the North Pacific Ocean: An Ongoing Legacy of Industrial Whaling?" *Proceedings of the National Academy of Sciences* 100, no. 21 (October 14, 2003): 223–28.

Stackpole, Edouard. *The Sea-Hunters: The New England Whalemen During Two Centuries 1635–1835.* Philadelphia: Lippincott, 1953.

Stanbury, Peter, and John Clegg. *A Field Guide to Aboriginal Rock Engravings with Special Reference to Those Around Sydney.* Sydney: Sydney University Press, 1990.

Stoett, Peter. *The International Politics of Whaling.* Vancouver: University of British Columbia Press, 1997.

Stoker, Sam, and Igor Krupnik. "Subsistence Whaling." In *The Bowhead Whale,* special pub. no. 2, edited by John J. Burns, J. Jerome Montague, and Cleveland J. Cowles, 579–629. Lawrence, KS: Society for Marine Mammalogy, 1993.

Stolzenburg, Will. *Where the Wild Things Were: Life, Death, and Ecological Wreckage in a Land of Vanishing Predators.* New York: Bloomsbury, 2008.

Thompson, Derek. "The Spectacular Rise and Fall of U.S. Whaling: An Innovation Story." *The Atlantic,* February 22, 2012.

Tønnessen, John, and Arne Odd Johnsen. *The History of Modern Whaling.* Berkeley: University of California Press, 1982.

* Van Dooren, Thom. *Flight Ways: Life and Loss at the Edge of Extinction.* New York: Columbia University Press, 2014.

Waterman, Thomas Talbot. *The Whaling Equipment of the Makah Indians.* Seattle: University of Washington Press, 1920.

Watkins, Graham, and Pete Oxford. *Galapagos: Both Sides of the Coin.* Morganville, NJ: Imagine Publishing, 2009.

Williams, Heathcote. *Whale Nation.* London: Jonathan Cape, 1988.

Yablokov, Alexey V. "Validity of Whaling Data." *Nature* 367, no. 6459 (January 13, 1994): 108.

* York, Richard. "Why Petroleum Did Not Save the Whales." *Socius: Sociological Research for a Dynamic World* 3 (January 2017): 1–13.

THE OOOO-ERS

Barash, David. "Why Did Humans Evolve to be So Fascinated with Other Animals?" *Aeon,* May 13, 2014.

Bartholomew, Kylie, and Rob Blackmore. "Humpback Whale Population Increasing 'Like Crazy,' Say Scientists." ABC Sunshine Coast, September 23, 2016 [audio].

Bearzi, Maddalena, and Craig B. Stanford. *Beautiful Minds: The Parallel Lives of Great Apes and Dolphins.* Cambridge, MA: Harvard University Press, 2008.

Bejder, Michelle, David W. Johnston, Joshua Smith, Ari Friedlaender, and Lars Bejder. "Embracing Conservation Success of Recovering Humpback Whale Populations: Evaluating the Case for Down-listing their Conservation Status in Australia." *Marine Policy* 66 (April 2016): 137–41.

* Berger, John. "Why Look at Animals." In *About Looking*. New York: Pantheon Books, 1980.

Berwald, Juli. *Spineless: The Science of Jellyfish and the Art of Growing a Backbone*. New York: Riverhead Books, 2017.

Brakes, Philippa, and Mark Peter Simmonds, eds. *Whales and Dolphins: Cognition, Culture, Conservation and Human Perceptions*. New York: Routledge, 2011.

British Antarctic Survey. "El Nino Events Affect Whale Breeding." *ScienceDaily*, January 10, 2006.

Brydon, Anne. *The Eye of the Guest: Icelandic Nationalist Discourse and the Whaling Issue*. Ottawa: Bibliothèque nationale du Canada, 1991 [Unpublished PhD thesis].

* Buell, Lawrence. "Global Commons as Resource and Icon." In *Writing for an Endangered World: Literature, Culture, and Environment in the U.S. and Beyond*, 196–223. Cambridge, MA: Belknap Press of Harvard University Press, 2001.

Cai, Wenju, Simon Borlace, Matthieu Lengaigne, Peter van Rensch, M. Collins, Gabriel A. Vecchi, Axel Timmermann, et al. "Increasing Frequency of Extreme El Niño Events Due to Greenhouse Warming." *Nature: Climate Change* 4 (January 2014): 111–16.

Calvez, Leigh. *The Breath of a Whale: The Science and Spirit of Pacific Ocean Giants*. Seattle: Sasquatch Books, 2019.

Carter, Neil. *The Politics of the Environment: Ideas, Activism, Policy*, 2nd ed. Cambridge, UK, and New York: Cambridge University Press, 2007.

Carwardine, Mark. *Whales, Dolphins, and Porpoises*. London: Dorling Kindersley, 1995.

Cherfas, Jeremy. *The Hunting of the Whale: A Tragedy That Must End*. London: Penguin, 1988.

Cisneros-Montemayor, Andrés, Rashid U. Sumaila, Kristin Kaschner, and Daniel Pauly. "The Global Potential for Whale Watching." *Marine Policy* 34, no. 6 (November 2010): 1273–78.

Clapham, Phil, and Colin Baxter. *Winged Leviathan: The Story of the Humpback Whale*. Grantown-on-Spey, Scotland: Colin Baxter Photography, 2013.

Clapham, Phillip J., Sharon B. Young, and Robert L. Brownell Jr. "Baleen Whales: Conservation Issues and the Status of the Most Endangered Populations." *Mammal Review* 29, no. 1 (March 1999): 37–62.

Corkeron, Peter J. "Whale Watching, Iconography, and Marine Conservation." *Conservation Biology* 18, no. 3 (June 2004): 847–49.

———. "How Shall We Watch Whales?" In *Gaining Ground: In Pursuit of Ecological Sustainability*, edited by David Lavigne. Guelph, Ontario: International Fund for Animal Welfare, 2006.

Corkeron, Peter J., and Richard Connor. "Why Do Baleen Whales Migrate?" *Marine Mammal Science* 15, no. 4 (October 1999): 1228–45.

Croll, Donald A., Baldo Marinovic, Scott Benson, Francisco P. Chavez, Nancy Black, Richard Ternullo, and Bernie Tershy. "From Wind to Whales: Trophic Links in a Coastal Upwelling System." *Marine Ecology Progress Series* 289 (March 2005): 117–30.

Cronon, William, ed. *Uncommon Ground: Rethinking the Human Place in Nature*. New York: W. W. Norton, 1996.

Davidson, Ana D., Alison G. Boyer, Hwahwan Kim, Sandra Pompa-Mansilla, Marcus J. Hamilton, Daniel P. Costa, Gerardo Ceballos, and James Brown. "Drivers and Hotspots of Extinction Risk in Marine Mammals." *Proceedings of the National Academy of Sciences* 109, no. 9 (February 2012): 3395–400.

Davidson, Helen. "Humpback Whales Make a Comeback in Australian Waters as Numbers Rebound." *Guardian*, July 27, 2015.

Day, David. *The Whale War*. San Francisco: Random House, 1987.

Despret, Vinciane. "Thinking Like a Rat." *Angelaki: Journal of Theoretical Humanities* 20, no. 2 (June 2015): 130.

* de Waal, Frans. *Are We Smart Enough to Know How Smart Animals Are?* New York: W. W. Norton, 2016.

Dingle, Hugh. *Migration: The Biology of Life on the Move*. Oxford, UK: Oxford University Press, 2014.

Donovan, Greg. "The International Whaling Commission: Given Its Past, Does It Have a Future?" In *Proceedings of the Symposium on "Whales: Biology—Threats—Conservation,"* edited by J. J. Symoens. Brussels: Royal Academy of Overseas Science, 1992.

Dorsey, Kurkpatrick. *Whales and Nations: Environmental Diplomacy on the High Seas*. Seattle: University of Washington Press, 2016.

Downhower, Jerry F., and Lawrence S. Blumer. "Calculating Just How Small a Whale Can Be." *Nature* 335 (1988): 675.

Dunlop, Rebecca, Douglas H. Cato, and Michael J. Noad. "Non-song Acoustic Communication in Migrating Humpback Whales (*Megaptera novaeangliae*)." *Marine Mammal Science* 24, no. 3 (July 2008): 613–29.

* Ehrenreich, Barbara. "The Animal Cure." *The Baffler* 19 (March 2012).

Epstein, Charlotte. "World Wide Whale: Globalisation and a Dialogue of Cultures?" *Cambridge Review of International Affairs* 16, no. 2 (2003): 309–22.

———. "The Making of Global Environmental Norms: Endangered Species Protection." *Global Environmental Politics* 6, no. 2 (May 2006): 32–54.

Esser-Miles, Carolin. "King of the Children of Pride: Symbolism, Physicality, and the Old English Whale." In *The Maritime World of the Anglo-Saxons*, edited by Stacy Klein, William Schipper, and Shannon Lewis-Simpson. Phoenix: ACMRS Arizona Center for Medieval and Renaissance, 2014.

Falk, Richard A. "Introduction: Preserving Whales in a World of Sovereign States." *Denver Journal of International Law and Policy* 17 (1989): 249–54.

Fretwell, Peter T., Jennifer A. Jackson, Mauricio J. Ulloa Encina, Vreni Häussermann, Maria J. Perez Alvarez, Carlos Olavarría, and Carolina S. Gutstein. "Using Remote Sensing to Detect Whale Strandings in Remote Areas: The Case of Sei Whales Mass Mortality in Chilean Patagonia." *PLOS ONE* 14, no. 11 (October 17, 2019): e0222498.

Freud, Sigmund. *The Standard Edition of the Complete Psychological Works of Sigmund Freud, Volume 21, 1927–1931: The Future of an Illusion, Civilization and Its Discontents, and Other Works*. London: Hogarth Press; Institute of Psychoanalysis, 1961.

Friends of the Earth. *Whale Manual '78*. Friends of the Earth Publication, 1978.

Gallo, Rubén. *Freud's Mexico: Into the Wilds of Psychoanalysis*. Cambridge, MA, and London: MIT Press, 2010.

Geib, Claudia. "Death by Killer Algae." *Hakai Magazine*, November 21, 2017.

Gero, Shane. "The Lost Cultures of Whales." *New York Times*, October 8, 2016.

Goldbogen, Jeremy. "The Ultimate Mouthful: Lunge Feeding in Rorqual Whales." *American Scientist* 98, no. 2 (March–April 2010): 124–31.

Goldbogen, Jeremy, David E. Cade, John Calambokidis, A. S. Friedlaender, Jean Potvin, P. S. Segre, and Alexander Werth. "How Baleen Whales Feed: The Biomechanics of Engulfment and Filtration." *Annual Review of Marine Science* 9, no. 1 (January 2017): 367–86.

Graeber, David. "What's the Point If We Can't Have Fun?" *The Baffler* 24 (January 2014).

Gulland, John. "The Management Regime for Living Resources." In *The Antarctic Legal Regime*, edited by Christopher C. Joyner and Sudhir K. Chopra. Dordrecht, Netherlands; Boston; and London: Springer, 1988.

Haas, Peter M., Robert O. Keohane, and Marc A. Levy, eds. *Institutions for the Earth: Sources of Effective International Environmental Protection*. Cambridge, MA, and London: MIT Press, 1993.

Häussermann, Verena, Carolina S. Gutstein, Michael Bedington, David Cassis, Carlos Olavarria, Andrew C. Dale, Ana M. Valenzuela-Toro, et al. "Largest Baleen Whale Mass Mortality During Strong El Niño Event Is Likely Related to Harmful Toxic Algal Bloom." *PeerJ* 5 (May 31, 2017): e3123.

* Heise, Ursula. *Imagining Extinction: The Cultural Meanings of Endangered Species*. Chicago: University of Chicago Press, 2016.

Hoyt, Erich, and Glen T. Hvenegaard. "A Review of Whale-Watching and Whaling with Applications for the Caribbean." *Coastal Management* 30, no. 4 (October 2002): 381–99.

Hunter, Robert. *The Greenpeace Chronicles*. London: Picador, 1980.

International Whaling Commission. "Report of the Sub-Committee on Whalewatching." *Journal of Cetacean Research and Management* 3 (2003).

Jackson, Jennifer A., Nathalie Patenaude, Emma Carroll, and C. Scott Baker. "How Few Whales Were There After Whaling? Inference from Contemporary mtDNA Diversity." *Molecular Ecology* 17, no. 1 (February 2008): 236–51.

Jackson, Jeremy. "Ecological Extinction and Evolution in the Brave New Ocean." *Proceedings of the National Academy of Sciences* 105, supp. 1 (August 12, 2008): 11458–65.

Kalland, Arne. "Management by Totemization: Whale Symbolism and the Anti-Whaling Campaign." *Arctic* 46, no. 2 (June 1993): 97–188.

Kawaguchi, S., Akio Ishida, R. King, Ben Raymond, N. Waller, Andrew Constable, Stephen Nicol, Masahide Wakita, and Atsushi Ishimatsu. "Risk Maps for Antarctic Krill Under Projected Southern Ocean Acidification." *Nature Climate Change* 3 (July 2013) 843–47.

Kimmerer, Robin Wall. "Nature Needs a New Pronoun: To Stop the Age of Extinction, Let's Start by Ditching 'It.'" *Yes Magazine*, Spring 2015.

Klein, Emily S., Simeon L. Hill, Jefferson T. Hinke, Tony Phillips, and George M. Watters. "Impacts of Rising Sea Temperature on Krill Increase Risks for Predators in the Scotia Sea." *PLOS ONE* 13, no. 1 (January 31, 2018): e0191011.

Lilly, John. *Lilly on Dolphins: Humans of the Sea*. Garden City, NY: Anchor Press, 1975.

* Lormier, Jamie. *Wildlife in the Anthropocene: Conservation After Nature*. Minneapolis: University of Minnesota Press, 2015.

MacLeod, Colin. "Global Climate Change, Range Changes and Potential Implications for the Conservation of Marine Cetaceans: A Review and Synthesis." *Endangered Species Research* 7, no. 2 (June 2009): 125–36.

Mann, Janet, ed. *Deep Thinkers: Inside the Minds of Whales, Dolphins, and Porpoises*. Chicago: University of Chicago Press, 2017.

Mapes, Lynda V. "Feds Propose Allowing Makah Tribe to Hunt Gray Whales Again." *Seattle Times*, April 4, 2019.

* Mathews, Freya. "From Biodiversity-based Conservation to an Ethic of Bio-proportionality." *Biological Conservation* 200 (August 2016): 140–48.

McCormick, John. *Reclaiming Paradise: The Global Environmental Movement*. Bloomington: Indiana University Press, 1989.

McIntyre, Joan. *Mind in the Waters: A Book to Celebrate the Consciousness of Whales and Dolphins*. San Francisco: Sierra Club Books, 1974.

McNally, Robert. *So Remorseless a Havoc: Of Dolphins, Whales & Men*. Boston: Little, Brown, 1981.

Meynecke, Jan-Olaf, Russell Richards, and Oz Sahin. "Whale Watch or No Watch: The Australian Whale Watching Tourism Industry and Climate Change." *Regional Environmental Change* 17, no. 2 (February 2017): 477–88.

Moeran, Brian. "The Cultural Construction of Value: 'Subsistence,' 'Commercial,' and Other Terms in the Debate about Whaling." *Maritime Anthropological Studies* 5, no. 2 (1992): 1–15.

* Morley, Simon A., David K. A. Barnes, and Michael J. Dunn. "Predicting Which Species Succeed in Climate-Forced Polar Seas." *Frontiers in Marine Science* 5 (January 17, 2019): 507.

Morton, Harry. *The Whale's Wake*. Dunedin: University of Otago Press, 1982.

Mowat, Farley. *A Whale for the Killing*. Boston: Little, Brown, 1972.

———. *Sea of Slaughter*. Boston: Atlantic Monthly Press, 1984.

Moyle, B. J., and M. Evans. "A Bioeconomic and Socio-Economic Analysis of Whale-Watching, With Attention Given to Associated Direct and Indirect Costs." Paper Presented to the IWC Scientific Committee, 2001.

National Oceanic and Atmospheric Administration. "Successful Conservation Efforts Pay Off for Humpback Whales: Division into Distinct Populations Paves the Way for Tailored Conservation Efforts," media release. September 6, 2016.

Nicol, Stephen. *The Curious Life of Krill: A Conservation Story from the Bottom of the World*. Washington, DC: Island Press, 2018.

Noad, Michael J., Eric Kniest, and Rebecca A. Dunlop. "Boom to Bust? Implications for the Continued Rapid Growth of the Eastern Australian Humpback Whale Population Despite Recovery." *Population Ecology* 61, no. 2 (April 2019): 198–209.

Norris, Scott. "Creatures of Culture: Making the Case for Cultural Systems in Whales and Dolphins." *BioScience* 52, no. 1 (January 2002): 9–14.

Nowacek, Douglas P., Ari S. Friedlaender, Patrick N. Halpin, Elliott L. Hazen, David W. Johnston, Andrew J. Read, Boris Espinasse, Meng Zhou, and Yiwu Zhu. "Super-Aggregations of Krill and Humpback Whales in Wilhelmina Bay, Antarctic Peninsula." *PLOS ONE* 6, no. 4 (April 27, 2011): e19173.

O'Connor, Simon, Roderick Campbell, Tristan Knowles, and Hernan Cortez. "Whale Watching Worldwide: Tourism Numbers, Expenditures and Expanding Economic Benefits," special report from the International Fund for Animal Welfare, Yarmouth, MA, June 2009.

Palacios, Manuela. "Inside the Whale: Configurations of An-other Female Subjectivity." *Women's Studies* 47, no. 2 (February 13, 2018): 160–72.

Parsons, E. C. M. "The Negative Impacts of Whale-Watching." In "Protecting Wild Dolphins and Whales: Current Crises, Strategies, and Future Projections," special issue, *Journal of Marine Sciences* (August 15, 2012).

Parsons, E. C. M, C. M. Fortuna, F. Ritter, N. A. Rose, M. P. Simmonds, M. Weinrich,

R. Williams, and S. Panigada. "Glossary of Whale Watching Terms." *Journal of Cetacean Research and Management* 8, supplement (2006): 249–51.

Payne, Roger. *Among Whales*. New York: Scribner, 1995.

Peace, Adrian. "Loving Leviathan: The Discourse of Whale-Watching in Australian Ecotourism." In *Animals in Person: Cultural Perspectives on Human-Animal Intimacies*, edited by John Knight. Oxford, UK, and New York: Berg, 2005.

———. "Rhetoric to the South, Reality to the North: Realigning the Conflict Over Whaling." *The Conversation*, December 12, 2011.

Pearce, Fred. *Green Warriors: The People and the Politics Behind the Environmental Revolution*. London: Bodley Head, 1991.

———. "Ghosts of Whales Past." *New Scientist* 205, no. 2746 (February 3, 2010): 40–42.

* Peters, John Durham. "Of Cetaceans and Ships; or, the Moorings of Our Being." In *The Marvelous Clouds: Toward a Philosophy of Elemental Media*. Chicago and London: Chicago University Press, 2015.

Peterson, M. J. "Whalers, Cetologists, Environmentalists, and the International Management of Whaling." *International Organization* 46, no. 1 (Winter 1992): 147–86.

Pulkkinen, Levi. "A Pod of Orcas Is Starving to Death: A Tribe Has a Radical Plan to Feed Them." *Guardian*, April 25, 2019.

Quammen, David. *The Song of the Dodo: Island Biogeography in an Age of Extinction*. New York: Scribner, 1996.

Rasmussen, Kristin, Daniel M. Palacios, John Calambokidis, Marco T. Saborío, Luciano Dalla Rosa, Eduardo R. Secchi, Gretchen H. Steiger, Judith M. Allen, and Gregory S. Stone. "Southern Hemisphere Humpback Whales Wintering Off Central America: Insights from Water Temperature into the Longest Mammalian Migration." *Biology Letters* 3, no. 3 (June 2007): 302.

Readfearn, Graham. "From Alaska to Australia, Anxious Observers Fear Mass Shearwater Deaths." *Guardian*, November 23, 2019.

Reilly, Steve, S. Hedley, J. Borberg, R. Hewitt, D. Thiele, J. Watkins, and M. Naganobu. "Biomass and Energy Transfer to Baleen Whales in the South Atlantic Sector of the Southern Ocean." *Deep Sea Research Part II: Tropical Studies in Oceanography* 51, no. 12–13 (June 2004): 1397–409.

Ripple, William J., Christopher Wolf, Thomas M. Newsome, Michael Hoffmann, Aaron J. Wirsing, and Douglas J. McCauley. "Extinction Risk Is Most Acute for the World's Largest and Smallest Vertebrates." *Proceedings of the National Academy of Sciences* 114, no. 40 (September 18, 2017): 10678-83.

Ris, Mats. "Why Look at Whales? Reflections on the Meaning of Whale Watching." In *11 Essays on Whales and Man*, edited by Georg Blichfeldt. Lofoten Reine, Norway: High North Alliance, 1991.

Roman, Joe, and Stephen Palumbi. "Whales Before Whaling in the North Atlantic." *Science* 301, no. 5632 (July 25, 2003): 508–10.

Rubenstein, Diane. "Hate Boat: Greenpeace, National Identity and Nuclear Criticism." *International/Intertextual Relations: Postmodern Readings of World Politics*, edited by James Der Derian and Michael Shapiro. Lexington, MA: Lexington Books, 1989.

Safina, Carl. *Beyond Words: What Animals Think and Feel*. New York: Henry Holt, 2015.

Sharma, Rajnish, Lisa L. Loseto, Sonja K. Ostertag, Matilde Tomaselli, Christina M. Bredtmann, Colleen Crill, Cristina Rodriguez-Pinacho, et al. "Qualitative Risk Assessment of Impact of *Toxoplasma gondii* on Health of Beluga Whales, *Delphinapterus leucas,* from the Eastern Beaufort Sea, Northwest Territories." *Arctic Science* 4, no. 3 (2018): 321–37.

Siebert, Charles. "Watching Whales Watching Us." *New York Times Magazine,* July 8, 2009.

Siegel, Volker, ed. *Biology and Ecology of Antarctic Krill.* Switzerland: Springer, 2016.

Silva, Santiago Piedra. "Whales Under Threat as Climate Change Impacts Migration." Phys.org, December 2, 2015.

Simmonds, Mark and Judith Hutchinson. *The Conservation of Whales and Dolphins: Science and Practice.* Chichester, UK, and New York: Wiley, 1996.

Simmons, I. G. *Interpreting Nature: Cultural Constructions of the Environment.* London and New York: Routledge, 1993.

Singer, Peter. "The Cow Who . . ." *Project Syndicate,* February 5, 2016.

Sumich, James. "Blowing." In *Encyclopedia of Marine Mammals,* edited by William Perrin, Bernd Würsig, and Johannes Thewissen. San Diego: Academic Press, 2002.

Tulloch, Vivitskaia J. D., Éva E. Plagányi, Christopher Brown, Anthony J. Richardson, and Richard Matear. "Future Recovery of Baleen Whales Is Imperiled by Climate Change." *Global Change Biology* 25, no. 4 (March 12, 2019): 1263–81.

Van Dolah, Frances, Gregory Doucette, Frances Gulland, Teri Rowles, and Gregory Bossart. "Impacts of Algal Toxins on Marine Mammals." In *Toxicology of Marine Mammals,* edited by Joseph G. Vos. London and New York: Taylor and Francis, 2003.

Van Duzer, Chet. *Sea Monsters on Medieval and Renaissance Maps.* London: British Library, 2013.

* Wallace-Wells, David. "Storytelling." In *The Uninhabitable Earth: Life After Warming,* 143–57. New York: Tim Duggan Books, 2019.

Ware, Colin, Ari S. Friedlaender, and Douglas P. Nowacek. "Shallow and Deep Lunge Feeding of Humpback Whales in Fjords of the West Antarctic Peninsula." *Marine Mammal Science* 27, no. 3 (July 2011): 587–605.

Waters, Hannah. "The Enchanting Sea Monsters on Medieval Maps." *Smithsonian Magazine,* October 15, 2013.

Werth, Alexander. "How Do Mysticetes Remove Prey Trapped in Baleen?" *Bulletin of the Museum of Comparative Zoology* 156 (January 2001): 189–203.

White, Thomas. *In Defense of Dolphins: The New Moral Frontier.* Oxford: John Wiley & Sons, 2008.

Whitehead, Hal. *Sperm Whales: Social Evolution in the Ocean.* Chicago and London: University of Chicago Press, 2003.

———. "Why Whales Leap." *Scientific American,* March 1985.

* Whitehead, Hal, and Luke Rendell. *The Cultural Life of Whales and Dolphins.* Chicago and London: University of Chicago Press, 2014.

Wilcove, David S. "Animal Migration: An Endangered Phenomenon?" *Issues in Science and Technology* XXIV, no. 3 (Spring 2008).

Wiley, David, Ari Friedlaender, Alessandro Bocconcelli, Danielle Cholewiak, Colin Ware, Mason Weinrich, and Michael Thompson. "Underwater Components of Humpback Whale Bubble-Net Feeding Behaviour." *Behaviour* 148, no. 5–6 (January 1, 2011): 575–602.

Woinarski, John C. Z., Andrew A. Burbidge, and Peter L. Harrison. "Ongoing Unraveling of a Continental Fauna: Decline and Extinction of Australian Mammals Since

European Settlement." *Proceedings of the National Academy of Sciences* 112, no. 15 (April 14, 2015): 4531–40.

Worm, Boris, and Robert T. Paine. "Humans as a Hyperkeystone Species." *Trends in Ecology & Evolution* 31, no. 8 (August 2016): 600–07.

Wright, Franz. "The Only Animal." In *Walking to Martha's Vineyard: Poems.* New York: Knopf, 2003.

Zelko, Frank. *Make It a Green Peace!: The Rise of Countercultural Environmentalism.* New York: Oxford University Press, 2013.

Zerbini, Alexandre N., Grant Adams, John Best, Phillip J. Clapham, Jennifer A. Jackson, and Andre E. Punt. "Assessing the Recovery of an Antarctic Predator from Historical Exploitation." *Royal Society Open Science* 6, no. 10 (October 1, 2019): 190368.

BLUE MUSEUM

In addition to the resources listed, this chapter relied on archival photographs and documents provided by Western Australian Museum, as well as microfiche of the *Bunbury Herald* and the *Sunday Times* accessed at the State Library of Western Australia. My thanks to both these public institutions, and to the librarians and archivists there.

Asma, Stephen. *Stuffed Animals and Pickled Heads: The Culture and Evolution of Natural History Museums.* Oxford, UK, and New York: Oxford University Press, 2001.

Bennett, Hywel, presenter. *Kingdom of the Ice Bear.* United Kingdom: BBC, 1985.

Berta, Annalisa. *Return to the Sea: The Life and Evolutionary Times of Marine Mammals.* Berkeley: University of California Press, 2012.

———. *The Rise of Marine Mammals: 50 Million Years of Evolution.* Baltimore: Johns Hopkins University Press, 2017.

Blunt, Wilfrid. *Linnaeus: The Complete Naturalist.* Princeton, NJ: Princeton University Press, 2002.

Branch, Trevor. "Abundance of Antarctic Blue Whales South of 60°S from Three Complete Circumpolar Sets of Surveys." International Whaling Commission, 2007.

Branch, Trevor, K. M. Stafford, D. M. Palacios, C. Allison, J. L. Bannister, C. L. K. Burton, E. Cabrera, et al. "Past and Present Distribution, Densities and Movements of Blue Whales *Balaenoptera musculus* in the Southern Hemisphere and Northern Indian Ocean." *Mammal Review* 37, no. 2 (June 12, 2007): 116–75.

Branch, Trevor A., Koji Matsuoka, and Tomio Miyashita. "Evidence for Increases in Antarctic Blue Whales Based on Bayesian Modelling." *Marine Mammal Science* 20, no. 4 (October 2004): 726–54.

Brown, Cecil. *Language and Living Things: Uniformities in Folk Classification and Naming.* New Brunswick, NJ: Rutgers University Press, 1984.

Burnett, Graham D. *Trying Leviathan: The Nineteenth-Century New York Court Case That Put the Whale on Trial and Challenged the Order of Nature.* Princeton, NJ: Princeton University Press, 2010.

Burnett, Graham D., and Sina Najafi. "Cutting the World at Its Joints: An Interview with Graham D. Burnett." *Cabinet Magazine*, no. 28 (Winter 2007–2008): 81–88.

DeCou, Christopher. "When Whales Were Fish." *Lateral Magazine*, October 8, 2018.

de Muizon, Christian. "Walking with Whales." *Nature* 413 (September 20, 2001): 259–60.

———. "Walrus-like Feeding Adaptation in a New Cetacean from the Pliocene of Peru." *Nature* 365 (October 21, 1993): 745–48.

Farber, Paul Lawrence. *Finding Order in Nature: The Naturalist Tradition from Linnaeus to E. O. Wilson*. Baltimore: Johns Hopkins University Press, 2000.

Frängsmyr, Tore, ed. *Linnaeus: The Man and His Work*. Berkeley: University of California Press, 1983.

Gatesy, John, and Maureen A. O'Leary. "Deciphering Whale Origins with Molecules and Fossils." *Trends in Ecology & Evolution* 16, no. 10 (October 1, 2001): 562–70.

George, John C., Jeffrey Bada, Judith Zeh, Laura Scott, Stephen E. Brown, Todd O'Hara, and Robert Suydam. "Age and Growth Estimates of Bowhead Whales (*Balaena mysticetus*) via Aspartic Acid Racemization." *Canadian Journal of Zoology* 77, no. 4 (September 1999): 571–80.

Gingerich, Philip. "Evolution of Whales from Land to Sea." *Proceedings of the American Philosophical Society* 156, no. 3 (September 2012): 309–23.

———. "Evolution of Whales from Land to Sea: Fossils and a Synthesis." In *Great Transformations: Vertebrate Evolution*, edited by Kenneth P. Dial, Neil Shubin, and Elizabeth L. Brainerd. Chicago: Chicago University Press, 2015.

Gingerich, Philip D., Nelson A. Wells, Don E. Russell, and S. M. A. Shah. "Origin of Whales in Epicontinental Remnant Seas: New Evidence from the Early Eocene of Pakistan." *Science* 220, no. 4595 (April 22, 1983): 403–06.

Goldbogen, Jeremy, D. E. Cade, John Calambokidis, Max Czapanskiy, James Fahlbusch, A. S. Friedlaender, William Gough, et al. "Extreme Bradycardia and Tachycardia in the World's Largest Animal." *Proceedings of the National Academy of Sciences* 116, no. 50 (December 10, 2019): 25329–32.

Gould, Stephen Jay. "Hooking Leviathan by Its Past." In *Dinosaur in a Haystack: Reflections in Natural History*. New York: Crown, 1997.

Greenblatt, Stephen. "Resonance and Wonder." In *Exhibiting Culture: The Poetics and Politics of Museum Display*, edited by Ivan Karp and Steven D. Lavine. Washington, DC: Smithsonian Institution Press, 1991.

Grönberg, Cecilia, and Jonas J. Magnusson. "The Gothenburg Leviathan." *Cabinet Magazine* 33 (Spring 2009).

Haag, Amanda Leigh. "Patented Harpoon Pins Down Whale Age." *Nature News*, June 19, 2007.

Haraway, Donna. "Teddy Bear Patriarchy: Taxidermy in the Garden of Eden, New York City, 1908–1936." *Social Text*, no. 11 (Winter 1984–1985): 20–64.

———. *When Species Meet*. Minneapolis: University of Minnesota Press, 2008.

Henning, Michelle. "Neurath's Whale." In *The Afterlives of Animals: A Museum Menagerie*, edited by Samuel J. M. M. Alberti. Charlottesville: University of Virginia Press, 2011.

Jamie, Kathleen. "The Hvalsalen." In *Sightlines: A Conversation with the Natural World*. New York: The Experiment, 2013.

Klinkenborg, Verlyn. "After the Great Quake, Living with Earth's Uncertainty." *Yale Environment 360*, March 28, 2011.

Linnaeus, Carl. *Systema Naturae*. Nieuwkoop: B. de Graff, 1964 (c. 1735).

Nguyen, Mai. "How Scientists Preserved a 440-Pound Blue Whale Heart." *Wired*, July 2, 2017.

Nin, Anaïs. *The Diary of Anaïs Nin: Volume Three, 1939–1944*, edited by Gunther Stuhlmann. New York: Harcourt Brace Jovanovich, 1969.

* Poliquin, Rachel. *The Breathless Zoo: Taxidermy and the Cultures of Longing*. University Park: Pennsylvania State University Press, 2012.

Prince, Sue Ann, ed. *Stuffing Birds, Pressing Plants, Shaping Knowledge: Natural History in North America, 1730–1860*. Philadelphia: American Philosophical Society, 2003.

* Pyenson, Nick. *Spying on Whales: The Past, Present, and Future of Earth's Most Awesome Creatures*. New York: Penguin Books, 2018.

———. "The Ecological Rise of Whales Chronicled by the Fossil Record." *Current Biology*, 27, no. 11 (June 5, 2017): R558–64.

Pyenson, Nick, and Geerat J. Vermeij. "The Rise of Ocean Giants: Maximum Body Size in Cenozoic Marine Mammals as an Indicator for Productivity in the Pacific and Atlantic Oceans." *Biology Letters* 12, no. 7 (July 1, 2016).

Romero, Aldemaro. "When Whales Became Mammals: The Scientific Journey of Cetaceans from Fish to Mammals in the History of Science." *IntechOpen*, November 7, 2012.

Rossi, Michael. "Modelling the Unknown: How to Make a Perfect Whale." *Endeavour* 32, no. 2 (June 2008): 58–63.

———. "Fabricating Authenticity: Modeling a Whale at the American Museum of Natural History, 1906–1974." *Isis* 101, no. 2 (June 2010): 338–61.

Rudwick, Martin. *Earth's Deep History: How It Was Discovered and Why It Matters*. Chicago and London: University of Chicago Press, 2014.

Samaran, Flore, Kathleen M. Stafford, Trevor A. Branch, Jason Gedamke, Jean-Yves Royer, Robert P. Dziak, and Christophe Guinet. "Seasonal and Geographic Variation of Southern Blue Whale Subspecies in the Indian Ocean." *PLOS ONE* 8, no. 8 (2013): e71561.

Small, George. *The Blue Whale*. New York and London: Columbia University Press, 1971.

* Switek, Brian. *Written in Stone: Evolution, the Fossil Record and Our Place in Nature*. New York: Bellevue Literary Press, 2010.

Thewissen, J. G. M., ed. *The Emergence of Whales: Evolutionary Patterns in the Origin of Cetacea*. New York: Plenum Press, 1998.

———. *The Walking Whales: From Land to Water in Eight Million Years*. Berkeley: University of California Press, 2014.

Thewissen, J., and Ellen M. Williams. "The Early Radiations of Cetacea (Mammalia): Evolutionary Pattern and Developmental Correlations." *Annual Review of Ecology and Systematics* 33 (2002): 73–90.

Zalasiewicz, Jan. *The Planet in a Pebble: A Journey into Earth's Deep History*. Oxford and New York: Oxford University Press, 2010.

Zimmer, Carl. *At the Water's Edge: Macroevolution and the Transformation of Life*. New York: Free Press, 1998.

CHARISMA

Acorn, John. "The Windshield Anecdote." *American Entomologist* 62, no. 4 (Winter 2016): 262–64.

Albert, Céline, Gloria M. Luque, and Franck Courchamp. "The Twenty Most Charismatic Species." *PLOS ONE* 13, no. 7 (July 9, 2018): e0199149.

Andrews, Candice Gaukel. "Your Travel Photos Are Helping Rhino Poachers." *Outside*, July 17, 2014.

Aragón, Oriana R., Margaret S. Clark, Rebecca L. Dyer, and John A. Bargh. "Dimorphous Expressions of Positive Emotion: Displays of Both Care and Aggression in Response to Cute Stimuli." *Psychological Science* 26, no. 3 (2015): 259–73.

Austin, Bryant. *Beautiful Whale*. New York: Abrams Books, 2013.

Australian Associated Press. "Migaloo's Red Rash Prompts Skin Cancer Fears for Albino Whale." *Guardian*, June 25, 2014.

"Autopsy Shows Killer Whale Nami Swallowed 180 lbs of Stones Before Death." *Orca Project*, February 2, 2011.

Barkham, Patrick. "Stone-Stacking: Cool for Instagram, Cruel for the Environment." *Guardian*, August 17, 2018.

* Bar-On, Yinon M., Rob Phillips, and Ron Milo. "The Biomass Distribution on Earth." *Proceedings of the National Academy of Sciences* 115, no. 25 (June 19, 2018): 6506–11.

Becker, Elizabeth. *Overbooked: The Exploding Business of Travel and Tourism*. New York: Simon & Schuster, 2013.

Benson, Etienne. *Wired Wilderness: Technologies of Tracking and the Making of Modern Wildlife*. Baltimore: Johns Hopkins University Press, 2010.

Bousé, Derek. "False Intimacy: Close-ups and Viewer Involvement in Wildlife Films." *Visual Studies* 18, no. 2 (October 2003): 123–32.

* Brower, Matthew. *Developing Animals: Wildlife and Early American Photography*. Minneapolis: University of Minnesota Press, 2011.

Byrld, Mette, and Nina Lykke. *Cosmodolphins: Feminist Cultural Studies of Technology, Animals and the Sacred*. London: Zed Books, 1999.

Courchamp Franck, Ivan Jaric, Céline Albert, Yves Meinard, William J. Ripple, and Guillaume Chapron. "The Paradoxical Extinction of the Most Charismatic Animals." *PLOS Biology* 16, no. 4 (April 12, 2018): e2003997.

Cowperthwaite, Gabriela, dir. *Blackfish*. United States: CNN Films/Magnolia Pictures, 2013.

Cummings, E. E. *E. E. Cummings: A Miscellany Revised*. New York: October House, 1965.

Downs C. A., Esti Kramarsky-Winter, Roee Segal, John Fauth, Sean Knutson, Omri Bronstein, Frederic R. Ciner, et al. "Toxicopathological Effects of the Sunscreen UV Filter, Oxybenzone (Benzophenone-3), on Coral Planulae and Cultured Primary Cells and Its Environmental Contamination in Hawaii and the U.S. Virgin Islands." *Archives of Environmental Contamination and Toxicology* 70, no. 2 (2016): 265–88.

Einarsson, Niels. "All Animals Are Equal but Some Are Cetaceans." In *Environmentalism: The View from Anthropology*, edited by Kay Milton. London and New York: Routledge, 1993.

* Falconer, Delia. "The Opposite of Glamour." *Sydney Review of Books*, July 28, 2017.

Fletcher, Robert. "Ecotourism After Nature: Anthropocene Tourism as a New Capitalist Fix." *Journal of Sustainable Tourism* 27, no. 4 (2019): 522–35.

———. *Romancing the Wild: Cultural Dimensions of Ecotourism*. Durham, NC: Duke University Press, 2014.

Foote, Andrew, Nagarjun Vijay, María C. Ávila-Arcos, Robin W. Baird, John W. Durban, Matteo Fumagalli, Richard A. Gibbs, et al. "Genome-culture Coevolution Promotes Rapid Divergence of Killer Whale Ecotypes." *Nature Communications* 7 (May 31, 2016): 11693.

Freedman, Eric. "Extinction Is Forever: A Quest for the Last Known Survivors." *Earth Island Journal*, Autumn 2011.

* *Global Assessment Report on Biodiversity and Ecosystem Services.* Intergovernmental Science-Policy Platform on Biodiversity and Ecosystem Services, 2019.

Grebowicz, Margret. *The National Park to Come.* Stanford, CA: Stanford University Press, 2015.

Gren, Martin, and Edward H. Huijbens, eds. *Tourism and the Anthropocene.* London: Routledge, 2015.

Grooten, M., and R. E. A. Almond, eds. *Living Planet Report—2018: Aiming Higher.* Gland, Switzerland: WWF, 2018.

Haigney, Sophie. "How Stone Stacking Wreaks Havoc on National Parks." *The New Yorker*, December 2, 2018.

Hall, C. Michael. "Degrowing Tourism: Décroissance, Sustainable Consumption and Steady-State Tourism." *Anatolia* 20, no. 1 (June 2009): 46–61.

Hallmann, Caspar A., Martin Sorg, Eelke Jongejans, Henk Siepel, Nick Hofland, Heinz Schwan, Werner Stenmans, et al. "More Than 75 Percent Decline Over 27 Years in Total Flying Insect Biomass in Protected Areas." *PLOS ONE* 12, no. 10 (October 18, 2017): e185809.

Hargrove, John, and Howard Chua-Eoan. *Beneath the Surface: Killer Whales, SeaWorld, and the Truth Beyond Blackfish.* New York: Palgrave Macmillan, 2015.

Harrison, Marissa. "Anthropomorphism, Empathy, and Perceived Communicative Ability Vary with Phylogenetic Relatedness to Humans." *Journal of Social, Evolutionary, and Cultural Psychology* 4, no. 1 (2010): 34–48.

Hecker, Bruce. "How Do Whales and Dolphins Sleep Without Drowning?" *Scientific American*, February 2, 1998.

Hoare, Philip. "Or the Whale," video. Address to University of Sydney, Sydney Environment Institute, July 26, 2017.

Jørgensen, Dolly. "Endling: The Power of the Last in an Extinction-Prone World." *Environmental Philosophy* 14, no. 1 (2017): 119–38.

Kalland, Arne. "Super-Whale: The Use of Myths and Symbols in Environmentalism." In *11 Essays on Whales and Man*, edited by Georg Blichfeldt. Lofoten Reine, Norway: High North Alliance, 1994.

* Kolbert, Elizabeth. *The Sixth Extinction: An Unnatural History.* New York: Henry Holt, 2014.

Leiren-Young, Mark. *The Killer Whale Who Changed the World.* Vancouver and Berkeley: Graystone Books, 2016.

Lewis, Helen. "Sense of an Endling." *New Statesman*, June 27, 2012.

Lippit, Akira Mizuta. *Electric Animal: Toward a Rhetoric of Wildlife.* Minneapolis: University of Minnesota Press, 2000.

Lister, Bradford, and Andres Garcia. "Climate-Driven Declines in Arthropod Abundance Restructure a Rainforest Food Web." *Proceedings of the National Academy of Sciences* 115, no. 44 (October 30, 2018): e10391–406.

Lutts, Ralph. *The Nature Fakers: Wildlife, Science & Sentiment.* Golden, CO: Fulcrum, 1990.

Lynas, Mark. *The God Species: How the Planet Can Survive the Age of Humans.* London: Fourth Estate, 2011.

Madrigal, Alexis C. "You're Eye-to-Eye With a Whale in the Ocean: What Does It See?" *The Atlantic*, March 28, 2013.

Marino, Lori, Daniel McShea, and Mark D. Uhen. "Origin and Evolution of Large Brains in Toothed Whales." *Anatomical Record* 281, no. 2 (December 2004): 1247–55.

McKibben, Bill. "The Problem with Wildlife Photography." *Doubletake* (Fall 1997): 50–56.

———. *The End of Nature*. New York: Anchor Books, 1989.

Miller, Patrick, Kagari Aoki, Luke E. Rendell, and Masao Amano. "Stereotypical Resting Behaviour of the Sperm Whale." *Current Biology* 18, no. 1 (January 8, 2008): R21–23.

Miralles, Aurélien, Michel Raymond, and Guillaume Lecointre. "Empathy and Compassion Toward Other Species Decrease with Evolutionary Divergence Time." *Scientific Reports* 9 (December 20, 2019): 19555.

Mormann, Florian Julien Dubois, Simon Kornblith, Milica Milosavljevic, Moran Cerf, Matias Ison, Naotsugu Tsuchiya, et al. "A Category-Specific Response to Animals in the Right Human Amygdala." *Nature Neuroscience* 14 (August 28, 2011): 1247–49.

Murzyn, Eva. "Do We Only Dream in Color? A Comparison of Reported Dream Color in Younger and Older Adults with Different Experiences of Black and White Media." *Consciousness and Cognition* 17, no. 4 (2008): 1228–37.

* Ngai, Sianne. "The Cuteness of the Avant-Garde." *Critical Inquiry* 31, no. 4 (Summer 2005): 811–47.

Paine, Stefani. *The World of the Arctic Whales: Belugas, Bowheads and Narwhals*. San Francisco: Sierra Club Books, 1995.

Passarello, Elena. *Animals Strike Curious Poses: Essays*. London: Jonathan Cape, 2017.

Pearce, Fred. "Global Extinction Rates: Why Do Estimates Vary So Wildly?" *Yale Environment 360*, August 17, 2015.

Peichl, Leo, Günther Behrmann, and Ronald Kröger. "For Whales and Seals the Ocean Is Not Blue: A Visual Pigment Loss in Marine Mammals." *European Journal of Neuroscience* 13, no. 8 (April 2001): 1520–28.

Reardon, Sara. "Do Dolphins Speak Whale in Their Sleep?" *Science Magazine*, January 20, 2012.

Régnier, Claire, Guillaume Achaz, Amaury Lambert, Robert H. Cowie, Philippe Bouchet, and Benoit Fontaine. "Mass Extinction in Poorly Known Taxa." *Proceedings of the National Academy of Sciences* 112, no. 25 (June 23, 2015): 7761–66.

* Reinert, Hugo. "Face of a Dead Bird: Notes on Grief, Spectrality and Wildlife Photography." *Rhizomes: Cultural Studies in Emerging Knowledge* 23 (2012).

———. "The Care of Migrants: Telemetry and the Fragile Wild." *Environmental Humanities* 3, no. 1 (May 1, 2013): 1–24.

Richards, Morgan. "Greening Wildlife Documentary." In *Environmental Conflict and the Media*, edited by Libby Lester and Brett Hutchins. New York: Peter Lang, 2013.

Safina, Carl. "Woo-woo; Whale Magic?" *National Geographic*, August 31, 2016.

Siebert, Charles. "The Story of One Whale Who Tried to Bridge the Linguistic Divide Between Animals and Humans." *Smithsonian Magazine*, June 2014.

Silko, Leslie Marmon. "Landscape, History, and the Pueblo Imagination." *Antaeus* 57 (Autumn 1986): 882–94.

* Simmonds, Charlotte, Annette McGivney, Patrick Reilly, Brian Maffly, Todd Wilkinson, Gabrielle Canon, Michael Wright, and Monte Whaley. "Crisis in Our National Parks, How Tourists Are Loving Nature to Death." *Guardian*, November 20, 2018.

Skibins, Jeffrey, Robert Powell, and Jeffrey C. Hallo. "Charisma and Conservation: Charismatic Megafauna's Influence on Safari and Zoo Tourists' Pro-Conservation Behaviors." *Biodiversity and Conservation* 22, no. 4 (April 2013): 959–82.

Stenger, Richard. "New Moby Dick? Boat Crasher a Rare White Whale." CNN, August 23, 2003.

Stien, Didier, Fanny Clergeaud, Alice M. S. Rodrigues, and Karine Lebaron. "Metabolomics Reveal That Octocrylene Accumulates in *Pocillopora damicornis* Tissues as Fatty Acid Conjugates and Triggers Coral Cell Mitochondrial Dysfunction." *Analytical Chemistry* 91, no. 1 (January 2, 2019): 990–95.

Trigger, David. "Whales, Whitefellas, and the Ambiguity of 'Nativeness': Reflections on the Emplacement of Australian Identities." *Island* 107 (2006): 25–36.

Vidal, John. "Stop scattering ashes, families are told." *Guardian*, January 18, 2009.

Walter, Benjamin. "Gloves." In *One-Way Street and Other Writings*. London: NLB, 1979.

Webster, Robert M., and Bruce Erickson. "The Last Word?," Correspondence. *Nature* 380 (April 4, 1996): 386.

* Wilson, Edward O. *Biophilia*. Cambridge, MA, and London: Harvard University Press, 1984.

———. *Half-Earth: Our Planet's Fight for Life*. New York: Liveright, 2016.

Yoon, Carol Kaesuk. *Naming Nature: The Clash Between Instinct and Science*. New York: W. W. Norton, 2009.

Yong, Ed. "The Last of Its Kind." *The Atlantic*, July 2019.

SOUNDING

Definitions of "sound" are taken from Webster's Dictionary, 1828 and 1913 editions. Definitions have been lightly edited for clarity and consistency.

Allen, Jenny, Ellen Garland, Rebecca Dunlop, and Michael Noad. "Cultural Revolutions Reduce Complexity in the Songs of Humpback Whales." *Proceedings of the Royal Society B: Biological Sciences* 285, no. 1891 (November 21, 2018).

Balcomb, K. C., and Diane Claridge. "A Mass Stranding of Cetaceans Caused by Naval Sonar in the Bahamas." *Bahamas Journal of Science* 8 (January 2001): 2–12.

Bataille, Georges. *The Unfinished System of Nonknowledge*. Minneapolis: University of Minnesota Press, 2001.

Bateson, Gregory. *Steps to an Ecology of Mind: Collected Essays in Anthropology, Psychiatry, Evolution, and Epistemology*. St Albans, UK: Paladin, 1972.

Biguenet, John. *Object Lesson: Silence*. London and New York: Bloomsbury Academic, 2015.

"Boat Noise Impacts Development, Survival of Sea Hares." *ScienceDaily*, with the University of Bristol, July 31, 2014.

Bosker, Bianca. "Why Is the World So Loud?" *The Atlantic*, November, 2019.

"Breaking the Silence: How Our Noise Pollution Is Harming Whales." International Fund for Animal Welfare Publication, 2013.

* Carson, Rachel. *Silent Spring*. London: Hamish Hamilton, 1962.

Cholewiak, Danielle M., Renata Santoro Sousa-Lima, and Salvatore Cerchio. "Humpback Whale Song Hierarchical Structure: Historical Context and Discussion of Current Classification Issues." *Marine Mammal Science* 29, no. 3 (July 2013): E312–32.

Clark, Christopher Willes, and Phillip J. Clapman. "Acoustic Monitoring on a Humpback Whale (*Megaptera novaeangliae*) Feeding Ground Shows Continual Singing Into Late Spring." *Proceedings of the Royal Society B: Biological Sciences* 271, no. 1543 (May 22, 2004): 1051–57.

Cousteau, Jacques, dir. *The Silent World*. France: FSJYC Productions, 1956.

Crane, Adam, and Maud Ferrari. "The Fishy Problem of Underwater Noise Pollution." *The Conversation*, April 9, 2018.

Darling, James D. "Humpback Whale Calls Detected in Tropical Ocean Basin Between Known Mexico and Hawaii Breeding Assemblies." *Journal of the Acoustical Society of America* 145, no. 6 (June 2019): EL534.

———. "Low Frequency, ca. 40 Hz, Pulse Trains Recorded in the Humpback Whale Assembly in Hawaii." *Journal of the Acoustical Society of America* 138, no. 5 (November 2015): EL452–58.

Darling, James D., Jo Marie V. Acebes, Oscar Frey, R. Jorge Urbán, and Manami Yamaguchi "Convergence and Divergence of Songs Suggests Ongoing, but Annually Variable, Mixing of Humpback Whale Populations Throughout the North Pacific." *Scientific Reports* 9, no. 7002 (May 7, 2019).

Darling, James D., Meagan E. Jones, and Charles P. Nicklin. "Humpback Whale Songs: Do They Organize Males During the Breeding Season?" *Behaviour* 143, no. 9 (September 2006): 1051–101.

Day, Ryan D., Robert D McCauley, Quinn P. Fitzgibbon, Klaas Hartmann, and Jayson M. Semmens. "Exposure to Seismic Air Gun Signals Causes Physiological Harm and Alters Behavior in the Scallop *Pecten fumatus*." *Proceedings of the National Academy of Sciences* 114, no. 40 (October 3, 2017): 8537–46.

de Quirós, Y. Bernaldo, A. Fernandez, R. W. Baird, R. L. Brownell, N. Aguilar de Soto, D. Allen, M. Arbelo, et al. "Advances in Research on the Impacts of Anti-Submarine Sonar on Beaked Whales." *Proceedings of the Royal Society B: Biological Sciences* 286, no. 1895 (January 30, 2019).

Donovan, Arthur; and Joseph Bonney. *The Box that Changed the World: Fifty Years of Container Shipping: An Illustrated History*. East Windsor, NJ: Commonwealth Business Media, 2006.

* Douglas, Mary. *Purity and Danger: An Analysis of the Concepts of Pollution and Taboo*. New York and London: Praeger, 1966.

Erbe, Christine, Rebecca Dunlop, K. Curt S. Jenner, Michelin N. M. Jenner, Robert D. McCauley, Iain Parnum, Miles Parsons, Tracey Rogers, and Chandra Salgado-Kent. "Review of Underwater and In-Air Sounds Emitted by Australian and Antarctic Marine Mammals." *Acoustics Australia* 45, no. 2 (2017): 179–241.

Ferrari, Maud C. O., et al. "School Is Out on Noisy Reefs: The Effect of Boat Noise on Predator Learning and Survival of Juvenile Coral Reef Fishes." *Proceedings of the Royal Society B: Biological Sciences* 285, no. 1871 (January 31, 2018).

Garland, Ellen C., Luke Rendell, Luca Lamoni, M. Michael Poole, and Michael J. Noad. "Song Hybridization Events During Revolutionary Song Change Provide Insights into Cultural Transmission in Humpback Whales" colloquium paper. *Proceedings of the National Academy of Sciences* 114, no. 30 (July 25, 2017): 7822–29.

Garland, Ellen C., Anne W. Goldizen, Melinda L. Rekdahl, Rochelle Constantine, Claire Garrigue, Nan Daeschler Hauser, M. Michael Poole, Jooke Robbins, and

Michael J. Noad. "Dynamic Horizontal Cultural Transmission of Humpback Whale Song at the Ocean Basin Scale." *Current Biology* 21, no. 8 (April 26, 2011): 687–91.

Gazioglu, Cem, Ahmet Edip Müftüglu, Volkan Demir, and Abdillah Aksu. "Connection between Ocean Acidification and Sound Propagation." *International Journal of Environment and Geoinformatics* 2, no. 2 (November 2015): 16–26.

Gende, Scott, Lawrence Vose, Jeff Baken, Christine M. Gabriele, Rich Preston, and A. Noble Hendrix. "Active Whale Avoidance by Large Ships: Components and Constraints of a Complementary Approach to Reducing Ship Strike Risk." *Frontiers in Marine Science* 6 (September 30, 2019): 592.

Gies, Erica. "An Earsplitting Threat Is Endangering the World's Rarest Killer Whales." *TakePart*, December 15, 2015.

Gol'din, Pavel. "'Antlers Inside': Are the Skull Structures of Beaked Whales (*Cetacea: Ziphiidae*) Used for Echoic Imaging and Visual Display?" *Biological Journal of the Linnean Society* 113, no. 2 (October 2014): 510–15.

Gray, Patricia, Bernie Krause, Jelle Atema, Roger Payne, Carol Krumhansl, and Luis Baptista. "The Music of Nature and the Nature of Music." *Science* 291, no. 5501 (January 5, 2001): 52–54.

* Grebowicz, Margret. *Object Lessons: Whale Song*. London and New York: Bloomsbury Academic, 2017.

Guinee, Linda, and Katharine Payne. "Rhyme-like Repetitions in Songs of Humpback Whales." *Ethology* 79, no. 4 (January–December 1988): 295–306.

Hamer, Ashley. "Sperm Whales Are Loud Enough to Burst Your Eardrums." *Curiosity*, July 22, 2016.

Jamison, Leslie. "52 Blue." In *Make It Scream, Make It Burn: Essays*. New York: Little, Brown, 2019.

Juchau, Mireille. "What Should We Send Into Space as a *New* Record of Humanity?" *Literary Hub*, April 22, 2019.

* Krause, Bernie. "Anatomy of the Soundscape." *Journal of the Audio Engineering Society* 56, no. 1/2 (January 2008): 73–80.

———. *Voices of the Wild: Animal Songs, Human Din and the Call to Save Natural Soundscapes*. New Haven, CT: Yale University Press, 2015.

Kroll, Gary. "Snarge." In *Future Remains: A Cabinet of Curiosities for the Anthropocene*, edited by Gregg Mitman, Marco Armiero, and Robert S. Emmett. Chicago: University of Chicago Press, 2018.

Kunc, Hansjoerg P., Gillian N. Lyons, Julia D. Sigwart, Kirsty E. McLaughlin, and Jonathan D. R. Houghton. "Anthropogenic Noise Affects Behavior across Sensory Modalities." *American Naturalist* 184, no. 4 (October 2014): E93–100.

* Leroy, Emmanuelle C., Jean-Yves Royer, Julien Bonnel, and Flore Samaran. "Long-Term and Seasonal Changes of Large Whale Call Frequency in the Southern Indian Ocean." *Journal of Geophysical Research: Oceans*, 123, no. 11 (November 2018): 8568–80.

MacKinnon, J. B. "It's Tough Being a Right Whale These Days." *The Atlantic*, July 30, 2018.

McCauley, Robert D. Ryan D. Day, Kerrie M. Swadling, Quinn P. Fitzgibbon, Reg A. Watson, and Jayson M. Semmens. "Widely Used Marine Seismic Survey Air Gun Operations Negatively Impact Zooplankton." *Nature Ecology & Evolution* 1 (June 22, 2017): 0195.

McGregor, Peter, ed. *Animal Communication Networks*. Cambridge, UK, and New York: Cambridge University Press, 2005.

McLendon, Russell. "Mysterious 'Ping' Reported in Arctic Ocean." MNN, November 4, 2016.

Menken, Alan, dir. *The Little Mermaid*. Milwaukee: H. Leonard Pub. Corp./Walt Disney, 1990.

Mercado, Eduardo, III, L. M. Herman, and A. A. Pack. "Stereotypical Sound Patterns in Humpback Whale Songs: Usage and Function." *Aquatic Mammals* 29, no. 1 (September 28, 2010): 37–52.

Moe, Aaron. *Zoopoetics: Animals and the Making of Poetry*. Plymouth, UK: Lexington Books, 2013.

Monbiot, George. *Feral: Searching for Enchantment on the Frontiers of Rewilding*. London: Allen Lane, 2013.

Mosher, Dave. "A Spacecraft Graveyard Exists in the Middle of the Ocean: Here's What's Down There." *Business Insider*, October 22, 2017.

Noad, Michael J., Douglas H. Cato, M. M. Bryden, Michelin N. Jenner, and K. Curt S. Jenner. "Cultural Revolution in Whale Songs." *Nature* 408, no. 6812 (November 30, 2000): 537.

Payne, Roger. "An Open Letter to the Youth of Japan." *Ocean Alliance*, 2005.

Payne, Roger, and Douglas Webb. "Orientation by Means of Long Range Acoustic Signalling in Baleen Whales." *New York Academy of Sciences* 188 (1971): 110–41.

Payne, Roger, and Scott McVay. "Songs of the Humpback Whales." *Science* 173 (1971), 585–97.

Pijanowski, Bryan C., Almo Farino, Stuart H. Gage, Sarah L. Dumyahn, and Bernie Krause. "What Is Soundscape Ecology? An Introduction and Overview of an Emerging New Science." *Landscape Ecology* 26, no. 9 (2011): 1213–32.

Podestà, Michela. "Beaked Whale Strandings in the Mediterranean Sea." Proceedings of the ECS Workshop, Beaked Whale Research. *ECS Special Publication Series No. 51* (2009).

Rankin, Shannon, and Jay Barlow. "Source of the North Pacific 'Boing' Sound Attributed to Minke Whales." *Journal of the Acoustical Society of America* 118, no. 5 (October 28, 2005): 3346–51.

Razafindrakoto, Y., S. Cerchio, T. Collins, H. Rosenbaum, and S. Ngouessono. "Similarity of Humpback Whale Song from Madagascar and Gabon Indicates Significant Contact between South Atlantic and Southwest Indian Ocean Populations." IWC Scientific Committee Paper SC/61/SH8, 15 (2009).

"Recordings That Made Waves: The Songs That Saved The Whales." *All Things Considered*, NPR, December 26, 2014 [audio].

Ridgeway, Sam, Donald Carder, Michelle Jeffries, and Mark Tood. "Spontaneous Human Speech Mimicry by a Cetacean." *Current Biology* 22, no. 20 (October 23, 2012): R860–61.

Rockwood, R. Cotton, John Calambokidis, and Jaime Jahncke. "High Mortality of Blue, Humpback and Fin Whales from Modeling of Vessel Collisions on the U.S. West Coast Suggests Population Impacts and Insufficient Protection." *PLOS ONE* 13, no. 7 (2017).

Rolland, R. M., Susan E. Parks, Kathleen E. Hunt, Manuel Castellote, Peter J. Corkeron, Douglas P. Nowacek, Samuel K. Wasser, and Scott D. Kraus. "Evidence that Ship Noise Increases Stress in Right Whales." *Proceedings of the Royal Society B: Biological Sciences* 279, no. 1737 (February 8, 2012): 2363–68.

* Rothenberg, David. *Thousand Mile Song: Whale Music in a Sea of Sound.* New York: Basic Books, 2008.

* Rozwadowski, Helen. *Vast Expanses: A History of the Oceans.* London: Reaktion Books, 2018.

Sagan, Carl. *Murmurs of Earth: The Voyager Interstellar Record.* New York: Random House, 1978.

Sandoe, P. "Do Whales Have Rights?" In *11 Essays on Whales and Man*, edited by Georg Blichfeldt. Lofoten Reine, Norway: High North Alliance, 1994.

Saxon, Wolfgang. "Christine Stevens, 84, a Friend to the Animals," obituary. *New York Times*, October 15, 2002.

Shen, Alice. "How Chinese Scientists Use Sperm Whale Sounds to Send Secret Messages for the Military." *South China Morning Post*, November 2, 2018.

Stimpert, Alison K., David N. Wiley, Whitlow Aw, and Mark P. Johnson. "'Megapclicks': Acoustic Click Trains and Buzzes Produced During Night-time Foraging of Humpback Whales (*Megaptera novaeangliae*)." *Biology Letters* 3, no. 5 (November 2007): 467–70.

Taylor, Stephanie, and Tony Walker. "North Atlantic Right Whales in Danger." *Science* 358, no. 6364 (November 2017): 730–31.

Tervo, Outi, Mads Fage Christoffersen, Susan Elizabeth Parks, and Reinhardt Møbjerg Kristensen. "Evidence for Simultaneous Sound Production in the Bowhead Whale (*Balaena mysticetus*)." *Journal of the Acoustical Society of America* 130, no. 4 (October 2011): 2257–62.

Tsujii, Koki, Tomonari Akamatsu, Ryosuke Okamoto, Kyoichi Mori, Yoko Mitani, and Naoya Umeda. "Change in Singing Behaviour of Humpback Whales Caused by Shipping Noise." *PLOS ONE* 13, no. 10 (October 24, 2018): e0204112.

Van Cise, Amy M., Sabre D. Mahaffy, Robin W. Baird, T. Aran Mooney, and Jay Barlow. "Song of My People: Dialect Differences Among Sympatric Groups of Short-Finned Pilot Whales in Hawai'i." *Behavioral Ecology and Sociobiology* 72 (December 2018): 1–13.

* Vanselow, Klaus Heinrich, Sven Jacobsen, Chris Hall, and Stefan Garthe. "Solar Storms May Trigger Sperm Whale Strandings: Explanation Approaches for Multiple Strandings in the North Sea in 2016," *International Journal of Astrobiology* 17, no. 4 (October 2018): 336–44.

Vidal, John. "Health Risks of Shipping Pollution Have Been 'Underestimated.'" *Guardian*, April 9, 2009.

Walker, M. M., J. L. Kirschvink, G. Ahmed, and A. E. Dizon. "Evidence That Fin Whales Respond to the Geomagnetic Field During Migration." *Journal of Experimental Biology* 171 (1992): 67–68.

Wallace, Samantha. "Underwater Compositions: Song Sharing Between Southern Ocean Humpback Whales." *PLOS Blog*, December 26, 2013.

Wallin, Nils Lennart, Björn Merker, and Steven Brown, eds. *The Origins of Music.* Cambridge, MA: MIT Press, 2000.

Williams, Rob. "Secret to a Sound Ocean." *The Walrus*, September 21, 2015 [audio].

Žižek, Slavoj. *The Pervert's Guide to Cinema*, directed by Sophie Fiennes. UK: Mischief Films, 2006.

SEA PIE

In evaluating the environmental impact of food waste and meat consumption, I have utilized research consolidated by Project Draw Down at www.drawdown.org.

Arch, Jakobina, and Paul Sutter. *Bringing Whales Ashore: Oceans and the Environment of Early Modern Japan*. Seattle: University of Washington Press, 2018.

Barsh, R. "Food Security, Food Hegemony and Charismatic Animals." *Toward a Sustainable Whaling Regime*, edited by Robert Friedheim. Seattle: University of Washington Press; Edmonton: Canadian Circumpolar Institute Press, 2001.

Bestor, Theodore C. *Tsukiji: The Fish Market at the Center of the World*. Berkeley and Los Angeles: University of California Press, 2004.

Bhattacharya, Shaoni. "Anti-Whalers Say Cruelty of Killing Requires Ban." *New Scientist*, March 9, 2004.

"Blood e-Commerce: Rakuten's Profits from the Slaughter of Elephants and Whales." Environmental Investigation Agency UK Publication, March 18, 2014.

Brasor, Philip, and Masako Tsubuku. "In 2019, How Hungry Is Japan for Whale Meat?" *Japan Times*, January 11, 2019.

de Toulouse-Lautrec, Henri, and Maurice Joyant. *The Art of Cuisine*. New York: Holt, Rinehart and Winston, 1966.

Douglas, Mary. *Food in the Social Order*. London and New York: Routledge, 2002.

Dudley, Paul. "An Essay Upon the Natural History of Whales, with Particular Account of the Ambergris Found in the Spermaceti Whale." *Philosophical Transactions of the Royal Society of London* (1725): 256–59.

* Goodyear, Dana. *Anything That Moves: Renegade Chefs, Fearless Eaters, and the Making of a New American Food Culture*. New York: Riverhead Books, 2013.

Greimel, H. "Most Japanese Support Commercial Whaling According to Survey." Associated Press, March 17, 2002.

Hiraguchi, T. "Prehistoric and Protohistoric Whaling and Diversity in Japanese Foods." In *Traditional Whaling Summit in Nagato*, edited by Institute for Cetacean Research. Tokyo and Nagato: Institute for Cetacean Research, 2003.

Hirata, Keiko. "Why Japan Supports Whaling." *Journal of International Wildlife Law & Policy* 8, no. 2–3 (2005): 129–49.

Hurst, Daniel. "Japanese Hunters Kill 120 Pregnant Minke Whales During Summer Months—Report." *Guardian*, May 30, 2018.

Institute for Cetacean Research. *Small Type Coastal Whaling in Japan: Report of an International Workshop*. Edmonton, Canada: Boreal Institute for Northern Studies, 1988.

_____. *Papers on Japanese Small Type Coastal Whaling Submitted by the Government of Japan to the International Whaling Commission, 1985–1995*. Tokyo: Institute for Cetacean Research, 1996.

Ishii, Atsushi, and Ayako Okubo. "An Alternative Explanation of Japan's Whaling Diplomacy in the Post-Moratorium Era." *Journal of Wildlife Law and Policy* 10, no. 1 (2007): 55–87.

Isihara, Akiko, and Junichi Yoshii. *A Survey of the Commercial Trade in Whale Meat Products*. Traffic Report, 2000.

Itoh, Mayumi. *The Japanese Culture of Mourning Whales: Whale Graves and Memorial Monuments in Japan*. Princeton, NJ: Palgrave Macmillan, 2018.

Kalland, Arne. "The Anti-Whaling Campaigns and Japanese Responses." In *Japanese Position on the Anti-Whaling Campaign*. Tokyo: Institute for Cetacean Research, 1998.

Kalland, Arne, and Brian Moeran. *Japanese Whaling: End of an Era?* London: Routledge, 2011.

Kang, Sue, and Marcus Phipps. *A Survey of Whale Meat Markets Along South Korea's Coast*. Traffic Report, 2000.

Kemp, Christopher. *Floating Gold: A Natural (and Unnatural) History of Ambergris*. Chicago and London: University of Chicago Press, 2012.

Kessler, Rebecca. "Written in Baleen." *Aeon*, September 28, 2016.

Komatsu, Masayuki, and S. Misaki. *The Truth Behind the Whaling Dispute*. Tokyo: Institute for Cetacean Research, 2001.

Leonard, Abigail. "In Japan, Few People Eat Whale Meat Anymore, But Whaling Remains Popular." Public Radio International, April 17, 2019 [audio].

McLeish, Todd. *Narwhals: Arctic Whales in a Melting World*. Seattle: University of Washington Press, 2014.

Misaki, S. *Whaling for the Twenty-first Century*. Tokyo: Institute for Cetacean Research, 1996.

Mithen, Steven, Chris Knight, and Camilla Power. "The Origins of Anthropomorphic Thinking." *Journal of the Royal Anthropological Institute* 4, no. 1 (March 1998): 129–32.

* Morikawa, Jun. *Whaling in Japan: Power, Politics and Diplomacy*. London: Hurst, 2009.

Mozingo, Joe. "Two Gold Standards." *Los Angeles Times*, November 1, 2008.

Palumbi, Stephen. "In the Market for Minke Whales." *Nature* 447 (May 16, 2007): 267–68.

Pollan, Michael. *In Defense of Food: An Eater's Manifesto*. New York: Penguin Press, 2008.

Serpell, James. A. "One Man's Meat: Further Thoughts on the Evolution of Animal Food Taboos." In *On the Human*, a project of the National Humanities Centre, November 27, 2011.

Shoemaker, Nancy. "Whale Meat in American History." *Environmental History* 10, no. 2 (April 2005): 269–94.

"Sixty Percent of Japanese Support Whale Hunt." Phys.org, April 22, 2014.

Smil, Vaclav, and Kazuhiko Kobayashi. *Japan's Dietary Transition and Its Impacts*. Cambridge, MA: MIT Press, 2012.

Tatar, Bradley. "The Safety of Bycatch: South Korean Responses to the Moratorium on Commercial Whaling." *Journal of Marine and Island Cultures* 3, no. 2 (December 2014): 89–97.

Vincent, Sam. *Blood and Guts: Dispatches from the Whale Wars*. Collingwood, Victoria: Black, 2014.

"Whale Meat Lunch to Boost New Food: Natural History Museum Presents War Substitute for Beef, Pork, and Mutton. Notables Try the Feast. Some Say It Tastes like Pot Roast, and Others That It Much Resembles Venison." *New York Times*, February 9, 1918.

"Whaling in Iceland Recommences and Byproducts Used for Medical Purposes." *Iceland Monitor*, April 17, 2018.

KITSCH INTERIOR

An image of a seahorse holding on to a Q-tip became briefly famous online in 2017: it was taken by Justin Hofman, and later titled "Sewerage Surfer." The picture listed as a finalist in that year's Wildlife Photographer of the Year competition.

Ackerman, Diane. *The Moon by Whale Light: And Other Adventures Among Bats, Penguins, Crocodilians, and Whales.* New York: Random House, 1991.

Alaimo, Stacey. "Oceanic Origins, Plastic Activism, and New Materialism at Sea." In *Material Ecocriticism*, edited by Serenella Iovino and Serpil Opperman. Bloomington: Indiana University Press, 2014.

Albeck-Ripka, Livia. "30 Vaquita Porpoises Are Left Alive: One Died in a Rescue Mission." *New York Times*, November 11, 2017.

Allen, Steve, Deonie Allen, Vernon R. Phoenix, Gaël Le Roux, Pilar Durántez Jiménez, Anaëlle Simonneau, Stéphane Binet, and Didier Galop. "Atmospheric Transport and Deposition of Microplastics in a Remote Mountain Catchment." *Nature Geoscience* 12 (April 15, 2019): 339–44.

"Beluga Whale Has Finally Left the Thames, Say Experts." *Telegraph*, May 13, 2019.

Benton, Tim. "Oceans of Garbage." *Nature* 352, no. 113 (July 11, 1991).

Brodeur, Paul. "In the Face of Doubt." *The New Yorker*, June 2, 1986.

Bryant, Peter, Christopher Lafferty, and Susan Lafferty. "Reoccupation of Laguna Guerrero Negro, Baja California, Mexico, by Gray Whales." In *The Gray Whale*, Eschrichtius robustus, edited by Mary Lou Jones, Steven L. Swartz, and Stephen Leatherwood. Orlando: Academic Press, 1984.

Catarino, Ana I., Valeria Macchia, William G. Sanderson, Richard C. Thompson, and Theodore B. Henry. "Low Levels of Microplastics (MP) in Wild Mussels Indicate that MP Ingestion by Humans is Minimal Compared to Exposure via Household Fibres Fallout During a Meal." *Environmental Pollution* 237 (June 2018): 675–84.

Chung, Emily. "Beluga Whales Adopt Lost Narwhal in St. Lawrence River." CBC News, September 13, 2018.

Custard, Ben. "Rare Bowhead Whale Spotted off Cornish Coast." *Countryfile*, May 19, 2016.

D'Agostino, Valeria C., Mónica S. Hoffmeyer, Gastón O. Almandoz, Viviana Sastre, and Mariana Degrati. "Potentially Toxic *Pseudo-Nitzschia* Species in Plankton and Fecal Samples of *Eubalaena australis* from Península Valdés Calving Ground, Argentina." *Journal of Sea Research* 106 (December 2015): 39–43.

Doward, Jamie. "How Did That Get There? Plastic Chunks on Arctic Ice Show How Far Pollution has Spread." *Guardian*, September 24, 2017.

Ellis, Richard. *The Empty Ocean: Plundering the World's Marine Life.* Washington, DC: Island Press, 2003.

Eriksen, Marcus, Laurent C. M. Lebreton, Henry S. Carson, Martin Thiel, Charles J. Moore, Jose C. Borerro, Francois Galgani, Peter G. Ryan, and Julia Reisser. "Plastic Pollution in the World's Oceans: More than 5 Trillion Plastic Pieces Weighing over 250,000 Tons Afloat at Sea." *PLOS ONE* 9, no. 12 (December 10, 2014): e111913.

Farrier, David. *Footprints: In Search of Future Fossils.* London: Farrar, Straus and Giroux, 2020.

Fazio, A., Marcelo Bertellotti, and Cecilia Villanueva. "Kelp Gulls Attack Southern Right Whales: A Conservation Concern?" *Marine Biology* 10 (September 2012).

Fossi, Maria Cristina, Cristina Panti, Cristiana Guerranti, Daniele Coppola, Matteo Giannetti, Letizia Marsili, and Roberta Minutoli. "Are Baleen Whales Exposed to the Threat of Microplastics? A Case Study of the Mediterranean Fin Whale (*Balaenoptera physalus*)." *Marine Pollution Bulletin* 64, no. 11 (November 2012): 2374–79.

Geyer, Roland, Jenna R. Jambeck, and Kara Lavender Law. "Production, Use, and Fate of All Plastics Ever Made." *Science Advances* 3, no. 7 (July 19, 2017): e1700782.

Gibbs, Susan, Chandra P. Salgado Kent, Boyan Slat, and Damien Morales. "Cetacean Sightings within the Great Pacific Garbage Patch." *Marine Biodiversity* 49 (April 9, 2019): 2021–27.

Goldfarb, Ben. "The Endling: Watching a Species Disappear in Real Time." *Pacific Standard Magazine*, updated September 7, 2018.

Gregory, Murray. "Environmental Implications of Plastic Debris in Marine Settings—Entanglement, Ingestion, Smothering, Hangers-on, Hitch-Hiking and Alien Invasions." *Philosophical Transactions of the Royal Society B* 364, no. 1526 (July 27, 2009): 2013–25.

Higdon, Jeff, and Steven Ferguson. "Loss of Arctic Sea Ice Causing Punctuated Change in Sightings of Killer Whales (*Orcinus orca*) over the Past Century." *Ecological Applications* 19, no. 5 (July 2009): 1365–75.

* Hohn, Donovan. *Moby-Duck: The True Story of 28,800 Bath Toys Lost at Sea and of the Beachcombers, Oceanographers, Environmentalists, and Fools, Including the Author, Who Went in Search of Them*. London: Viking, 2011.

Hoy, C. M. "The 'White-Flag' Dolphin of Tung Ting Lake." *China Journal of Arts and Sciences* 1 (1923): 154–57.

Jamieson, Alan J., Tamas Malkocs, Stuart B. Piertney, Toyonobu Fujii, and Zulin Zhangl. "Bioaccumulation of Persistent Organic Pollutants in the Deepest Ocean Fauna." *Nature Ecology Evolution* 1 (February 13, 2017): 0051.

Jamieson, Alan J., L. S. R. Brooks, W. D. K. Reid, S. B. Piertney, B. E. Narayanaswamy, and T. D. Linley. "Microplastics and Synthetic Particles Ingested by Deep-Sea Amphipods in Six of the Deepest Marine Ecosystems on Earth." *Royal Society Open Science* 6, no. 2 (February 1, 2019).

Keartes, Sarah. "For the First Time Ever, a Narwhal has Stranded on Belgium's Shores." *Earth Touch News Network*, April 29, 2016.

Kelly, Brendan P., Andrew Whiteley, and David Tallmon. "The Arctic Melting Pot." *Nature* 468 (December 15, 2010): 891.

Kershaw, Peter, ed. "Sources, Fate and Effects of Microplastics in the Marine Environment: A Global Assessment." Joint Group of Experts on the Scientific Aspects of Marine Environmental Protection, for the International Maritime Organization, 2015.

* Kormann, Carolyn. "Where Does All the Plastic Go?" *The New Yorker*, September 16, 2019.

Lebreton, Laurent, Boyan Slat, Francesco F. Ferrari, and Bruno Sainte-Rose. "Evidence that the Great Pacific Garbage Patch Is Rapidly Accumulating Plastic." *Scientific Reports* 8 (March 22, 2018): 4666.

Liu, Mengting, Shibo Lu, Yang Song, Lili Lei, Jiani Hu, Weiwei Lv, Wenzong Zhou, et al. "Microplastic and Mesoplastic Pollution in Farmland Soils in Suburbs of Shanghai, China." *Environmental Pollution* 242, part A (November 2018): 855–62.

Marón, Carina F., Lucas Beltramino, Matias Di Martino, Andrew Chirife, Jon Seger, Marcela Uhart, Mariano Sironi, and Victoria J. Rowntree. "Increased Wounding of Southern Right Whale (*Eubalaena australis*) Calves by Kelp Gulls (*Larus dominicanus*) at Península Valdés, Argentina." *PLOS ONE* 10, no. 11 (October 21, 2015): e0139291.

McDonnell, Tim. "A Strange New Gene Pool of Animals Is Brewing in the Arctic." *Nautilus*, December 11, 2014.

McGrath, Matt. "Whale Killing: DNA Shows Iceland Whale Was Rare Hybrid." BBC News, July 20, 2018.

Moore, Charles, and Cassandra Phillips. *Plastic Ocean: How a Sea Captain's Chance Discovery Launched a Determined Quest to Save the Oceans.* New York: Avery, 2011.

Moore, Thomas. "Sky Ocean Rescue: A Plastic Whale." Sky News UK, June 23, 2017.

Neilson, Alasdair. "Considering the Importance of Metaphors for Marine Conservation." *Marine Policy* 97 (November 2018): 239–43.

Noakes, S. E., Nick Pyenson, and G. McFall. "Late Pleistocene Gray Whales (*Eschrichtius robustus*) Offshore Georgia USA, and the Antiquity of Gray Whale Migration in the North Atlantic Ocean." *Palaeogeography, Palaeoclimatology, Palaeoecology* 392 (December 15, 2013): 502–09.

Orwell, George. "Inside the Whale." In *Inside the Whale and Other Essays.* Harmondsworth, UK: Penguin, 1940.

Plumer, Brad. "We Dump 8 Million Tons of Plastic into the Ocean Each Year. Where Does It All Go?" *Vox*, October 21, 2015.

Price, Jennifer. "A Brief Natural History of the Plastic Pink Flamingo." In *Flight Maps: Adventures with Nature In Modern America.* New York: Basic Books, 2000.

Pyenson, Nick. "Ballad of the Last Porpoise." *Smithsonian Magazine*, November 2017, 29–33.

Roser, Max. "Oil Spills." *Our World in Data*, April 2017.

Santora, Jarrod A., Nathan J. Mantua, Isaac D. Schroeder, John C. Field, Elliott L. Hazen, Steven J. Bograd, William J. Sydeman, et al. "Habitat Compression and Ecosystem Shifts as Potential Links Between Marine Heatwave and Record Whale Entanglements." *Nature: Communications* 11 (January 27, 2020): 536.

Savoca, Matthew. "The Oceans Are Full of Plastic, but Why Do Seabirds Eat It?" *The Conversation*, November 9, 2016.

Scheinin, Aviad P., Dan Kerem, Colin D. MacLeod, Manel Gazo, Carla Alvarez Chicote, and Manuel Castellote. "Gray Whale (*Eschrichtius robustus*) in the Mediterranean Sea: Anomalous Event or Early Sign of Climate-Driven Distribution Change." *Marine Biodiversity Records* 4 (December 2011): e28.

Sebille, Erik van, Chris Wilcox, Laurent Lebreton, Nikolai Maximenko, Britta Denise Hardesty, Jan A. van Franeker, Marcus Eriksen, et al. "A Global Inventory of Small Floating Plastic Debris." *Environmental Research Letters* 10, no. 12 (December 8, 2015).

Semeena, Valiyaveetil S., and G. Lammel. "The Significance of the Grasshopper Effect on the Atmospheric Distribution of Persistent Organic Substances." *Geophysical Research Letters* 32, no. 7 (April 2005).

Struzik, Ed. "Arctic Roamers: The Move of Southern Species into Far North." *Yale Environment 360*, February 14, 2011.

Tranströmer, Tomas. *Selected Poems, 1954–1986.* New York: Ecco Press, 2011.

Turvey, Samuel T., Robert L. Pitman, Barbara L. Taylor, Jay Barlow, Tomonari Akamatsu, Leigh A. Barrett, Xiujiang Zhao, et al. "First Human-Caused Extinction of a Cetacean Species?" *Biology Letters* 3, no. 5 (October 22, 2007): 537–40.

Weinberger, Eliot. *An Elemental Thing.* New York: New Directions Books, 2007.

Wilcox, Chris, et al. "Threat of Plastic Pollution to Seabirds Is Global, Pervasive, and Increasing." *Proceedings of the National Academy of Sciences* 112, no. 38 (September 22, 2015): 11899–904.

SCANTLING

Adamowsky, Natascha. *The Mysterious Science of the Sea 1775–1943*. London and New York: Routledge, 2015.

Behrman, Cynthia Fansler. *Victorian Myths of the Sea*. Athens: Ohio University Press, 1977.

Brown, Chandros Michael. "A Natural History of the Gloucester Sea Serpent: Knowledge, Power and Culture of Science in Antebellum America." *American Quarterly* XLII, no. 3 (September 1990): 402–36.

Corbin, Alain. *The Lure of the Sea: The Discovery of the Seaside in the Western World 1750–1840*. Berkeley: University of California Press, 1994.

Croll, Donald A., Raphael Kudela, and Bernie R. Tershy. "Ecosystem Impact of the Decline of Large Whales in the North Pacific (Great Whales as Consumers)." In *Whales, Whaling, and Ocean Ecosystems*, edited by James Estes. Berkeley, California: University of California Press, 2007.

Darwin, Charles. *On the Origin of Species by Means of Natural Selection, or Preservation of Favored Races in the Struggle for Life*. London: John Murray, 1859.

Dillard, Annie. *Pilgrim at Tinker Creek*. Maine: Thorndike Press, 1974.

Ellis, Richard. *Monsters of the Sea*. Westminster, Maryland: Knopf, 1994.

Feltman, Rachel. "Scientists Found a New Whale Species Hiding in Plain Sight—Including in a High School Gym." *Washington Post*, July 28, 2016.

* France, Robert. "Historicity of Sea Turtles Misidentified as Sea Monsters: A Case for the Early Entanglement of Marine Chelonians in Pre-plastic Fishing Nets and Maritime Debris." *Coriolis: Interdisciplinary Journal of Maritime Studies* 6, no. 2 (2016).

_____. "Reinterpreting Nineteenth-Century Accounts of Whales Battling Sea Serpents as an Illation of Early Entanglement in Pre-Plastic Fishing Gear or Maritime Debris." *International Journal of Maritime History* 28, no. 4 (November 25, 2016): 686–714.

* Hamilton-Paterson, James. *Seven Tenths: The Sea and its Thresholds*. London: Faber, 1992.

Harlan, Richard. *Fauna Americana: Being a Description of the Mammiferous Animals Inhabiting North America*. Philadelphia: Finley, 1825.

Heuvelmans, Bernard. *In the Wake of the Sea Serpents*. New York: Hill and Wang, 1965.

"The Hydrarchos!! Or, Leviathan of the Antediluvian world!" Advertisement for exhibition of the Hydrarchos, the skeleton of a "sea serpent," at Niblo's Garden, New York City, 1845.

Iwasa-Arai, Tammy, Salvatore Siciliano, Cristiana S. Serejo, and Ghennie T. Rodríguez-Rey. "Life History Told by a Whale-Louse: A Possible Interaction of a Southern Right Whale *Eubalaena australis* Calf with Humpback Whales *Megaptera novaeangliae*." *Helgoland Marine Research* 71 (2017): 6.

Kaliszewska, Zofia, Jon Seger, Victoria J. Rowntree, Susan G. Barco, Rafael Benegas, Peter B. Best, Moira W. Brown, et al. "Population Histories of Right Whales (Cetacea: *Eubalaena*) Inferred from Mitochondrial Sequence Diversities and Divergences of their Whale Lice (Amphipoda: *Cyamus*)." *Molecular Ecology* 14, no. 11 (October 2005): 3439–56.

Kennedy, Merrit. "Mysterious and Known as the 'Raven,' Scientists Identify New Whale Species." *The Two Way*, NPR, July 27, 2016 [audio].

Kingdon, Amorina. "The Stories Whale Lice Tell." *Hakai Magazine*, January 2, 2019.

Loxton, Daniel, and Donald R. Prothero. *Abominable Science: Origins of the Yeti, Nessie, and Other Famous Cryptids*. New York: Columbia University Press, 2013.

Lucas, Frederic Augustus. *Animals of the Past: An Account of Some of the Creatures of the Ancient World*. New York: American Museum of Natural History, 1913.

McClain, Craig R., Meghan A. Balk, Mark C. Benfield, Trevor A. Branch, Catherine Chen, James Cosgrove, Alistair D. M. Dove, et al. "Sizing Ocean Giants: Patterns of Intraspecific Size Variation in Marine Megafauna." *PeerJ* 3 (January 13, 2015).

Merwin, W. S. *The Lice: Poems*. London: Hart-Davis, 1969.

Morin, Phillip A., C. Scott Baker, Reid S. Brewer, Alexander M. Burdin, Merel L. Dalebout, James P. Dines, Ivan Fedutin, et al. "Genetic Structure of the Beaked Whale Genus *Berardius* in the North Pacific, with Genetic Evidence for a New Species." *Marine Mammal Science* 33, no. 1 (January 2017): 96–111.

National Oceanic and Atmospheric Administration (NOAA) Marine Debris Program. *Report on the Entanglement of Marine Species in Marine Debris with an Emphasis on Species in the United States*, 2014.

Netting, Jesse Forte. "Whale Lice Offer Links to Past." *Discover*, January 17, 2006.

* Olalquiaga, Celeste. *Artificial Kingdom: A Treasury of Kitsch*. New York: Pantheon Books, 1998.

Oudemans, Anthonie Cornelis. *The Great Sea-Serpent: An Historical and Critical Treatise*. Leiden: E. J. Brill, 1892.

Pascual, Santiago, and Elvira Abollo. "Whaleworms as a Tag to Map Zones of Heavy-Metal Pollution." *Trends in Parasitology* 21, no. 5 (May 2005): 204–06.

Paxton, C. G. M., E. Knatterud, and S. L. Hedley "Cetaceans, Sex and Sea Serpents: An Analysis of the Egede Accounts of 'a Most Dreadful Monster' Seen off the Coast of Greenland in 1734." *Archives of Natural History* 32, no. 1 (April 2005): 1–9.

Pierce, Sidney K., Steven E. Massey, Nicholas E. Curtis, and Gerald N. Smith. "Microscopic, Biochemical, and Molecular Characteristics of the Chilean Blob, and a Comparison with the Remains of Other Sea Monsters—Nothing but Whales." *Biological Bulletin* 206, no. 3 (June 2004): 125–33.

Rieppel, Lukas. "Albert Koch's Hydrarchos Craze: Credibility, Identity, and Authenticity in Nineteenth-Century Natural History." In *Science Museums in Transition: Cultures of Display in Nineteenth Century Britain and America*, edited by Carin Berkowitz and Bernard Lightman. Pittsburgh: University of Pittsburgh Press, 2017.

Ritvo, Harriet. *The Platypus and the Mermaid, and Other Figments of the Classifying Imagination*. Cambridge, MA: Harvard University Press, 1997.

Rotschafer, Paula A. "Serpentine Imagery in Nineteenth-Century Prints," unpublished thesis, University of Nebraska, School of Art, Art History & Design, 2014.

Rozwadowski, Helen M., and David K. van Keuren, eds. *The Machine in Neptune's Garden: Historical Perspectives on Technology and the Marine Environment*. Canton, MA: Science History, 2004.

Ryder, Richard D. *Animal Revolution: Changing Attitudes Towards Speciesism*. Oxford, UK: Berg, 2000.

Schorr, Gregory, Erin A. Falcone, David J. Moretti, and Russel D. Andrews. "First

Long-Term Behavioral Records from Cuvier's Beaked Whales (*Ziphius cavirostris*) Record-Breaking Dives." *PLOS ONE* 9, no. 3 (March 26, 2014): e92633.

Schulz, Kathryn. "Fantastic Beasts and How to Rank Them." *The New Yorker*, October 2017.

Seger, John, and V. J. Rowntree. "Whale Lice." In *Encyclopedia of Marine Mammals*, 3rd ed., edited by Bernd Würsig, J. G. M. Thewissen, and Kit M. Kovacs. London: Elsevier, Academic Press, 2018.

Simon, Matt. "Kraken and Owl Whales: Take a Dip with History's Most Terrifying Sea Monsters." *Wired*, September 13, 2013.

Smith, Craig. "Bigger Is Better: The Role of Whales as Detritus in Marine Ecosystems." In *Whales, Whaling and Ocean Ecosystems*, edited by James Estes. Berkeley, California: University of California Press, 2007.

Snively, Eric, Julia M. Fahlke, and Robert C. Walsh. "Bone-Breaking Bite Force of *Basilosaurus isis* (Mammalia, Cetacea) from the Late Eocene of Egypt, Estimated by Finite Element Analysis." *PLOS ONE* 10, no. 2 (February 25, 2015): e0118380.

Snyder, Gary. *A Place in Space: Ethics, Aesthetics, and Watersheds: New and Selected Prose.* Washington, DC: Counterpoint, 1995.

Soini, Wayne. *Gloucester's Sea Serpent.* Charleston, SC, and London: History Press, 2010.

Steinberg, Philip. *The Social Construction of the Ocean.* Cambridge, UK: Cambridge University Press, 2001.

Taylor, Larry D., Aaron O'Dea, Timoth J. Bralower, and Seth Finnegan. "Isotopes from Fossil Coronulid Barnacle Shells Record Evidence of Migration in Multiple Pleistocene Whale Populations." *Proceedings of the National Academy of Sciences* 116, no. 15 (April 9, 2019): 7377–81.

Thompson, Kirsten, C. Scott Baker, Anton van Helden, Selina Patel, Craig Miller, and Rochelle Constantine. "The World's Rarest Whale." *Current Biology* 22, no. 21 (November 2012): R905–06.

University of Utah News Centre. "Secrets of the Whale Riders: Lice Show How Endangered Cetaceans Evolved." September 14, 2005.

Walker, Matt. "Rare Whale Gathering Sighted." BBC Earth News, November 4, 2009.

Whiting, Candace Calloway. "Stuck on Whales and Dolphins, Remoras Are Not as Creepy as They Look." *HuffPost*, November 18, 2013.

Wilson, Edward O. *The Meaning of Human Existence.* New York: W. W. Norton, 2014.

Wood, Chelsea, and Pieter Johnson. "A World without Parasites: Exploring the Hidden Ecology of Infection." *Frontiers in Ecology and the Environment* 13, no. 8 (October 2015): 425–34.

* Zimmer, Carl. *Parasite Rex: Inside the Bizarre World of Nature's Most Dangerous Creatures.* New York: Free Press, 2000.

DEEP END

Bull, Jacob, ed. *Animal Movements—Moving Animals: Essays on Direction, Velocity and Agency in Human-Animal Encounters.* Uppsala: Uppsala University Press, 2011.

Crouch, Ian. "What Do We Do With This Whale?" *The New Yorker*, May 2, 2014.

Ingold, Tim. "Life Beyond the Edge of Nature? Or, the Mirage of Society." In *The Mark of the Social: Discovery or Invention*, edited by John D. Greenwood. New York: Rowman & Littlefield, 1997.

Mooallem, Jon. *Wild Ones: A Sometimes Dismaying, Weirdly Reassuring Story About Looking at People Looking at Animals in America.* New York: Penguin Press, 2013.

Index